FUNDAMENTALS OF PHOTOCHEMISTRY

FUNDAMENTALS OF PHOTOCHEMISTRY

K K Rohatgi-Mukherjee
Jadavpur University, Calcutta

A HALSTED PRESS BOOK

JOHN WILEY & SONS
New York Toronto

Published in the U.S.A., Canada and
Latin America by Halsted Press,
a division of John Wiley & Sons, Inc., New York

Library of Congress Cataloging in Publication Data

Rohatgi-Mukherjee, K K
Fundamentals of photochemistry.

"A Halsted Press book."
Bibliography: p.
Includes index.
1. Photochemistry. I. Title.

QD708.2.R63 1978 541'.35 78-12088
ISBN 0-470-26547-7

Printed in India by Prabhat Press, Meerut.

To

FATHER

who cared for education

MOTHER

who still encourages

Foreword

Not many years ago photochemistry was a flimsy subject, devoid of system and ill-equipped with apparatus. Theory now provides an effective way of understanding the interaction of light with atoms and molecules, and measuring devices of extremely high sensitivity and accuracy are now available. New light sources, photomultipliers, electronics, chromatography, etc. have entirely transformed the subject. The molecular bases of plant growth, vision, photobiological effects and upper air chemistry are some of the areas now being rapidly explored. Specialists in many fields find photochemistry essential in their work. This book gives an overall description of the subject, showing how it has been built up to its present state. It should prove an invaluable aid to students, lecturers and researchers who wish to form a coherent understanding of the whole field.

E. J. Bowen

Oxford
March 1977

Preface

"नूनं जनाः सूर्येण प्रसूताः ।"
(All that exists was born from the Sun)
—*Brhad-devata*, I: 61

In the last ten years photochemistry has seen a tremendous upsurge of interest and activity. A great deal of fundamental knowledge about the excited states has come to light as a result of the advent of tunable and high intensity laser beams. The field is developing so fast that any knowledge gained becomes outdated before it is fully comprehended. In the circumstances, perhaps another textbook is justified.

This book is written as a university level textbook, suitable for graduate, postgraduate and research students in the field of photochemistry, photophysics and photobiology. During the long years of teaching photochemistry at the graduate and postgraduate levels, I have always found it difficult to recommend a single textbook to the students. My first introduction to photochemistry was through Bowen's *Chemical Aspects of Light* which very lucidly explained the interactions between radiation and matter and their consequences and which has influenced me the most although photochemistry has travelled a long way since then. I have freely taken the help of books and monographs which are now available on the subject. All these books are listed in the beginning of the bibliography. J.B. Birks' *Photophysics of Aromatic Molecules*, N.J. Turro's *Molecular Photochemistry*, J.P. Simons' *Photochemistry and Spectroscopy* and A.A. Lamola and N.J. Turro (ed) *Organic Photochemistry and Energy Transfer* are some of the books from which I have drawn heavily. To these should be added the many review articles which have been of great help. I have adapted diagrams from some of these articles which have been acknowledged.

As the title implies, the book emphasizes the fundamental aspects of photochemistry. The first section introduces the subject by enumerating the relevance of photochemistry. Since the vocabulary of photochemistry is that of spectroscopy, the second section in which is discussed energy level schemes and symmetry properties, is like a refresher course. In the third section the actual mechanism of light absorption is taken up in detail because the probability of absorption forms the basis of photochemistry. A proper understanding of the process is essential before one can appreciate photochemistry. The next three sections present the properties of the electronically excited states and the fundamentals of photophysical processes. The primary photochemical processes form a separate section because chemical reactions in the excited states present certain new concepts. The rest of the book is mainly concerned with the application of the knowledge so gained to some typical photochemical reactions. Some current topics which are being actively pursued and are of great relevance have been presented in section nine. The last section discusses the latest tools and techniques for the determination of various photophysical and photochemical parameters. An attempt has been made, as far as possible, to explain the concepts by simple examples. A summary is given at the end of each of the first six sections which deal mainly with the fundamental aspects.

My thanks are due to the University Grants Commission for approving the project for writing this book and for providing necessary funds and facilities, and to the National Book Trust for subsidizing the book. I take this opportunity to acknowledge with thanks the help and suggestions that I have received from various quarters. I am deeply indebted to my teacher Dr. E.J. Bowen, FRS, Oxford University, for going through the entire manuscript with a 'fine-toothed comb' as he puts it, for suggestions and criticisms and for writing a Foreword to this book. Only because of his encouragement could I confidently embark upon a project of such magnitude. I also thank professors C.N.R. Rao, M.R. Padhye and H.J. Arnikar for their valuable comments. Mention must be made of S.K. Chakraborty, A.K. Gupta, P.K. Bhattacharya, S.K. Ash, U. Samanta, S. Basu and Shyamsree Gupta, who have helped me in various ways. To the scholar-poet professor P. Lal I owe a special debt for suggesting a beautiful couplet from the Vedas, pronouncing the glory of the *Sun*—the soul of the world.

Words fail to express the patience with which my husband, Dr. S.K. Mukherjee bore my writing bouts at the cost of my household duties. His constant encouragement gave me the moral and mental support which I needed in large measure in course of this arduous task.

Calcutta K.K. ROHATGI–MUKHERJEE

Contents

ONE

Introducing Photochemistry

1.1 IMPORTANCE OF PHOTOCHEMISTRY

Photochemistry is concerned with reactions which are initiated by electronically excited molecules. Such molecules are produced by the absorption of suitable radiation in the visible and near ultraviolet region of the spectrum. Photochemistry is basic to the world we live in. With sun as the central figure, the origin of life itself must have been a photochemical act. In the primitive earth conditions radiation from the sun was the only source of energy. Simple gaseous molecules like methane, ammonia and carbon dioxide must have reacted photochemically to synthesize complex organic molecules like proteins and nucleic acids. Through the ages, nature has perfected her machinery for the utilization of solar radiant energy for all photobiological phenomena and providing food for the propagation of life itself. Photobiology, the photochemistry of biological reactions, is a rapidly developing subject and helps the understanding of phenomena like photosynthesis, phototaxis, photoperiodism, photodynamic action, vision and mutagenic effects of light. In doing so it tries to integrate knowledge of physics, chemistry and biology.

The relevance of photochemistry also lies in its varied applications in science and technology. Synthetic organic photochemistry has provided methods for the manufacture of many chemicals which could not be produced by dark reactions. Moreover, greater efficiency and selectivity of these methods have an added advantage. Some examples of industrially viable photochemical syntheses may be mentioned here: (i) synthesis of vitamin D_2 from ergosterol isolated from certain yeasts, (ii) synthesis of cubanes which are antiviral agents, (iii) industrial synthesis of caprolactam, the monomer for Nylon 6, (iv) manufacture of cleaning solvents, insecticides and halogenated aromatics (used as synthetic intermediates) by photochlorination, and (v) synthesis of antioxidants by photosulphonation.

Photoinitiated polymerization and photopolymerization are used in photography, lithoprinting and manufacture of printed circuits for the electronic industry. The deleterious effect of sunlight on coloured cotton fabrics is of everyday experience, the worst sufferers being window curtains. The light absorbed by dyes used for colouring the fabric initiates oxidative chain reaction in cellulose fibres. This causes the tendering of cotton. Similar depolymerizing action is observed in plastic materials. Researches are going on to find suitable colourless chemicals which when added to dyed materials or plastics will take over the excitation energy and divert it to nondestructive pathways. These are known as energy degraders or photostabilizers, e.g., *o*-hydroxybenzophenones.

The photophysical phenomena of fluorescence and phosphorescence have found varied applications in fluorescent tube lights, X-ray and TV screens, as luminescent dials for watches, as 'optical brighteners' in white dress materials, as paints in advertisement hoardings which show enhanced brilliance by utilizing fluorescence, for detection of cracks in metal work, for tracing the course of river through caves, as microanalytical reagents, and so on.

Certain chemicals change their colour, that is, their absorption characteristics, when exposed to suitable radiation and reverse when the irradiation source is removed. These are known as photochromic materials. A well known example is the spiropyrans. Their use in photochromic sunglasses is obvious. But they have found application in information storage and self-developing self-erasing films in digital computers also. It is said that a company experimenting on such photochromic memory used UV light for writing the information, green light for reading it and blue light for erasing it. Unfortunately organic substances usually lack the stability for very large numbers of reversals.

Another revolutionary application of electronically excited molecular systems is in laser technology. Lasers are intense sources of monochromatic and coherent radiation. From their early development in 1960 they have found wide fields of application. They have provided powerful tools for the study of diverse phenomena ranging from moonquakes to picosecond processes of nonradiative decay of excitational energy in molecules. The intense and powerful beam of coherent radiation capable of concentration to a tiny point is used for eye surgery, cutting metals, boring diamonds, as military range finders and detectors, and many such applications. The advent of tunable dye lasers has increased the possibility of their application in science and technology.

A further impetus to the study of photochemical reaction has been provided by the energy crisis. This has initiated researches into the conversion and storage of solar energy, processes which plants carry out so efficiently. Solar energy provides a readily available source of energy, especially in those countries which lie between the tropics of cancer and capricorn. In these areas, the daily incident energy per square kilometre is equivalent to 3000 tonnes of coal. If suitable photochemical reactions are discovered and devices for proper utilization of this abundant source of energy perfected, half the world's energy problem might be solved. Solar batteries working on the principle of photovoltaic effects is one such device. For basic researches in these fields, the understanding of various photophysical and photochemical processes is essential. The fundamental study of excited states of molecules is exciting by itself. Short-lived energy states with nano and pico-second reaction kinetics have led to the proper understanding of chemical reactions, modes of energy transfer and the intricate structure of matter. Flash photolysis and pulsed laser photolysis are newer tools for the study of higher energy states. Now it is possible to excite individual vibronic levels or isotopically substituted compounds by using appropriate beams from tunable dye lasers.

1.2 LAWS OF PHOTOCHEMISTRY

Prior to 1817, photochemical changes such as photofading of coloured materials, photosynthesis in plants, blackening of silver halides, etc. was observed and studied qualitatively. The quantitative approach to photochemistry was initiated by Grotthus and Draper in the beginning of the nineteenth century. It was realized that all the incident light was not effective in bringing about a chemical change and the *first law of photochemistry*, now known as *Grotthus-Draper law* was formulated:

Only that light which is absorbed by a system can cause chemical change.

The probability or rate of absorption is given by the Lambert-Beer Law. The *Lambert law* states that *the fraction of incident radiation absorbed by a transparent medium is independent of the intensity of incident radiation and that each successive layer of the medium absorbs an equal fraction of incident radiation.* The *Beer law* states that *the amount of radiation absorbed is proportional to the number of molecules absorbing the radiation, that is the concentration C of the absorbing species.* The two are combined and expressed as

$$\frac{dI}{I} = \alpha_\nu C dl \tag{1.1}$$

where α_ν is the proportionality constant. The quantity Cdl, measures the amount of solute per unit area of the layer, dl being the thickness of the layer. Since

$$C = \frac{\text{mole}}{\text{volume}} = \frac{\text{mole}}{\text{area} \times \text{thickness}}$$

Therefore, $$Cdl = \frac{\text{mole}}{\text{area}}$$

On integrating equation (1.1) within the boundary conditions, we get (i) $I = I_0$, when $l = 0$, and (ii) $I = I$, when $l = l$, we have

$$\ln \frac{I_0}{I} = \alpha_\nu Cl \tag{1.2}$$

α_ν, known as absorption coefficient, is a function of frequency or wavelength of radiation. The final form is expressed in the decadic logarithm,

$$\log \frac{I_0}{I} = \epsilon_\nu Cl \tag{1.3}$$

where $\epsilon_\nu = \alpha_\nu/2.303$, is called the *molar extinction coefficient* and is a function of frequency ν, the concentration is expressed in moles per litre and l is the optical path length in cm. The SI units of c, l and ϵ are $mol\ dm^{-3}$, mm and $m^2 mol^{-1}$ respectively. I_0 and I are the incident and transmitted intensity respectively (Figure 1.1). The quantity $\log I_0/I$ is commonly known as the optical density OD or absorbance A. A plot of ϵ_ν (or its logarithm) vs wavelength or wavenumber gives rise to familiar absorption bands. Since

$$I = I_0 10^{-\epsilon_\nu Cl} \tag{1.4}$$

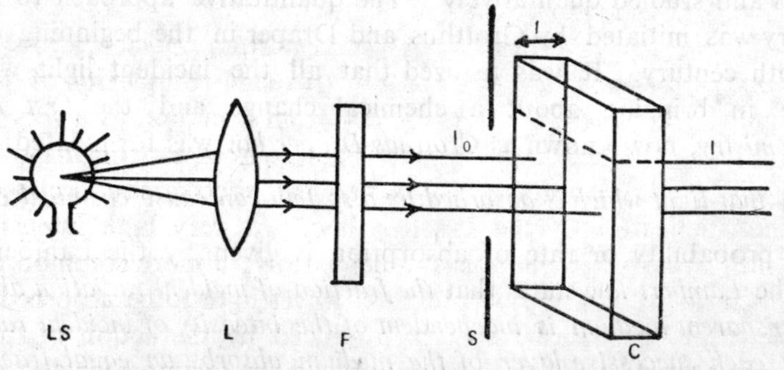

Figure 1.1 Optical arrangement for a photochemical reaction by a collimated beam of radiation of cross-sectional area A. LS=light source, L=lens, F=filter, S=collimating shield, C=reaction cell, l=optical path length, Io=incident light intensity, I=transmitted light intensity.

the amount of light absorbed I_a, by the system is

$$I_a = I_0 - I = I_0 - I_0\,10^{-\epsilon_\nu Cl}$$
$$= I_0(1 - 10^{-\epsilon_\nu Cl}) \qquad (1.5)$$

For more than one absorbing components, optical density is $\sum_i \epsilon_{\nu_i} C_i$, where ϵ_{ν_i} is the molar absorptivity at frequency ν_i for the ith component whose concentration is C_i, assuming path length to be unity. Hence the measured OD is

$$OD = OD_1 + OD_2 + OD_3 + \ldots \qquad (1.6)$$

The *second law of photochemistry* was first enunciated by Stark (1908) and later by Einstein (1912). The *Stark-Einstein law* states that:

One quantum of light is absorbed per molecule of absorbing and reacting substance that disappears.

Work by Warburg and Bodenstein (1912–1925) clarified earlier confusions between photon absorption and observed chemical change. Molecules which absorb photons become physically 'excited', and this must be distinguished from becoming chemically 'active'. Excited molecules may lose their energy in nonchemical ways, or alternatively may trigger off thermal reactions of large chemical yield. The socalled 'law', therefore, rarely holds in its strict sense, but rather provides essential information about the primary photochemical act.

To express the efficiency of a photochemical reaction, the quantity *quantum efficiency* ϕ is defined as

$$\phi_{\text{reaction}} = \frac{\text{number of molecules decomposed or formed}}{\text{number of quanta absorbed}} \qquad (1.7)$$

The concept of quantum yield or quantum efficiency was first introduced by Einstein. Because of the frequent complexity of photoreactions, quantum yields as observed vary from a million to a very small fraction of unity. When high intensity light sources as from flash lamps or lasers are used 'biphotonic' photochemical effects may occur which modify the application of the Einstein law. At very high intensities a molecule may absorb two photons simultaneously; a more common effect, however, is for a second photon of longer wavelength to be absorbed by a metastable (triplet or radical) species produced by the action of the first photon. The nature of the photo-products and the quantum yields are here dependent on the light intensity. The concept of quantum yield can be extended to any act, physical or chemical, following light absorption. It provides a mode of account-keeping for partition of absorbed quanta into various pathways.

$$\phi_{\text{process}} = \frac{\text{number of molecules undergoing that process}}{\text{number of quanta absorbed}}$$

$$= \frac{\text{rate of the process}}{\text{rate of absorption}} \tag{1.8}$$

1.3 PHOTOCHEMISTRY AND SPECTROSCOPY

Since the primary photoprocess is absorption of a photon to create a photoexcited molecule, photochemistry and spectroscopy are intimately related. Quantum mechanics has played a vital part in describing the energy states of molecules.

For any chemical reaction, energy is required in two ways: (i) as energy of activation ΔE_a, and (ii) as enthalpy or heat of reaction ΔH. The need for energy of activation arises because on close approach, the charge clouds of the two reacting partners repel each other. The reactants must have sufficient energy to overcome this energy barrier for fruitful interaction. The enthalpy of reaction is the net heat change associated with the breaking and making of bonds leading to reaction products. In thermal or dark reactions, the energy of activation is supplied as heat energy. In photochemical reactions, the energy barrier is bypassed due to electronic excitation and one of the products may appear in the excited state.

The bond dissociation energy per mole for most of the molecules lie between 150 kJ and 600 kJ. These energies are available from Avogadro's number of photons of wavelengths lying between 800 nm and 200 nm respectively, which correspond to the visible and near ultraviolet regions of the electromagnetic spectrum. The same range of energies is required for electronic transitions in most atoms and molecules. For example, anthracene has an absorption band with a maximum at wavelength 365 nm. This signifies that a photon of this wavelength is absorbed by the anthracene molecule to promote it from the ground energy state E_1, to upper energy state E_2. From Bohr's relationship, the energy equivalent of a photon of this wavelength is calculated as

$$E_{365} = E_2 - E_1 = h\nu \tag{1.9}$$

where, h = Planck's constant and ν is the frequency of absorbed radiation. When expressed in wavenumber in reciprocal centimetre (cm^{-1}) or wavelength in nanometre (nm) and substituting the values for h and c (the velocity of light), we get

$$E_{365} = h\nu = hc\bar{\nu} \qquad (\nu = c\bar{\nu}) \tag{1.10}$$

$$= \frac{hc}{\lambda} \qquad (\bar{\nu} = 1/\lambda) \tag{1.11}$$

$$= \frac{6.62 \times 10^{-27}\ \text{erg s} \times 3.00 \times 10^{10}\ \text{cm s}^{-1}}{365 \times 10^{-7}\ \text{cm}}$$

$$= 5.44 \times 10^{-12}\ \text{erg photon}^{-1}$$

1.4 UNITS AND DIMENSIONS

According to the modern convention, measurable quantities are expressed in SI (System Internationalé) units and replace the centimetre-gram-second (cgs) system. In this system, the unit of length is a metre (m), the unit of mass is kilogram (kg) and the unit of time is second (s). All the other units are derived from these fundamental units. The unit of thermal energy, *calorie*, is replaced by *joule* ($1\ J = 10^7$ erg) to rationalize the definition of thermal energy. Thus, Planck's constant

$$h = 6.62 \times 10^{-34}\ \text{J s};$$

velocity of light

$$c = 3.00 \times 10^{8}\ \text{m s}^{-1};$$

the wavelength of radiation λ is expressed in nanometres ($1\ \text{nm} = 10^{-9}$ m). Therefore in the SI units:

$$E_{365} = \frac{6.62 \times 10^{-34}\ \text{J s} \times 3.00 \times 10^{8}\ \text{m s}^{-1}}{365 \times 10^{-9}\ \text{m}}$$

$$= 5.44 \times 10^{-19}\ \text{J photon}^{-1}.$$

This quantum of energy is contained in a photon of wavelength 365 nm. An Avogadro number of photons is called an *einstein*. The amount of energy absorbed to promote one mole of anthracene molecules to the first excited electronic state will be

$$= 5.44 \times 10^{-19}\ \text{J photon}^{-1} \times 6.02 \times 10^{23}\ \text{photon mol}^{-1}$$

$$= 3.27 \times 10^{5}\ \text{J mol}^{-1}$$

$$= 327\ \text{kJ (kilojoule) mol}^{-1}$$

This amount of energy is contained in one mole or one einstein of photons of wavelength 365 nm.

The energy of an einstein of radiation of wavelength λ (expressed in nm) can be calculated from the simplified expression

$$\frac{1.196 \times 10^{8}}{\lambda}\ \text{kJ einstein}^{-1} \tag{1.12}$$

Rate of absorption is expressed in einstein per unit area per second

$$I_a = \frac{1.196 \times 10^{8}}{\lambda}\ \text{einstein m}^{-2}\ \text{s}^{-1} \tag{1.13}$$

The energy of radiation is quite often expressed in terms of kilo-calorie per mole (kcal/mole), (1 calorie = 4·186 J). Sometimes, merely cm^{-1}, the unit of wavenumber is used to express energy. The proportionality constant *hc*, is implied therein. The unit of electron-volt (eV) is used for single atom or molecule events. A chemical potential of one volt signifies an energy of one electron volt per molecule.

Some values for the energy of radiation in the visible and ultraviolet regions are given in Table 1.1.

TABLE 1.1

Energy of electromagnetic photon in the visible and uv regions expressed in different units

Region	Approx. wave-length range nm	Wavenumber cm^{-1}	Energy mol^{-1} kJ	kcal	eV
	200	50,000	598	142.9	6.20
Ultraviolet					
	400	25,000	299	71.4	3.10
Violet					
	450	22,222	266	63.5	2.76
Blue					
	500	20,000	239	57.1	2.48
Green					
	570	17,544	209	49.9	2.16
Yellow					
	590	16,949	203	48.5	2.10
Orange					
	620	16,129	192	45.9	2.0
Red					
	750	13,333	159	38.0	1.6

$$\lambda\,(\text{nm}) = \frac{c}{\nu} = \frac{1}{\bar{\nu}}\ \text{cm}^{-1} = 10^{-9}\ \text{m}$$

$$1\ \text{cal} = 4.186\ \text{J}$$

$$1\ \text{eV} = 1.6 \times 10^{-19}\ \text{J}$$

$$1\ \text{cm}^{-1}\ \text{mol}^{-1} = 2.859\ \text{cal}\ \ \text{mol}^{-1}$$

$$= 0.0135\ \text{kJ mol}^{-1}$$

$$1\ \text{eV mol}^{-1} = 23.06\ \text{kcal mol}^{-1}$$

$$= 96.39\ \text{kJ mol}^{-1}$$

The intensity of incident flux from light sources is in general defined in terms of *power*, i.e. *watt per unit cross-section* (watt = J s^{-1}). Since *power* is energy per unit time and each photon has energy associated with it, intensity I can be expressed in number of quanta $m^{-2}\ s^{-1}$. We have, $E = nh\bar{\nu}c$ and

$$\text{Power} = \frac{\text{watt}}{\text{m}^2} = \frac{\text{J}}{\text{m}^2\ \text{s}}$$

$$I = \frac{\text{n}}{\text{m}^2\ \text{s}} = \frac{E}{h\bar{\nu}c} \cdot \frac{\text{J}}{\text{m}^2\ \text{s}}$$

$$= \frac{\text{watt}}{h\,c\bar{\nu}} = 5.03 \times 10^{24} \times \lambda\,(\text{nm}) \times \text{power (watt)}$$

Also $I = \dfrac{\text{einstein}}{\text{m}^2\ \text{s}} = 8.36 \times \lambda\ (\text{nm}) \times \text{power (watt)}$

For example, a helium-argon laser with a power of 2×10^{-3} W at 632.8 nm will emit 6.37×10^{15} quanta $s^{-1}\ m^{-2}$ or 1.66×10^{-8} einstein $s^{-1}\ m^{-2}$.

If the area of the reaction vessel exposed to the radiation is A, *the rate of incidence* is given as the intensity I times the area A.

1.5 THERMAL EMISSION AND PHOTOLUMINESCENCE

Atoms and molecules absorb only specific frequencies of radiation dictated by their electronic configurations. Under suitable conditions they also emit some of these frequencies. A perfect absorber is defined as one which absorbs all the radiation falling on it and, under steady state conditions, emits all frequencies with unit efficiency. Such an absorber is called a *black body*. When a system is in thermal equilibrium with its environment rates of absorption and emission are equal (Kirchhoff's law). This equilibrium is disturbed if energy from another source flows in. Molecules electronically excited by light are not in thermal equilibrium with their neighbours.

The total energy E, of all wavelengths radiated per m^2 per second by a black body at temperature TK is given by the Stefan-Boltzmann law

$$E = \sigma T^4 \tag{1.14}$$

where the Stefan's constant

$$\sigma = 5.699 \times 10^{-8} \text{ J m}^{-2} \text{ deg}^{-4} \text{ s}^{-1}$$

From Planck's radiation law, the energy per m^3 of radiation or radiation density ρ in an enclosure having wavelength between λ and $\lambda + d\lambda$ is $\rho_\lambda \, d\lambda$, that is

$$\rho_\lambda \, d\lambda = \frac{8\pi hc}{\lambda^5} \frac{d\lambda}{e^{hc/\lambda kT} - 1} = \frac{C_1}{\lambda^5}\left(\frac{d\lambda}{e^{C_2/\lambda T} - 1}\right) \tag{1.15}$$

where $C_1 = 4.992 \times 10^{-24}$ J m^{-1}, $C_2 = 1.439 \times 10^{-2}$ m deg and k = Boltzmann constant $= 1.38 \times 10^{-23}$ J molecule^{-1}.

The corresponding radiation density within frequency range ν and $\nu + d\nu$ is

$$\rho_\nu \, d\nu = \frac{8\pi h\nu^3}{c^3} \frac{d\nu}{e^{h\nu/kT} - 1} \tag{1.16}$$

On multiplying the expression (1.15) by $c/4$ where c is the velocity of light, the expression for energy density can be converted into energy flux E, the energy emitted in units of J per second per unit area within unit wavelength interval at wavelength λ (expressed in nm) by an ideal black body of surface area A at TK. Hence

$$dE = E_\lambda d\lambda = \frac{c}{4} \frac{8\pi hc \times 10^{-7} \times A}{\lambda^5 (e^{hc/\lambda kT} - 1)} \, d\lambda \text{ Js}^{-1} \text{ nm}^{-1}$$

$$\frac{dE}{d\lambda} = \frac{2\pi hc^2 \times 10^{-7} \times A}{\lambda^5 (e^{hc/\lambda kT} - 1)} \text{ Js}^{-1} \text{ nm}^{-1} \tag{1.17}$$

$$= \frac{3.74 \times 10^{-7} \times A}{\lambda^5 (e^{hc/\lambda kT} - 1)} \text{ watt nm}^{-1}$$

To express in units of quanta $m^{-3}\ s^{-1}$, Planck's equation is divided by the energy of one quantum $h\nu$:

$$Q_\nu\ (\text{quanta/m}^3\text{s}) = \frac{\rho_\nu\ (\text{J/m}^3)}{h\nu} = \frac{8\pi}{c^3}\left(\frac{\nu^2}{e^{h\nu/kT}-1}\right) \qquad (1.18)$$

where Q_ν is quantum density per unit frequency interval per second. The rate of emission per unit area per unit wavenumber interval is obtained by dividing by $c/4$.

Planck's equation applies strictly to the emission into space at absolute zero, but for wavelengths in the visible and ultraviolet region from incandescent sources, this is substantially the same as emission into space at room temperature. For low temperatures and frequencies in the optical range $e^{h\nu/kT} \gg 1$ the following simplification can be made:

$$\rho_\nu = \frac{8\pi h\nu^3}{c^3} \exp^{-h\nu/kT} \qquad (1.19)$$

$$Q_\nu = \frac{8\pi\nu^2}{c^3} \exp^{-h\nu/kT} \qquad (1.20)$$

Light emitted from a black body solely as a result of high temperature as in electric bulb is known as *incandescence* or *thermal* radiation. The quality and quantity of thermal radiation is a function of temperature only. The wavelength of most strongly emitted radiation in the continuous spectrum from black body is given by Wien's Displacement Law, $\lambda_{max} T = b$, (where b is Wien's constant $= 2.898 \times 10^{-3}$ m deg).

On the other hand, the quality and quantity of emission from an electronically excited molecule, as in fluorescent tube lamps, are not basically functions of temperature. *Photoluminescence* is known as cold light. It is characteristic of the absorbing system.

Summary

1. Photochemistry is the chemistry of the electronically excited molecules.
2. The first law of photochemistry states: *Only that light which is absorbed by a system can cause chemical change* (Grotthus–Draper Law).
3. The probability or rate of absorption is given by Lambert–Beer law: *Fractional light absorption is proportional to concentration C in mol l^{-1} and the thickness dl of the absorbing system*

$$\frac{dI}{I} = \alpha_\nu\, C\, dl$$

where α_ν is the proportionality constant.
In the integrated form

$$\log \frac{I_0}{I} = \epsilon_\nu\, Cl = \text{optical density}$$

where $\epsilon_\nu = \alpha_\nu/2.303$ is called the molar extinction coefficient and is a function of frequency or wavelength.

4. The second law states: *One quantum of light is absorbed per molecule of absorbing and reacting substances that disappear.*
5. The efficiency of a photochemical reaction is expressed in terms of a quantity called quantum yield ϕ, defined as

$$\phi = \frac{\text{number of molecules decomposed or formed per unit time}}{\text{number of quanta absorbed per unit time}}$$

6. Photochemistry and spectroscopy are related intimately.
7. Quality and quantity of thermal emission is a function of temperature only.
8. Quality and quantity of photo-luminiscence is characteristic of the absorbing system.

TWO

Nature of Light and Nature of Matter

2.1 INTERACTION BETWEEN LIGHT AND MATTER

The interaction between light and matter is the basis of all life in this world. Even our knowledge of the physical world is based on such interactions, because to understand matter we have to make use of light, and to understand light we must involve matter. Here, by light we mean the complete spectrum of the electromagnetic radiation (Figure 2.1) from radioactive rays to radio waves; hence light and radiation have been used synonymously.

For example, we use X-rays to elucidate the structure of molecules in their crystalline state, and take the help of various types of spectroscopic methods for the understanding of the intricate architecture of atoms and molecules (Figure 2.1). On the other hand, if we wish to study the nature of light, we must let it fall on matter which reflects, transmits, scatters or absorbs it and thus allowing us to understand its behaviour. A beam of light in a dark room will not be visible to us unless it is scattered by dust particles floating in the air. A microscope will view a particle only when incident light is scattered by it into the aperture of the objective. All light measuring devices are based on such interactions. In some of these interactions light behaves as a particle and in some others its behaviour is akin to a wave motion. Therefore, to obtain a basic understanding of the interaction of light with matter, we must first

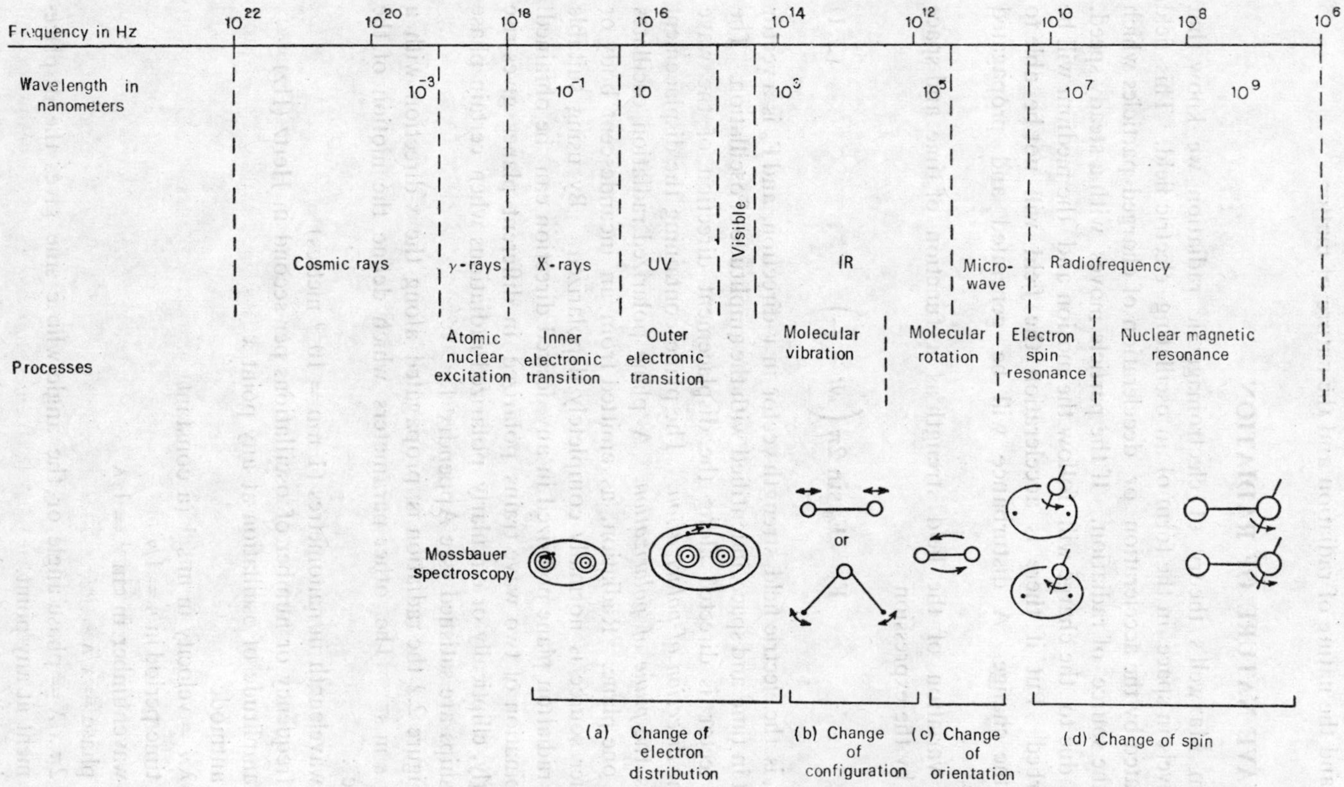

Figure 2.1 The complete spectrum of electromagnetic radiation, types of processes initiated and the corresponding spectroscopic terminology. (a) Change of electron distribution—x-ray, *uv* and visible electron spectroscopy; (b) change of configuration—vibration-rotation spectroscopy; (c) change of orientation-microwave spectroscopy; (d) change of spin—eleetron spin resonance (ESR) and nuclear magnetic resonance (NMR) spectroscopy.

understand the nature of radiation and the nature of matter.

2.2 WAVE NATURE OF RADIATION

From Maxwell's theory of electromagnetic radiation we know that light travels in space in the form of an oscillating electric field. This field is generated by the acceleration or deceleration of charged particles which act as the source of radiation. If the particle moves with a steady speed, the field due to the charge will follow the motion and the medium will be undisturbed. But if there is acceleration the field will not be able to follow the change. A disturbance will be generated and propagated in space.

The variation of the field strength as a function of time and space is given by the expression

$$\mathbf{E}_y = \mathbf{E}_0 \sin 2\pi \left(\nu t - \frac{x}{\lambda} \right) \quad (2.1)$$

where $\mathbf{E}_y$ is the electric field strength vector in y-direction, and $\mathbf{E}_o$ is a vector constant in time and space, indentified with the amplitude of oscillation. The electric vector is directed along the displacement direction of the wave called the *direction of polarization*. The plane containing the displacement vector is the *plane of polarization*. A plane polarized radiation oscillates only in one plane. Radiation, as emitted from an incandescent body or any other source is normally completely depolarized. By using suitable devices, radiation plane polarized in any desired direction can be obtained. A combination of two wave trains polarized in different planes gives rise to linearly, elliptically or circularly polarized radiations when certain phase relationships are satisfied (see Appendix 1).

In Figure 2.2 the radiation is propagated along the x-direction with a velocity c m s^{-1}. The other parameters which define the motion of the wave are :

λ = wavelength in nanometres (1 nm = 10^{-9} metres)
ν = frequency or number of oscillations per second in Hertz (Hz)
A = amplitude of oscillation at any point x
A_{max} = antinode
$c = \lambda \nu$ = velocity in m s^{-1} (a constant)
T = time period in s = $1/\nu$
$\bar{\nu}$ = wavenumber in cm^{-1} = $1/\lambda$
ϕ = phase = x/λ
$2\pi\phi = 2\pi x/\lambda$ = phase angle or the angle whose sine gives the displacement at any point.
I = intensity at any point x
= square of displacement at that point
node = point where the amplitude is zero.

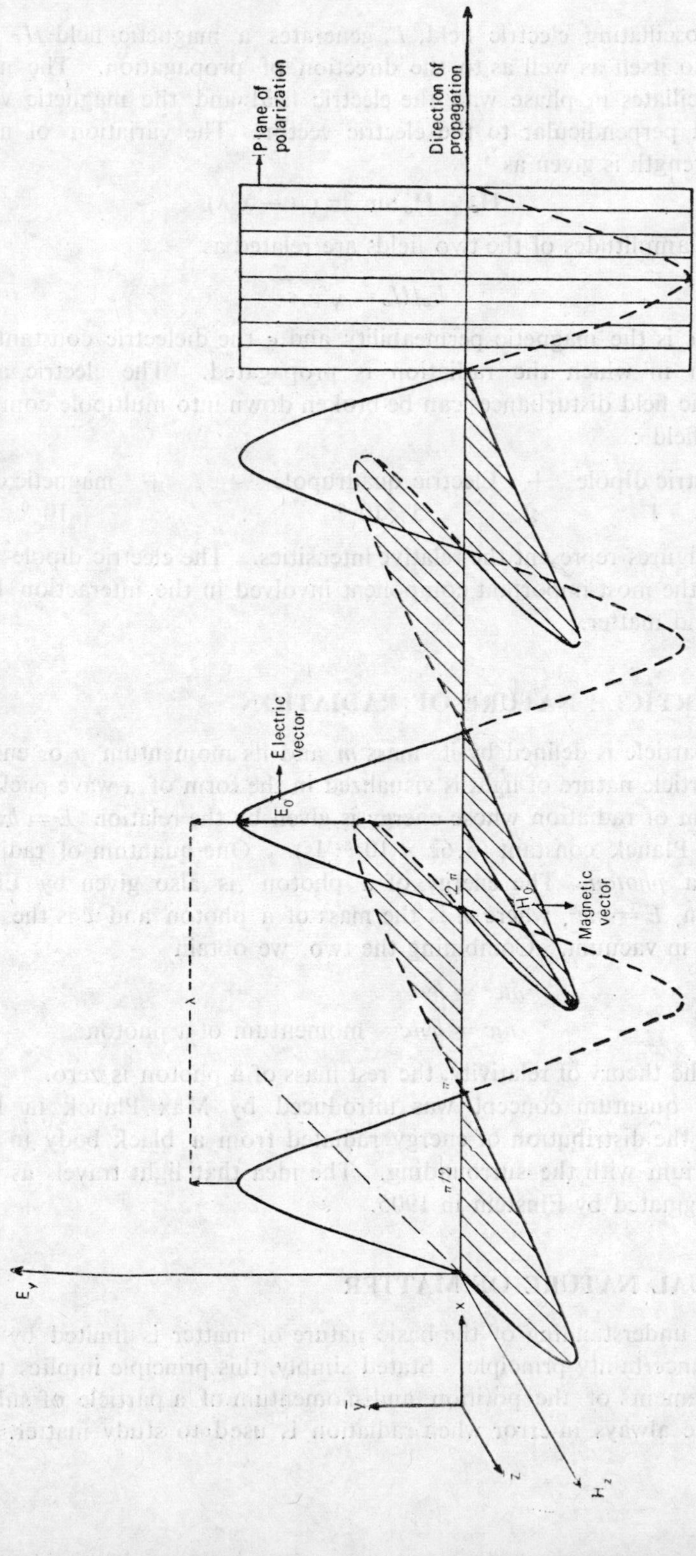

Figure 2. 2 Nature and propagation of plane polarized electromagnetic radiation.

An oscillating electric field, E, generates a magnetic field H, at right angles to itself as well as to the direction of propagation. The magnetic field oscillates in phase with the electric field and the magnetic vector is directed perpendicular to the electric vector. The variation of magnetic field strength is given as

$$\mathbf{H}_z = \mathbf{H}_o \sin 2\pi (\nu t - x/\lambda) \qquad (2.2)$$

The amplitudes of the two fields are related as

$$E_o/H_o = \sqrt{\mu/\epsilon} \qquad (2.3)$$

where μ is the magnetic permeability and ϵ the dielectric constant of the medium in which the radiation is propagated. The electric and the magnetic field disturbances can be broken down into multipole components of the field :

Electric dipole	+	Electric quadrupole	+.. +	magnetic dipole
1	:	5×10^{-6}	:	10^{-6}

where figures represent the relative intensities. The electric dipole component is the most important component involved in the interaction between light and matter.

2.3 PARTICLE NATURE OF RADIATION

A particle is defined by its mass m and its momentum p or energy E. The particle nature of light is visualized in the form of a wave packet or a quantum of radiation whose energy is given by the relation $E = h\nu$, where h is the Planck constant (6.62×10^{-34} Js). One quantum of radiation is called a *photon*. The energy of a photon is also given by Einstein's equation, $E = mc^2$, where m is the mass of a photon and c is the velocity of light in vacuum. Combining the two, we obtain

$$mc^2 = h\nu \qquad (2.4)$$

or

$$mc = h\nu/c = \text{momentum of a photon}$$

From the theory of relativity, the rest mass of a photon is zero.

The quantum concept was introduced by Max Planck in 1900 to explain the distribution of energy radiated from a black body in thermal equilibrium with the surrounding. The idea that light travels as *photons* was originated by Einstein in 1905.

2.4 DUAL NATURE OF MATTER

Our understanding of the basic nature of matter is limited by Heisenberg's uncertainty principle. Stated simply, this principle implies that our measurements of the position and momentum of a particle of subatomic mass are always in error when radiation is used to study matter. If Δx

is the error in the location of the particle, in the same experiment, Δp is the inherent error in the measurement of its momentum such that the product $\Delta x \, \Delta p \simeq h$ where h is Planck's constant. If Δx is made small, Δp becomes large and vice versa. Similarly, energy E, and time t, form a conjugate pair and $\Delta E \, \Delta t \simeq h$.

This principle has profound influence on our study of the structure of matter. Bohr's concept of well defined *orbits* is invalidated and the only way to express the dynamics of electron motion in atoms is in terms of probability distribution functions called *orbitals.* The necessity for such a probability distribution function immediately suggests the notion of a three dimensional standing wave. In 1924, de Broglie emphasized the dual nature of matter and obtained an expression similar to that of the light wave in which de Broglie's wavelength λ for the electron wave is related to the momentum p of the particle by Planck's constant h: Expressing p in terms of energy of the system

$$\lambda = \frac{h}{\sqrt{2m(E-V)}}$$

where E is the total energy and V is the potential energy.

An expression for describing such a wave motion was obtained by Schrödinger in 1925. The Schrödinger equation is a second order differential equation which can be solved to obtain the total energy E of a dynamic system when expressed as a sum of kinetic and potential energies:

$$-\frac{h^2}{8\pi^2 m}\left[\frac{d^2}{dx^2}+\frac{d^2}{dy^2}+\frac{d^2}{dz^2}\right]\Psi + V\Psi = E\Psi \tag{2.5}$$

$$\left[-\frac{h^2}{8\pi^2 m}\nabla^2 + \hat{V}\right]\Psi = E\Psi \tag{2.6}$$

$$\mathcal{H}\Psi = E\Psi \tag{2.7}$$

where

$$\nabla^2 \equiv \frac{d^2}{dx^2}+\frac{d^2}{dy^2}+\frac{d^2}{dz^2} = \text{Laplacian operator (del)}^2$$

$$\hat{V} \equiv V_x + V_y + V_z = \text{Potential energy operator}$$

$$\mathcal{H} \equiv -\frac{h^2}{8\pi^2 m}\nabla^2 + \hat{V} = \text{Hamiltonian operator}$$

In this equation Ψ is known as the *eigenfunction* and E the *eigenvalue* or the characteristic energy value corresponding to this function. On solving this equation, a number of values for Ψ such as $\Psi_0, \Psi_1, \Psi_2, \ldots, \Psi_n$ are obtained. Corresponding to each eigenfunction, there are the characteristic energies $E_0, E_1, E_2, \ldots, E_n$. Thus an electron constrained to move in a potential field can have only definite energy values for its motion. It is governed by the condition that the motion shall be described by a standing wave.

2.5 ELECTRONIC ENERGY STATES OF ATOMS

The hydrogen atom has a single electron confined to the neighbourhood of the nucleus by a potential field V, given by $-e^2/r$. The solution of the appropriate Schrödinger equation becomes possible if the equation is expressed in polar coordinates r, θ and ϕ (Figure 2.3), since in that case it can be resolved into three independent equations each containing only one variable:

$$\Psi(r, \theta, \phi) = R(r)\,\Theta(\theta)\,\Phi(\phi)$$

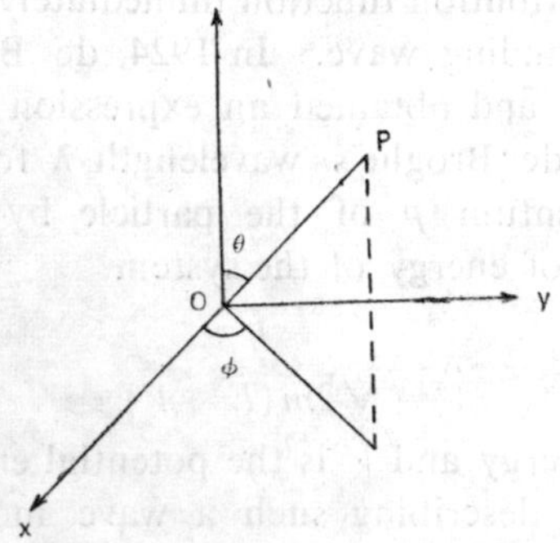

Figure 2.3 Polar coordinates r, θ and ϕ and factorization of total wavefunction, $\Psi(r, \theta, \phi) = R(r).\theta(\theta)\,\Phi(\phi)$

The solutions of $R(r)\,\theta(\theta)\,\Phi(\phi)$ equations introduce the three quantum numbers n, l, and m which have integral values as required by the quantum theory. These quantum numbers are:

n: the principal quantum number indicates the energy state of the system and is related to the dimensions of the orbital. An orbital of number n will have n nodes including one at infinity. n can have values from 1, 2, 3..., ∞.

l: the azimuthal or orbital angular momentum quantum number arises due to the motion of the electron in its orbital. Angular momentum is a vector quantity whose value is given by $\sqrt{l(l+1)}\ h/2\pi$. This quantum number is related to the geometry or the shape of the orbital and is denoted by the symbols s, p, d, f, etc. corresponding to l values of 0, 1, 2, 3 respectively. The possible values of l are governed by the principal quantum number n, such that the maximum value is $(n-1)$, e.g.

when $n = 1$, $l = 0$

$n = 2$, $l = 0, 1$

$n = n$ $l = 0, 1, 2, ..., (n-1)$

For the H-atom, the set of l values for a given n orbital have the same energy, i.e., they are degenerate. The angular momenta are eigenvalues of the angular momentum operator $\hat{L}^2$:

$$\hat{L}^2 \Psi_{n, l \pm m} = l(l+1) \frac{h^2}{4\pi^2} \Psi_{n, l \pm m} \tag{2.8}$$

m: the magnetic quantum number defines the orientation of the orbital in space and is effective in presence of an externally applied magnetic field (Zeeman effect). It corresponds to the component of the angular mamentum (L_z) in the direction of the field. In the absence of the field, each orbital of given values of n and l is $(2l+l)$ – fold degenerate and can have values $0, \pm 1, \pm 2, \ldots, \pm l$. The magnetic momentum quantum numbers are eigenvalues of the operator $\hat{L}_z$ such that

$$\hat{L}_z \Psi = \pm m \frac{h}{2\pi} \Psi \tag{2.9}$$

The pictorial representation of radial and spherical distribution functions for values of $n = 1, 2, 3$ are shown in Figure 2.4.

Another very fundamental concept which has to be introduced foı systems with more than one electron is the spin quantum number s.

s: the spin quantum number arises due to spinning of an electron on an axis defined by an existing magnetic field. This generates an angular momentum which is a vector quantity of magnitude $\sqrt{s(s+1)}\, h/2\pi$; s can have only two values $+\frac{1}{2}$ and $-\frac{1}{2}$. Since it is assumed to be an intrinsic property of the electron, the concept of spin quantum number cannot be deduced from the Schrödinger equation. It was introduced empirically by Uhlenbeck and Goudsmit to explain the doublet structure in the emission spectra of alkali metals. Implication of spin quantum number is emphasized by the *Pauli exclusion principle* which states: *No two electrons can have the same value for all the four quantum numbers (n, l, m and s). If n, l, m values are the same, then the two electrons must differ in their spin.*

For example, for a normal He atom with two electrons having quantum numbers $n = 1$, $l = 0$, $m = 0$, one electron shall be in $s = +\frac{1}{2}$ state and the other in $s = -\frac{1}{2}$ state. On the other hand, for an electronically excited He atom, since now the two electrons reside in two different energy states, i.e. n values differ the two electrons may have the same spin values.

The wave function for a spinning electron is, therefore, written as

$$\Phi = \Psi_{n, l, m} \cdot \sigma_s \tag{2.10}$$

where Ψ is the space dependent function and σ depends on the spin coordinates only. Normally, $\sigma(+\frac{1}{2})$ is designated as α and $\sigma(-\frac{1}{2})$ as β functions.

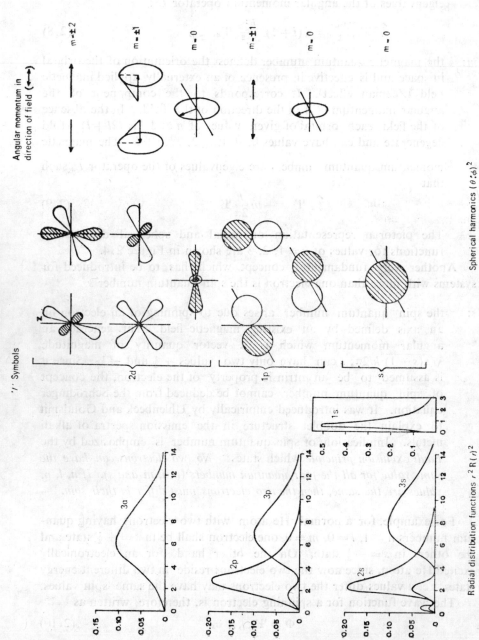

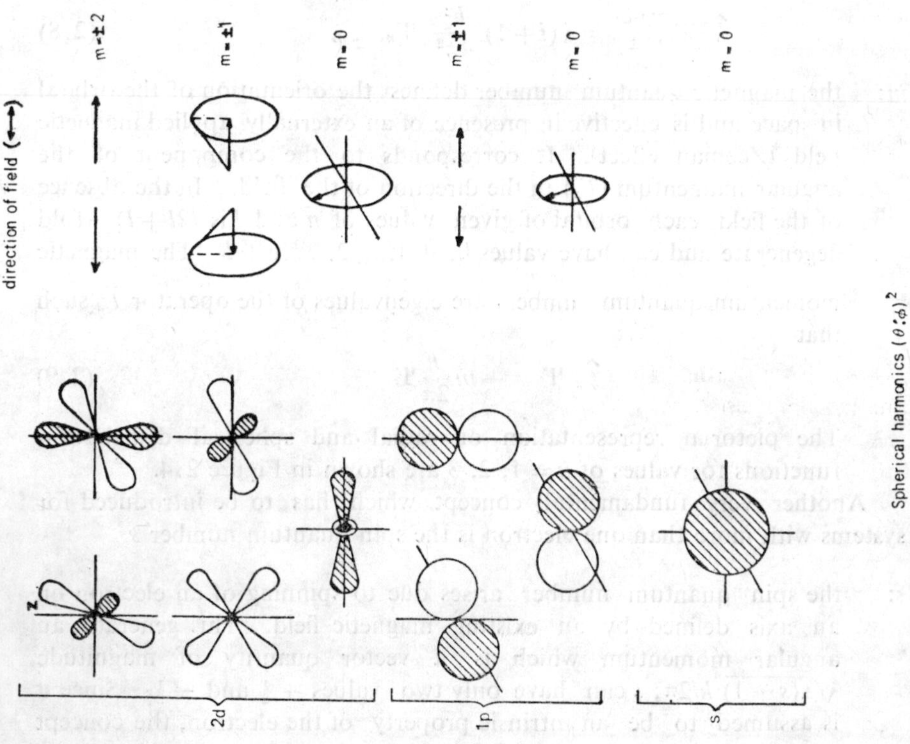

Figure 2.4 Radial and spherical distribution functions for atomic orbital

2.5.1 Interaction of Spin and Orbital Angular Momenta

The spin and orbital angular momenta of the electron are expected to interact with each other. The resultant angular momentum can be predicted by the vector addition rule and can have as many possibilities as the quantized orientations of the orbitals given by the values of m. With only one electron in the unfilled energy shell, the orbital angular momentum, $\sqrt{l(l+1)}\,h/2\pi$ and the spin angular momentum, $\sqrt{s(s+1)}\,h/2\pi$, vectorially add to give the resultant as $\sqrt{j(j+1)}\,h/2\pi$, where j is the total angular momentum quantum number. For the one-electron system, there can be only two values of j for any given l state: $j = l + \frac{1}{2}$ or $l - \frac{1}{2}$, except for $l = 0$, i.e. for an s electron the absolute value of $j = \frac{1}{2}$. For sodium atom with one electron in the valence shell, possible j value are 1/2, 3/2, 5/2, etc.

When the number of electrons is more than one, there are more than one possibilities for such interactions. In a completed shell or a subshell the contributions of individual electrons cancel each other and the total angular momentum is zero. For two electrons in an unfilled shell where the orbital angular momenta are denoted by l_1 and l_2 and spin angular momenta by s_1 and s_2, the possible interactions are:

(1) l_1 with l_2 to give L, and s_1 with s_2 to give S, followed by interaction between L and S to give J.

(2) l_1 with s_1 to give j_1 and l_2 with s_2 to give j_2, followed by interaction between j_1 and j_2 to give J.

The interactions of type (1) are known as *L-S coupling* or *Russell-Saunders* coupling. From the vector addition rule and the constraint that the values must differ by one in the unit of $h/2\pi$, the possible values of L and S are:

$$L = (l_1 + l_2), (l_1 + l_2 - 1), \ldots, |l_1 - l_2| \quad (2.11)$$

$$S = (s_1 + s_2), (s_1 + s_2 - 1), \ldots, |s_1 - s_2| \quad (2.12)$$

Therefore, $$J = L + S, L + S - 1, \ldots, |L - S| \quad (2.13)$$

For N number of electrons, the vector addition is carried out one by one to get the total L. In the same way the total S is obtained. The two are finally coupled to get the total J. There will be $2S + 1$ values of J when $L > S$ and $2L + 1$ values when $L < S$.

For multielectron atoms the symbols for L values are: S $(L = 0)$, P $(L = 1)$, D $(L = 2)$, F $(L = 3)$ similar to lower case symbols, s, p, d, f for one-electron atoms. The symbol S when $L = 0$, or s, when $l = 0$ should not be confused with the spin quantum number S for multielectron system and s for individual spin quantum number.

The type (2) spin orbit interaction is known as the $j - j$ coupling. For each electron, j can have values $(l + s) > j > (l - s)$. These j values further couple to give total J. The j-j coupling is observed for heavier atoms ($Z > 30$).

$$j_i = (l_i + s_i), (l_i + s_i - 1), (l_i + s_i - 2), \ldots, | l_i - s_i | \qquad (2.14)$$

$$J = (j_1 + j_2), (j_1 + j_2 - 1), \ldots, | j_1 - j_2 | \qquad (2.15)$$

where l_i and s_i are one electron orbital and spin angular momentum numbers respectively for the ith electron and i can be 1, 2, 3, 4, etc. In those cases where the j-j coupling is observed, the energy state may be represented by J value alone as this is now a good quantum number.

Spin-orbit interactions cause splitting of the energy states into $(2S + 1)$ values. These are known as the *multiplicity* of a given energy state. For one-electron atoms only doublet states are possible; for two-electron atoms, singlet and triplet states arise; for three-electron atoms, doublets and quartets can occur; for four-electron atoms singlet, triplet and quintet states are generated and so on. Odd number of electrons give rise to even multiplicity, whereas even number of electrons give rise to odd multiplicity. The complete description of the energy state of an atom is represented by the term symbol,

$$n\,^{2S+1}L_J$$

For example, 6^3P_1 (six triplet P one) state of mercury signifies that the total energy of the state corresponds to $n = 6$; the orbital angular momentum is $L = 1$; the multiplicity is three; hence it is a triplet energy state and the spins of the two valence electrons must be parallel ($S = 1$) and the particular value of J is 1 ($J = 1$). Since a normal mercury atom, has a pair of electrons with opposed spin in the S orbital, this must be an excited energy state, where a $6S$ electron is promoted to a $6P$ state.

In the same way, the sodium atoms can be promoted to the doublet levels $3\,^2P_{1/2}$ or $3\,^2P_{3/2}$ which are split due to spin-orbit coupling. When such atoms return to the ground state, the two closely spaced lines are observed in the emission spectrum. These are the well known D lines of sodium (Figure 2.5).

The various term values that are obtained by vector addition of orbital and spin angular momenta are energetically different. A suitable guide for energy level scheme is provided by *Hund's rules* as follows:

RULE 1: For terms resulting from equivalent electrons, those with the highest multiplicity will be the most stable.

RULE 2: Among the levels having the same electron configuration and the same multiplicity, the most stable state is the one with the largest L value.

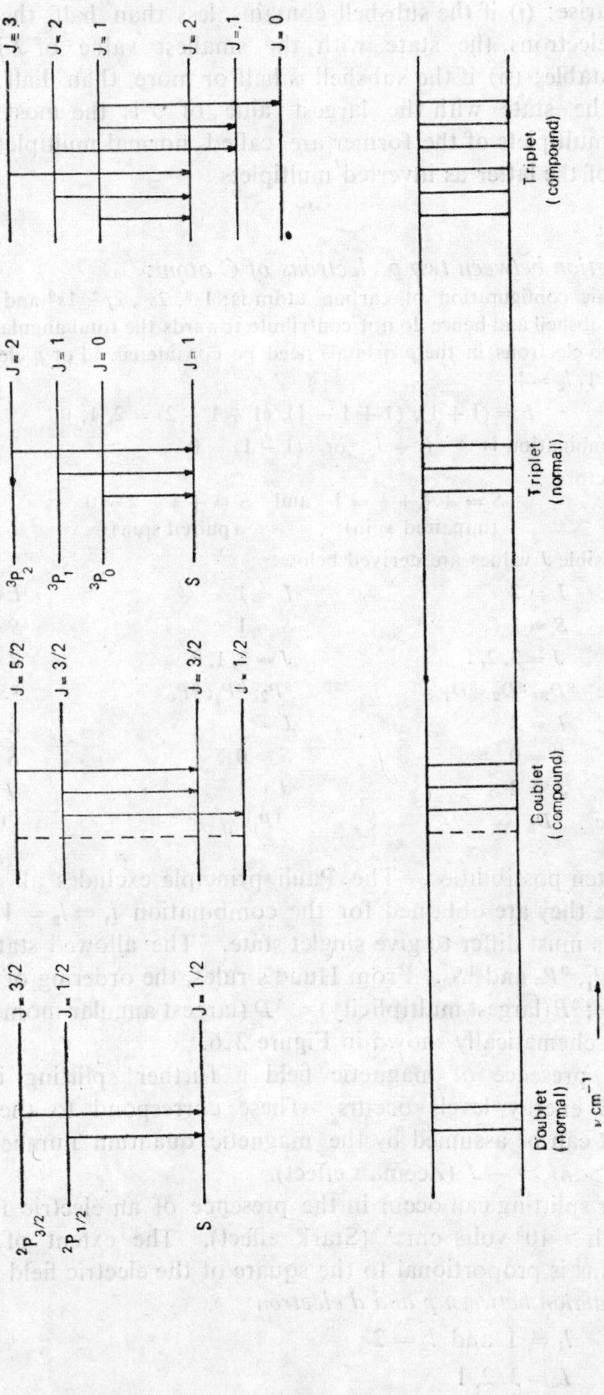

Figure 2.5 Origin of doublets and triplets in the atomic emission spectra.

RULE 3: In the case of states with given L and S values two situations arise: (i) if the subshell contains less than half the number of electrons, the state with the smallest value of J is the most stable; (ii) if the subshell is half or more than half filled, then the state with the largest value of J is the most stable. The multiplets of the former are called normal multiplets and those of the latter as inverted multiplets.

Examples

(*i*) *Interaction between two p electrons of C atom*:

The electronic configuration of carbon atom is: $1s^2$, $2s^2$, $2p^2$. $1s^2$ and $2s^2$ form the completed subshell and hence do not contribute towards the total angular momentum. Only the two electrons in the p orbitals need be considered. For p electrons, $l = 1$. Hence, $l_1 = 1, l_2 = 1$.

$$L = (1 + 1),\ (1 + 1 - 1),\ (1 + 1 - 2) = 2, 1, 0$$

The last combination is $l_1 - l_2$ or $(1 - 1) = 0$

For spin vectors,

$$S = +\tfrac{1}{2} + \tfrac{1}{2} = 1 \quad \text{and} \quad S = +\tfrac{1}{2} - \tfrac{1}{2} = 0$$

(unpaired spin) (paired spin)

The possible J values are derived below:

	$L = 2$	$L = 1$	$L = 0$
	$S = 1$	$S = 1$	$S = 1$
	$J = 3, 2, 1$	$J = 2, 1, 0$	$J = 1$
Energy state	${}^3D_3, {}^3D_2, {}^3D_1$	${}^3P_2, {}^3P_1, {}^3P_0$	${}^3S_1$
	$L = 2$	$L = 1$	$L = 0$
	$S = 0$	$S = 0$	$S = 0$
	$J = 2$	$J = 1$	$J = 0$
Energy state	${}^1D_2$	${}^1P_1$	${}^1S_0$

There are ten possibilities. The Pauli principle excludes all the three ${}^3D$ states since they are obtained for the combination $l_1 = l_2 = 1\,(L = 2)$ and hence spins must differ to give singlet state. The allowed states for C are ${}^1D_2$, ${}^3P_2$, ${}^3P_1$, ${}^3P_0$ and ${}^1S_0$. From Hund's rules, the ordering of these energy level will be: ${}^3P$ (largest multiplicity) $<$ ${}^1D$ (largest angular momentum) $<$ ${}^1S$. These are schematically shown in Figure 2.6.

In the presence of magnetic field a further splitting into $(2J + l)$ equispaced energy levels occurs. These correspond to the number of values that can be assumed by the magnetic quantum number M ranging from $+J > M > -J$ (Zeeman effect).

Similar splitting can occur in the presence of an electric field typically of strength $> 10^5$ volts cm^{-1} (Stark effect). The extent of splitting of energy terms is proportional to the square of the electric field strength.

(*ii*) *Interaction between p and d electron*

$$l_1 = 1 \text{ and } l_2 = 2$$

$$L = 3, 2, 1$$

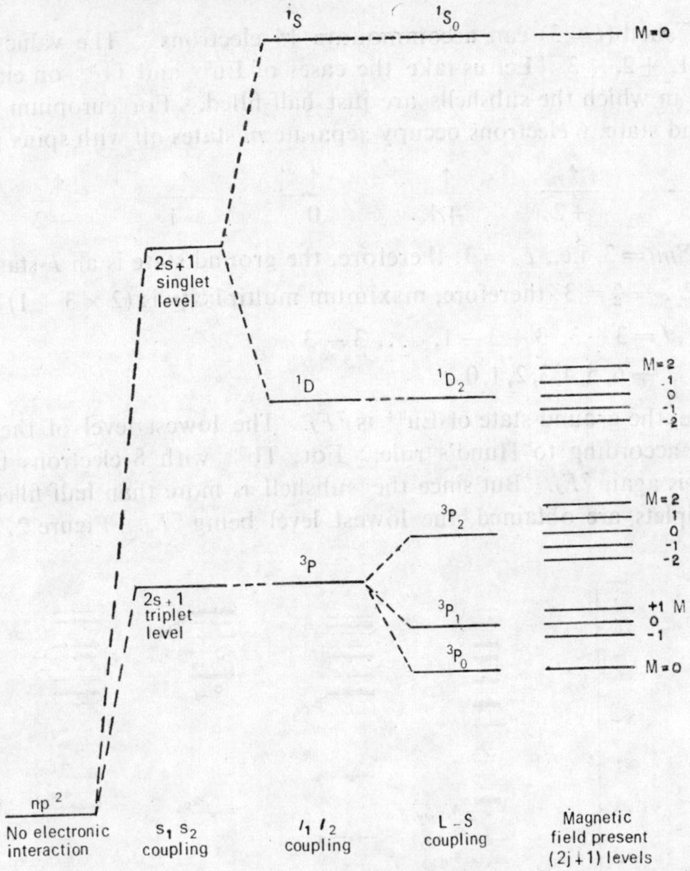

Figure 2.6 Energy levels for the electron configuration $(np)^2$, e.g. carbon, illustrating spin-orbit coupling and Hund's rules. (Adapted from Eyring, Walter and Kimball, *Quantum Chemistry*, Wiley, New York, 1946.)

$S = 1, 0$

$J = 4, 3, 2$, when $L = 3$ and $S = 1$

(*iii*) *Interaction between two d electrons*

$l_1 = 2, l_2 = 2$

$L = 4, 3, 2, 1, 0$

$S = 1, 0$

$J = 5, 4, 3$, when $L = 4$ and $S = 1$

(*iv*) *The energy levels of rare earth ions La^{3+}.* In rare earths or lanthanide ions, the *f* electronic shell is being gradually built up. The number of *f* electrons for the first nine members of the series is given as:

	Ce^{3+}	Pr^{3+}	Nd^{3+}	Pm^{3+}	Sm^{3+}	Eu^{3+}	Gd^{3+}	Tb^{3+}	Dy^{3+}
No. of *f* electrons	1	2	3	4	5	6	7	8	9

An f shell ($l = 3$) can accommodate 14 electrons. The values of m are 0, ± 1, ± 2, ± 3. Let us take the cases of Eu^{3+} and Tb^{3+} on either side of Gd^{3+} in which the subshells are just half-filled. For europium ion in the ground state, 6 electrons occupy separate m states all with spins parallel:

↑	↑	↑	↑	↑	↑	
+3	+2	+1	0	−1	−2	−3

$\Sigma m_l = 3$, i.e., $L_z = 3$; therefore, the ground state is an F-state.

$\Sigma\, s = \frac{6}{2} = 3$; therefore, maximum multiplicity is $(2 \times 3 + 1) = 7$.

And $J = 3 + 3,\ 3 + 3 - 1, \ldots, 3 - 3$

$= 6, 5, 4, 3, 2, 1, 0$

Hence, the ground state of Eu^{3+} is ^7F_J. The lowest level of the multiplet is $^7F_0$ according to Hund's rule. For, Tb^{3+} with 8 electrons the ground state is again ^7F_J. But since the subshell is more than half-filled, inverted multiplets are obtained, the lowest level being $^7F_6$. Figure 2.7 gives the

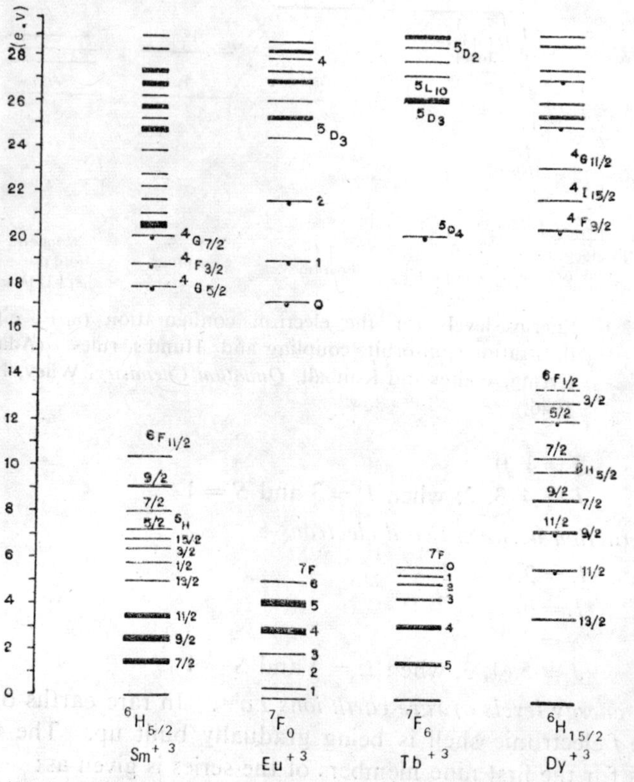

Figure 2.7 Energy levels of trivalent rare earth ions. (A.P.B. Sinha—"Fluorescence and Laser Action in Rare Earth Chelates" in Spectroscopy in Inorganic Chemistry Ed. CNR Rao and JR Ferraro.)

energy level schemes for Sm^{3+}, Eu^{3+} and Tb^{3+}. These ions are paramagnetic.

2.5.2 Inverted Multiplets

Oxygen atom with p^4 effective electron configuration has terms similar to those of carbon with p^2 effective configuration. But since the subshell is more than half-filled for oxygen, the multiplet manifold is inverted $^3P_2$, $^3P_1$, $^3P_0$. For sodium atom, $3^2P_{1/2}$ level lies below $3^2P_{3/2}$ but for chlorine atom the order is reversed. The case for Tb^{3+} is already mentioned above.

2.6 THE SELECTION RULE

The transition between the possible electronic energy states is governed by certain selection rules initially derived empirically. These are:
In an electronic transition:

(i) there is no restriction on changes in n; $\Delta n =$ any value
(ii) S can combine with its own value; $\Delta S = 0$
(iii) L can vary by 0 or ± 1 unit; $\Delta L = 0, \pm 1$
(iv) J can vary by 0 or ± 1 except that $J = 0$ to $J = 0$ transition is not allowed; $\Delta J = 0, \pm 1$ (except $0 \not\rightarrow 0$)

A basis for these empirical observations is provided by quantum mechanics according to which an odd term can combine with an even term and vice versa. This selection rule is known as *Laporte's rule*. Quantum mechanical justification for this rule is given in the next chapter. A convenient mode of representing these selection rules is the Grotian diagram for the energy states of an atom. Such a diagram for Hg atom is given in Figure 2.8.

The allowed transitions are between adjacent columns of energy states. The singlet and triplet manifolds are separated as they are forbidden by spin selection rules. Under certain conditions they do occur with reduced efficiency, as for example, the transitions between 6^1S and 6^3P states of mercury. They are indicated by dashed lines in the diagram. The wavelength associated with each transition is indicated in Å units.

2.7 DIATOMIC AND POLYATOMIC MOLECULES

Molecules differ from atoms in having more than one nuclei. These nuclei can vibrate with respect to each other and can also rotate around the molecular axes. Since vibrational and rotational energies are also quantized, they give rise to discrete energy levels which can be calculated from the Schrödinger equation. The differences in quantized energy levels for vibrational energy and those for rotational energy are respectively smaller by nearly 10^2 and 10^4 times than those for the electronic energy

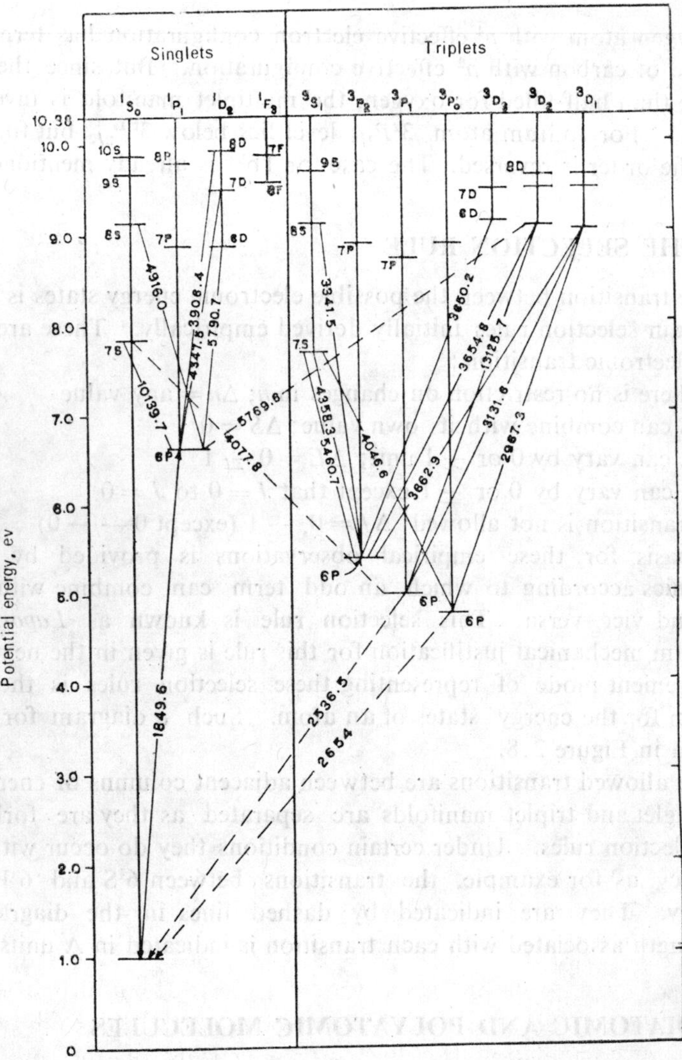

Figure 2.8 Grotian diagram for Hg atom. Wavelengths are in Å units.

levels. Therefore, the changes associated with rotational transitions only are observed in the far infrared region and those with vibration and rotation in the near infrared. The electronic transitions require energies in the visible and ultraviolet regions of the electromagnetic radiation and are accompanied by simultaneous changes in the vibrational and rotational quantum numbers.

In principle, it should be possible to obtain the electronic energy levels of the molecules as a solution of the Schrödinger equation, if inter-electronic and internuclear cross-coulombic terms are included in the potential energy for the Hamiltonian. But the equation can be solved only if it can be broken up into equations which are functions of one variable at a time. A simplifying feature is that because of the much larger mass of the nucleus the motion of the electrons can be treated as independent of that of the nucleus. This is known as the *Born-Oppenheimer approximation*. Even with this simplification, the exact solution has been possible for the simplest of molecules, that is, the hydrogen molecule ion, H_2^+ only, and with some approximations for the H_2 molecule.

The variation of total energy of the system on approach of two atoms towards each other to form a diatomic molecule when plotted as a function of internuclear distance R is given in Figure 2.9.

At the equilibrium distance r_e, the electrostatic attraction terms balance the repulsion terms. This equilibrium distance is identified as the *bond length* of the molecule and the curve is known as the *potential energy diagram*. If no attractive interaction is possible, then no bond formation is predicted and the potential energy curve shows no minimum.

The wave functions for the molecular systems are described in terms of the atomic orbitals of the constituent atoms. The molecular orbitals or MOs are obtained as algebraic summation or linear combination of atomic orbitals (LCAO) with suitable weighting factors (LCAO—MO method)

$$\Psi_{MO} = C_1\phi_1 + C_2\phi_2 + C_3\phi_3 + \dots \tag{2.16}$$
$$= \Sigma C_\nu \phi_\nu$$

where C_ν is the coefficient of the νth atom whose atomic orbital is described by the function ϕ_ν. For the simplest molecule. H_2^+, there are two possible modes of combination for the two 1s orbitals of the two H atoms, A and B:

$$\Psi_+ = C_1\phi_A + C_2\phi_B \qquad \text{(in-phase)}$$
$$\Psi_- = C_3\phi_A - C_4\phi_B \qquad \text{(out-of-phase)}$$

For the homopolar diatomic case, $C_1 = C_2 = C_3 = C_4$, the corresponding energy levels will be equally above and below the atomic energy level. For heteropolar molecules, the splitting will be unequal.

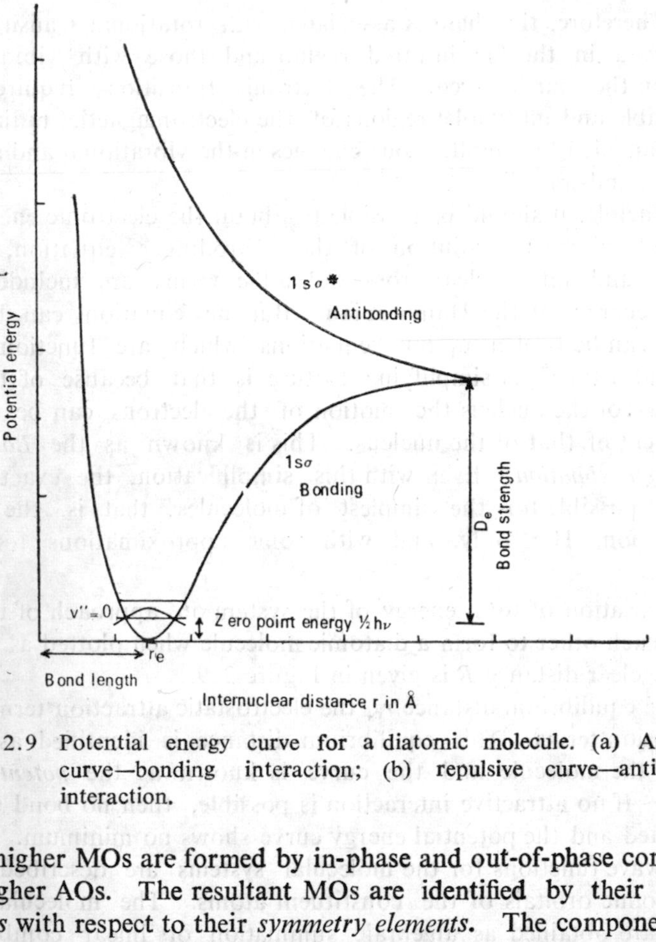

Figure 2.9 Potential energy curve for a diatomic molecule. (a) Attractive curve—bonding interaction; (b) repulsive curve—antibonding interaction.

The higher MOs are formed by in-phase and out-of-phase combination of the higher AOs. The resultant MOs are identified by their *symmetry properties* with respect to their *symmetry elements*. The component of the angular momentum in the direction of the bond is now more important and for a single electron, is designated by λ. It can have values 0, ± 1, ± 2, etc. represented by σ, π, δ, etc. λ has the same meaning as the quantum number m in the atomic case as can be shown by compressing the two hydrogen nuclei to the extent that they coalesce to give ^2He, helium atom with mass 2. Figure 2.10 illustrates the construction of MOs by bonding and antibonding combinations of s and p atomic orbitals. A σ orbital is symmetric with respect to *reflection* on a plane passing through the molecular axis. A π orbital is antisymmetric to this operation and has a *nodal plane* perpendicular to the bond axis. The letters g and u stand for *gerade* or symmetric and *ungerade* or antisymmetric function respectively under the operation of *inversion* at the point of symmetry located at the bond axis. For a g orbital, the wave function does not change sign if a point x, y, z on the left side of the inversion centre is transferred to the right side. For a u orbital, the wave function changes sign, $\Psi_{(x)} = -\Psi_{(-x)}$

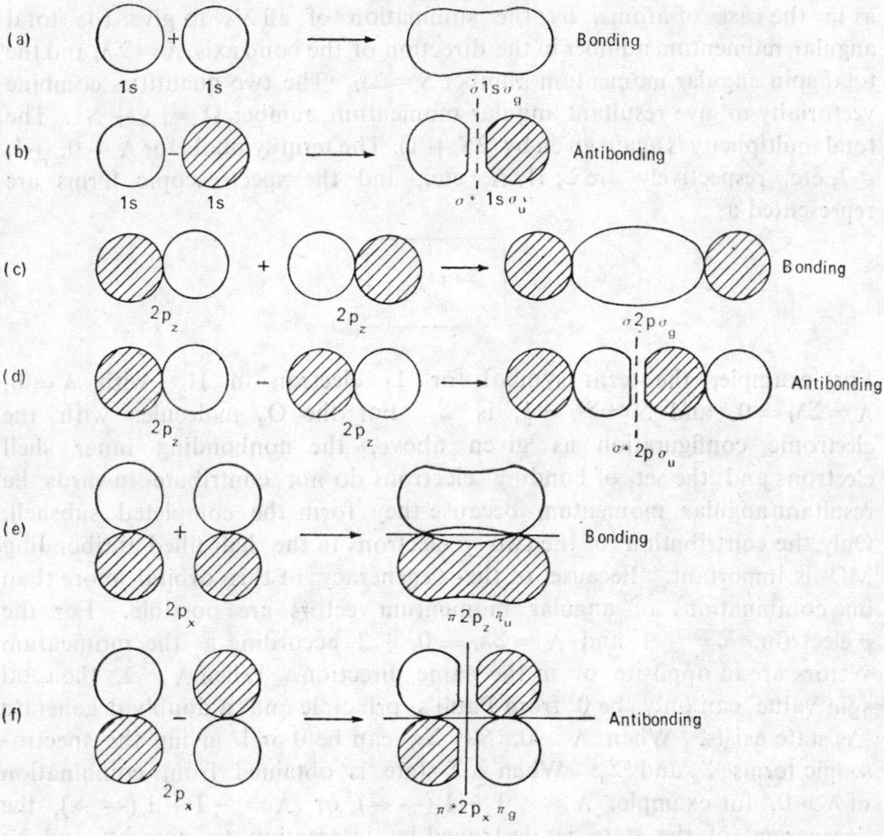

Figure 2.10 Formation of molecular orbitals (MOs) from atomic orbitals (AOs).

when the coordinates are $-x, -y, -z$. These subscripts are useful for centrosymmetric systems only such as homonuclear diatomic molecules. As in the case of atoms, the available electrons for the molecules are gradually fed into these energy levels obeying Pauli's principle and Hund's rule to obtain the complete electronic configuration of the molecule. For oxygen molecule with 16 electrons, the ground state can be represented in terms of nonbonding inner electrons, bonding and antibonding electrons:

$$\overbrace{(1s\sigma_g)^2\,(1s\sigma_u)^2\,(2s\sigma_g)^2}^{\text{nonbonding}}\;\overbrace{(2s\sigma_u)^2\,(2p\sigma_g)^2\,(2p_{x,y}\pi_u)^4}^{\text{bonding}}\;\overbrace{(2p_{x,y}\pi_g)^2}^{\text{antibonding}}$$

Mulliken notation: $K \quad K \quad z\sigma_g^2 \quad y\sigma_u^2 \quad x\sigma_g^2 \quad w\pi_u^4 \quad v\pi_g^2$

2.8 SPECTROSCOPIC TERMS FOR ELECTRONIC STATES

The spectroscopic term symbols for the molecular case can be obtained, as in the case of atoms, by the summation of all λ's to give the total angular momentum number in the direction of the bond axis, $\Lambda = \Sigma\lambda_i$ and the total spin angular momentum number $S = \Sigma s_i$. The two quantities combine vectorially to give resultant angular momentum number $\Omega = |\Lambda + S|$. The total multiplicity is again given by $(2S + 1)$. The term symbols for $\Lambda = 0, \pm 1, \pm 2$, etc. respectively are Σ, Π, Δ, etc., and the spectroscopic terms are represented as

$$^{2S+1}\Lambda_{\Omega}$$

For example, the term symbol for $1s$ electron in H_2^+ with $\lambda = 0$, $\Lambda = \Sigma\lambda_i = 0$, and $S = \Sigma s_i = \frac{1}{2}$, is $^2\Sigma$. For the O_2 molecule, with the electronic configuration as given above, the nonbonding inner shell electrons and the set of bonding electrons do not contribute towards the resultant angular momentum because they form the completed subshell. Only the contribution of the two p electrons in the half-filled antibonding MO is important. Because of the degeneracy of this orbital more than one combinations of angular momentum vectors are possible. For the p electron, $\lambda = \pm 1$ and $\Lambda = \Sigma\lambda_i = 0, \pm 2$ according as the momentum vectors are in opposite or in the same direction. When $\Lambda = 2$, the total spin value can only be 0 from Pauli's principle and a doubly degenerate $^1\Delta_g$ state exists. When $\Lambda = 0$, $S = \Sigma s_i$ can be 0 or 1 giving the spectroscopic terms $^1\Sigma_g$ and $^3\Sigma_g$. When a Σ state is obtained from combination of $\lambda > 0$, for example, $\Lambda = +1 - 1\ (\rightarrow\leftarrow)$ or $\Lambda = -1 + 1\ (\leftarrow\rightarrow)$, the degeneracy of the state is destroyed by interaction to give Σ^+ and Σ^- states respectively. A (+) sign indicates that the MO is symmetric with respect to the *operation of reflection* from a plane containing the molecular axis whereas the (−) sign indicates that it is antisymmetric. From the requirement that the total wavefunction including spin should be antisymmetric (antisymmetrization principle), only $^3\Sigma_g^-$ and $^1\Sigma_g^+$ states exist. From Hund's rule the ground state of O_2 will be $^3\Sigma_g^-$. The other two states $^1\Delta_g$ and $^1\Sigma_g^+$ are, respectively, 96 kJ (22.5 kcal mol^{-1}) and 163 kJ (37.7 kcal mol^{-1}) above it.

For MOs, the principal quantum number n has no meaning. According to convention, the ground state is denoted by the symbol $\widetilde{X}$, higher excited states of the same multiplicity as the ground state by A, B, C, etc. and those of different multiplicity by small letters a, b, c, etc. Thus, the ground state of oxygen in $\widetilde{X}^3\Sigma_g^-$ and the higher excited states are $a^1\Delta_g$ and $b^1\Sigma_g^+$. The unpaired electrons in the ground state account for the paramagnetic property of the oxygen molecule. The role of excited singlet oxygen $^1\Delta_g$ and $^1\Sigma_g^+$ in thermal and photochemical oxidation by molecular oxygen is being gradually realized.

In general, the electronic state of a molecule can be obtained from the direct product of the symmetry of the occupied orbitals. A doubly filled shell will always have Σ symmetry. Other combinations are given in Table 2.1.

TABLE 2.1

Direct product rule for assigning molecular symmetry from orbital symmetry

Orbital symmetry	$\lambda_1+\lambda_2=\Lambda$	Molecular symmetry
$\sigma\ \sigma$	$0+0=0$	$\Sigma+$
$\sigma\ \pi$	$0\pm1=1$	π
$\sigma\ \delta$	$0\pm2=2$	Δ
$\pi\ \pi$	$\pm1\pm1=0\pm2$	$\Sigma+$, Σ^-, Δ
$\pi\ \delta$	$\pm1\pm2=\pm1, \pm3$	Π, Φ
$\delta\ \delta$	$\pm2\pm2=0, \pm4$	$\Sigma+$, Σ^-, Γ

when the molecule is centrosymmetric, the *g* or *u* character of the product is given by: (i) $g \times g = u \times u = g$ and (ii) $g \times u = u \times g = u$.

At this point it is important to distinguish between the terms *electronic state* and *electronic orbitals*. *An electronic orbital* is defined as *that volume element of space in which there is a high probability* (99.9%) *of finding an electron.* It is calculated from the one-electron wave function and is assumed to be independent of all other electrons in the molecule. *Electronic states* signify *the properties of all the electrons in all of the orbitals.* Since the interaction between electrons is quite significant, the transition of an electron from one orbital to another will result in a change in the *electronic state* of the molecule. Therefore, it is important to consider the states of the molecule involved in such electron promotion. For example, consider the O_2 molecule again. It has 4 electrons in the $2p_{x,y}\ \pi_u$ orbital. If one electron is excited from $2p\pi_u$ to a partially filled $2p\ \pi_g$ orbital, each orbital will possess an odd electron. The possible electronic state associated with this configuration will now have the symmetries $\pi_u \times \pi_g = \Sigma^+_u, \Sigma^-_u, \Delta_u$. Since the two odd electrons now occupy separate orbitals, Hund's rule permits both the singlet and triplet states of the above symmetries and $^1\Sigma^+_u\ ^1\Sigma^-_u$, $^1\Delta_u$, $^3\Sigma^+_u$, $^3\Sigma^-_u$, $^3\Delta_u$ all are possible. The strong ultraviolet absorption of oxygen, which marks the onset of 'vacuum ultraviolet region' of the spectrum, Schumann-Runge continuum is associated with the electronic transition

$$B\ ^3\Sigma^-_u \leftarrow \tilde{X}\ ^3\Sigma^-_g$$

Because of the importance of O_2 molecule in our environmental photochemistry, various energy states and the corresponding potential functions are given in Figure 2.11.

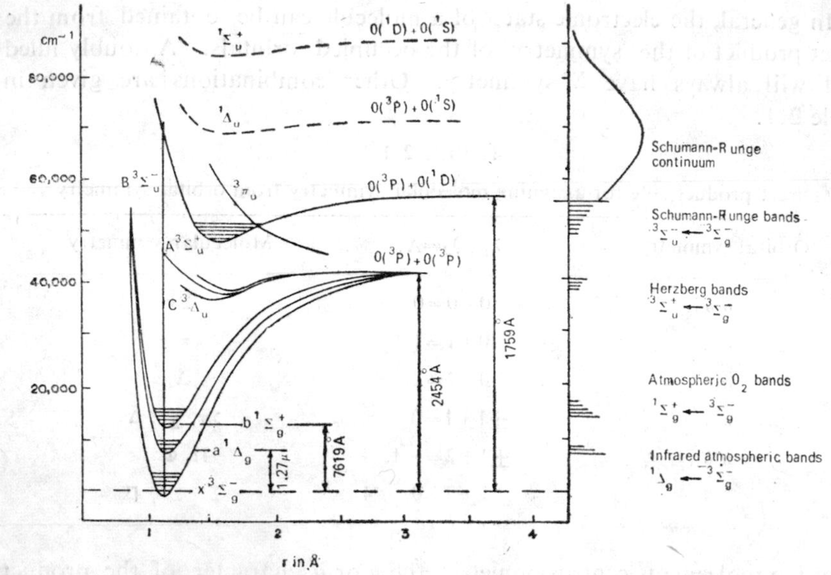

Figure 2.11 Potential energy diagrams for molecular oxygen electronic energy states and the absorption spectrum of oxygen molecule.

2.9 ORBITAL SYMMETRY AND MOLECULAR SYMMETRY

As already evident from the previous section, symmetry properties of a molecule are of utmost importance in understanding its chemical and physical behaviour in general, and spectroscopy and photochemistry in particular. The selection rules which govern the transition between the energy states of atoms and molecules can be established from considerations of the behaviour of atoms or molecules under certain symmetry operations. For each type of symmetry, there is a group of operations and, therefore, they can be treated by *group theory*, a branch of mathematics.

A *symmetry operation* is one which leaves the framework of a molecule unchanged, such that an observer who has not watched the operation cannot tell that an operation has been carried out on the molecule (of course one presupposes the structure of the molecule from other experimental sources). The geometry of the molecule is governed by the geometry of the orbitals used by the constituent atoms to form the molecule. There are five kinds of symmetry operations which are necessary for classifying a point group.

(i) Rotations about an axis of symmetry: C_p.

(ii) Reflection in a plane of symmetry: σ.

(iii) Inversion through a centre of symmetry: i.

(iv) Rotation about an axis followed by reflection in a plane perpendicular to it (also called improper rotation): S.

(v) Identity operation or leaving the molecule unchanged: *I*.

The axes, planes and centre of symmetry are known as the *elements of symmetry*. All these elements intersect at one point, the centre of gravity of the molecule, which does not change during these operations. Hence the designation *point symmetry*, in contrast to translational symmetry observed in crystals. Let us take the simple molecule, say, water to understand some of these terminologies (Figure 2.12).

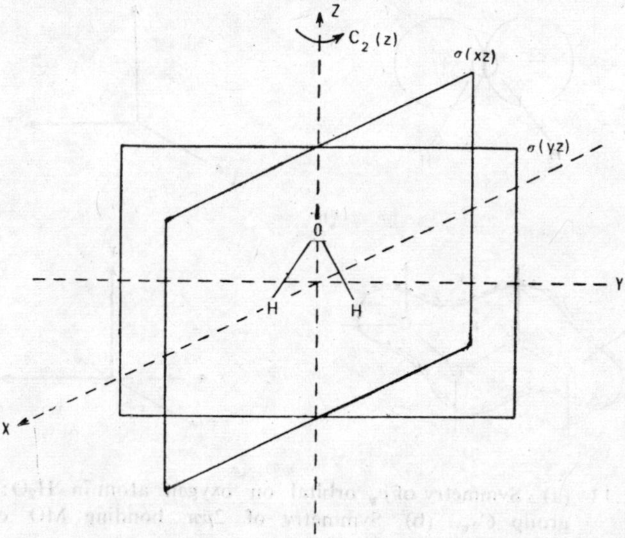

Figure 2.12 Elements of symmetry for H_2O molecule.

The water molecule has a two-fold ($p=2$) rotation axis along the *z*-direction. On complete rotation of the molecule through 360°, the molecule has indistinguishable geometry at two positions, 360°/2 and 360°. It has two planes of mirror symmetry, σ_{yz} passing through the plane of the molecule and the other σ_{xz}, bisecting the HOH bond angle. These three operations together with the identity operation *I* form the point group C_{2v} to which the water molecule belongs.

In the molecular orbital theory and electronic spectroscopy we are interested in the electronic wave functions of the molecules. Since each of the symmetry operations of the point group carries the molecule into a physically equivalent configuration, any physically observable property of the molecule must remain unchanged by the symmetry operation. Energy of the molecule is one such property and the Hamiltonian must be unchanged by any symmetry operation of the point group. This is only possible if the symmetry operator has values ± 1. Hence, the only possible wave functions of the molecules are those which are either symmetric or antisymmetric towards the symmetry operations of the group, provided the wave functions are nondegenerate. The symmetric and antisymmetric behaviours are usually denoted by + 1 and − 1 respectively, and are called the *character* of the motion with respect to the symmetry operation.

Let us examine the behaviour of p_y orbital in water under the symmetry operation of the point group C_{2v} (Figure 2.13a). Rotation around the z-axis changes sign of the wave function, hence under C_2^z, p_y orbital is

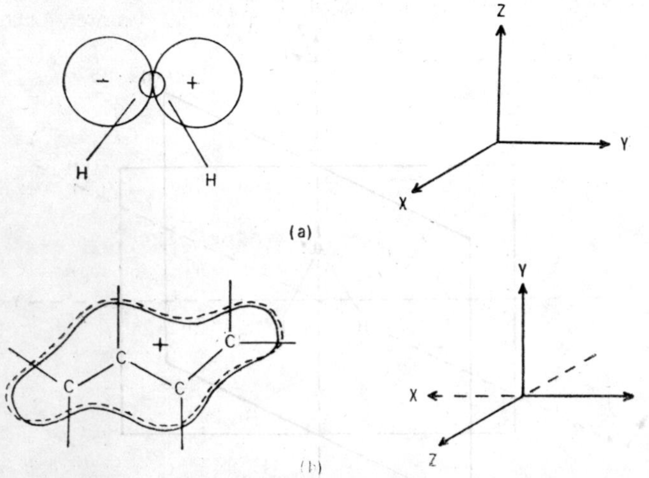

Figure 2.13 (a) Symmetry of p_y orbital on oxygen atom in H_2O: symmetry group C_{2v}. (b) Symmetry of $2p\pi$ bonding MO of *s*-trans-butadiene: symmetry C_{2h}.

antisymmetric and has the character − 1. Similarly, reflection on σ_v^{zx} changes the sign and hence is − 1, whereas σ_v^{yz} transforms the orbital into itself and hence has the character + 1. The identity operation also leaves the orbital unchanged and hence is + 1. On the other hand, p_z orbital transforms as + 1, + 1, + 1, + 1 for the operations I, C_2^z, σ_v^{xz} and σ_v^{yz}. Other possible combinations are + 1, − 1, + 1, − 1 as for p_x orbital and + 1, + 1, − 1, − 1. All this information can be put down in a tabular form, called the *character table*, for the point group C_{2v}.

TABLE 2.2

The character table for the point group C_{2v}

C_{2v}	I	C_2^z	σ_v^{xz}	σ_v^{yz}	T	R
a_1, A_1	+1	+1	+1	+1	T_z	
a_2, A_2	+1	+1	−1	−1		R_z
b_1, B_1	+1	−1	+1	−1	T_x	R_y
b_2, B_2	+1	−1	−1	+1	T_y	R_x

T=translational transformation; R=rotational transformation.

In column 1, by convention small letters are representations for electron orbital symmetries and capital letters for molecular symmetries. These four distinct behaviour patterns are called *symmetry species*. Thus symmetries of oxygen atomic orbitals are

$$2s\,(a_1),\ 2p_x\,(b_1),\ 2p_y\,(b_2),\ 2p_z\,(a_1)$$

The set of MOs are generated, taking into consideration the symmetry of the molecule and the atomic orbitals used for their formation. The net symmetry is obtained by the direct product of the symmetry species of the occupied orbitals. Thus

$$a_1 \times a_1 = (+1)^2\,(+1)^2\,(+1)^2\,(+1)^2 = a_1$$
$$b_1 \times b_1 = (+1)^2\,(-1)^2\,(+1)^2\,(-1)^2 = a_1$$
$$b_1 \times a_1 = (+1)^2\,(+1)\,(-1)\,(+1)^2\,(+1)\,(-1) = b_1$$

All doubly occupied orbitals have a_1 symmetry and if all the orbitals have paired electrons for any given molecular configuration, the state is totally symmetric and belongs to the species A_1. Species A is symmetric with respect to rotation about the z-axis and species B anti-symmetric. If there are more than one A and B species they are further given numerical subscripts.

For centrosymmetric systems with a centre of inversion i, subscripts g (symmetric) and u (antisymmetric) are also used to designate the behaviour with respect to the operation of inversion. The molecule *trans*-butadiene belongs to the point group C_{2h} (Figure 2.13b). Under this point group the symmetry operations are I, C_2^z, σ_h and i, and the following symmetry species can be generated:

TABLE 2.3

The character table for the point group C_{2h}

C_{2h}	I	C_2^z	σ_h	i
A_g	+1	+1	+1	+1
A_u	+1	+1	−1	−1
B_g	+1	−1	−1	+1
B_u	+1	−1	+1	−1

The π-orbital system of butadiene has a node in the plane of the molecule in the bonding combination and also contains σ_h plane (σ_{xy}) horizontal to the z-axis. In Figure 2.13b the molecular plane is the plane of the paper. If we consider the two $p\pi$-lobes above and below the paper and the *z*-axis perpendicular to the plane of paper, the $p\pi$-orbital of butadiene is found to belong to the symmetry species A_u.

The doubly degenerate single electron MOs are designated by the symbol *e* and the triply degenerate by *t*. Corresponding molecular symmetry species are termed *E* and *T*, respectively. Other important symmetry groups are T_d and D_{6h}.

2.10 NOTATION FOR EXCITED STATES OF ORGANIC MOLECULES

Representation by overall symmetry of the molecule is the most useful way of designating the energy states of a polyatomic molecule. Classification by the quantized component of the orbital angular momentum along the line of centres, Σ, Π, Δ is possible for linear molecules only. When details of the electronic structure of states are unknown or not necessary, the most common method is to denote them by their multiplicities, *S* (singlet) or *T* (triplet). The ground state is denoted as S_0 and higher excited states as S_1, S_2, S_3, T_1, T_2, etc. (Figure 2.14).

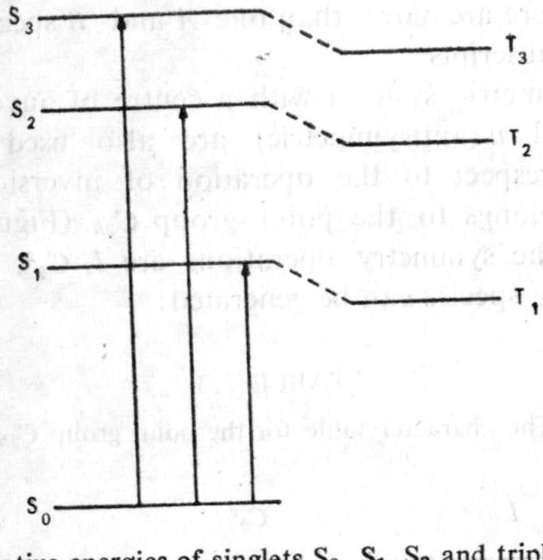

Figure 2.14 Relative energies of singlets S_0, S_1, S_2 and triplets T_1, T_2 states of a typical organic molecule.

M. Kasha has evolved still another system of notation in which electronic states are expressed in terms of the initial and final orbitals involved in a transition. This form of description is less precise than the symmetry notation but is very convenient for photochemical purposes, specially for designating energy levels of polyatomic organic compounds. In general, four types of molecules can be identified.

(i) Saturated molecules with σ-MOs only, e.g. the paraffin hydrocarbons, B_2H_6.

(ii) Saturated molecules with σ and nonbonding *n*-MOs, e.g. H_2O, NH_3, CH_3I.

(iii) Unsaturated molecules with σ and π-MOs, e.g., CO_2, C_2H_4, aromatic hydrocarbons.

(iv) Unsaturated molecules with σ, π and *n*-MOs, e.g., aldehydes, ketones, pyridine, amines and other heterocyclic compounds containing O, N, S, etc.

The σ-MOs are bonding type with axial symmetry. They are formed by overlap of *s* or p_z orbitals or hybrid orbitals like *sp*, sp^2, sp^3, etc. along the direction of the bond. They form strong single bonds with two spin paired electrons localized between the combining atoms.

The π-MOs have a nodal plane in the plane of the molecule. They are formed by the overlap of p_x, p_y or hybridized orbitals of similar symmetry. They participate in double (as in ethylene) or triple (as in acetylene) bonds using a pair of electrons with antiparallel spins. Some of these MOs are delocalized over a number of atoms as in molecules containing conjugated system of double bonds, e.g. butadiene and benzene. The π-bonds are less strong than σ-bonds but give rigidity to the molecule. σ* and π*-MOs are antibonding equivalents of σ and π-MOs, respectively. The *n*-MOs are, in general, pure atomic orbitals and do not take part in construction of MO, hence described as *n* or *nonbonding*. The pair of electrons occupying these orbitals is called *lone pair electrons* or *nonbonding electrons*. The lone pair electrons have far-reaching importance in H-bond formation which determines, for example, unusual properties of water and is responsible for the native structure of biomolecules such as deoxyribonucleic acid (DNA). They are also involved in the formation of coordinate dative bonds.

The characteristics of the lone pair electrons can vary if their nodal plane has a suitable geometry to conjugate with the π orbitals of the rest of the conjugated molecule. Since now lone pair orbitals are no longer nonbonding, in such cases they are designated as *l*-orbitals. Types of "lone pair" electrons in heteroatomic molecules described by Kasha are given in Figure 2.15.

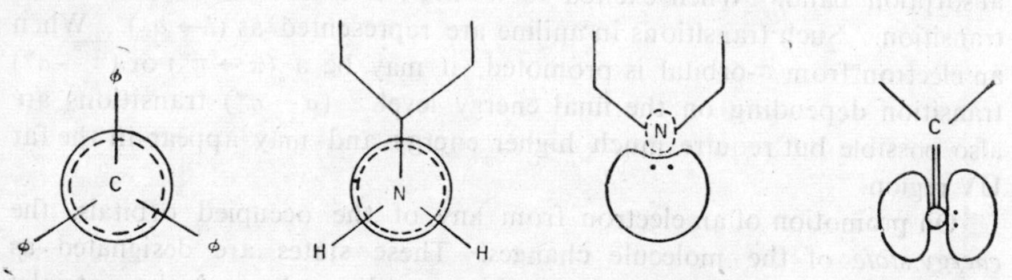

Figure 2.15 Types of lone pair electrons. (a) *l*-orbital (lone pair); (b) *n*-orbital (nonbonding).

A clear picture of all these different types of MOs can be obtained from the discussion of the formaldehyde molecule: $H_2C{=}O$. C uses sp^2 hybrid orbitals to form σ-bonds with 1*s* orbital of 2 hydrogen atoms and $2p_z$ orbital of oxygen (2*s* AO is supposed to be localized on oxygen). Pure p_x AOs of C and O form π-bonds. $2p_y$ of O is nonbonding. The formation of >C=O bond can be represented by the energy level diagram (Figure 2.16). The six electrons, four contributed by oxygen and two by carbon are accommodated in the three lower energy levels of σ, π and *n* character.

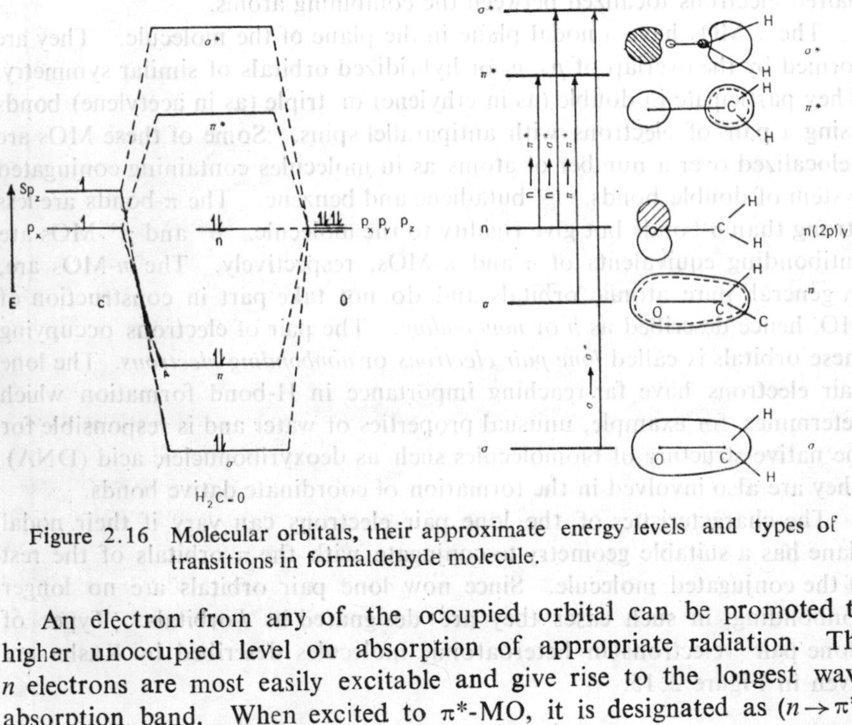

Figure 2.16 Molecular orbitals, their approximate energy levels and types of transitions in formaldehyde molecule.

An electron from any of the occupied orbital can be promoted to higher unoccupied level on absorption of appropriate radiation. The *n* electrons are most easily excitable and give rise to the longest wave absorption band. When excited to π*-MO, it is designated as $(n \rightarrow \pi^*)$ transition. Such transitions in aniline are represented as $(l \rightarrow a_\pi)$. When an electron from π-orbital is promoted, it may be a $(\pi \rightarrow \pi^*)$ or $(\pi \rightarrow \sigma^*)$ transition depending on the final energy level. $(\sigma \rightarrow \sigma^*)$ transitions are also possible but require much higher energy and may appear in the far UV region.

On promotion of an electron from any of the occupied orbitals, the *energy state* of the molecule changes. These states are designated as (n, π^*), (π, π^*), (n, σ^*), (σ, σ^*), (σ, π^*), etc. and may have singlet or triplet character. In the ground state when all the bonding MOs are doubly occupied, the Pauli principle predicts only the singlet state. Once the electrons are orbitally decoupled on excitation, spin restrictions are lifted. Both *S* singlet (spin paired) and *T* triplet (spin parallel) states are possible. Thus, the possible energy states are $^1(n, \pi^*)$, $^3(n, \pi^*)$, $^1(\pi, \pi^*)$, $^3(\pi, \pi^*)$, etc.

whose energies are dictated by the nature of the molecule. According to Hund's rule, the triplet state is the lowest energy excited state. State diagrams are represented with all the singlet levels expressed as horizontal lines one above the other and the triplet levels are drawn slightly shifted in space, maintaining the order of energy values. The ground state is arbitrarily assigned a zero value for the energy. Such a diagram is also known as Jablonski diagram (Figure 2.17) and is useful in representing various photophysical processes that may occur after the initial act of absorption of radiation.

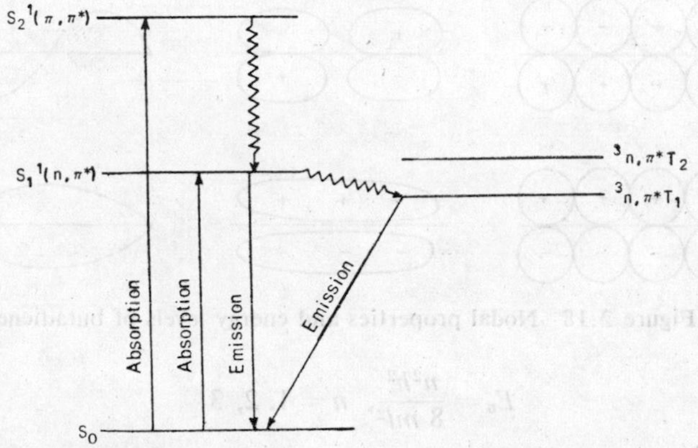

Figure 2.17 Jablonski diagram ior (π, π^*) and (n, π^*), singlet and triplet states.

The (n, π^*) state plays a very important role in the photochemistry of carbonyl compounds and many heterocyclic systems. In conjugated hydrocarbons ($\pi \rightarrow \pi^*$) transitions are most important and give intense characteristic absorption bands. Because of some overlap forbidden character, the ($n \rightarrow \pi^*$) transitions have low probability and hence weak absorption bands.

2.10.1 Unsaturated Molecules with Conjugated System of Double Bond

In simple conjugated hydrocarbons, carbon utilizes sp^2 hybrid orbitals to form σ-bonds and the pure p_x orbital to give the π-MOs. Since the σ-skeleton of the hydrocarbon is perpendicular to the wave functions of π-MO, only p_x AOs need be considered for the formation of π-MOs of interest for photochemists. Let us consider the case of butadiene with p_x AO contributed by 4 carbon atoms. The possible combinations are given in Figure 2.18. The energy increases with the number of nodes so that $E_1 < E_2 < E_3 < E_4$.

Immediately, one can observe the similarity of the wave functions thus obtained with that for a free particle-in-a-box. The energy values can be approximately calculated from the expression

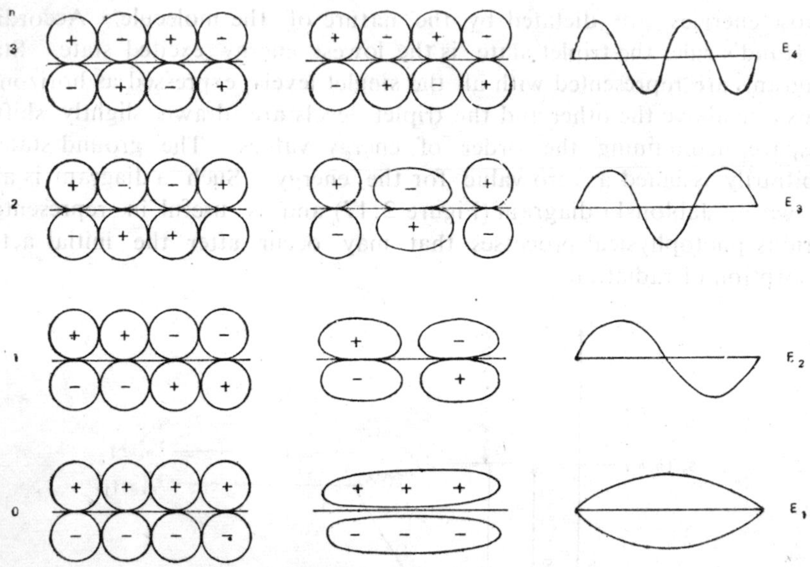

Figure 2.18 Nodal properties and energy levels of butadiene.

$$E_n = \frac{n^2h^2}{8\ ml^2}, \quad n = 1, 2, 3 \tag{2.17}$$

if l the length of the molecule can be estimated from the $-\mathrm{C}=\mathrm{C}-$ bond lengths. In the ground state of butadiene, the 4 electrons will occupy the lower two orbitals. The energy requirement for the lowest energy transition is given by:

$$\Delta E = (3^2 - 2^2)\frac{h^2}{8\ ml^2}$$

Both singlet and triplet states are generated by the orbital promotion of an electron. $\pi \rightarrow \pi^*$ transitions are totally allowed. These energy values can also be calculated from Hückel molecular orbital (HMO) method. For benzene, the free electron perimeter model has been found to be useful. The energy levels and nodal properties of benzene molecule are given in Figure 2.19.

Pyridine is isoelectronic with benzene in which C – H is replaced by N. The two electrons which formed the σ-bonds in – CH are now localized in a nonbonding hybrid orbital centred on the nitrogen atom. The substitution of N lowers the overall symmetry of the molecule from D_{6h} to C_{2v} and the degenerate e_{1g} and e_{2u} orbitals are split into a_1 and b_2 type orbitals. The MOs of pyridine are given in Figure 2.20.

2.11 ENERGY LEVELS FOR INORGANIC COMPLEXES

The MOs for inorganic complexes are obtained by combining AOs for

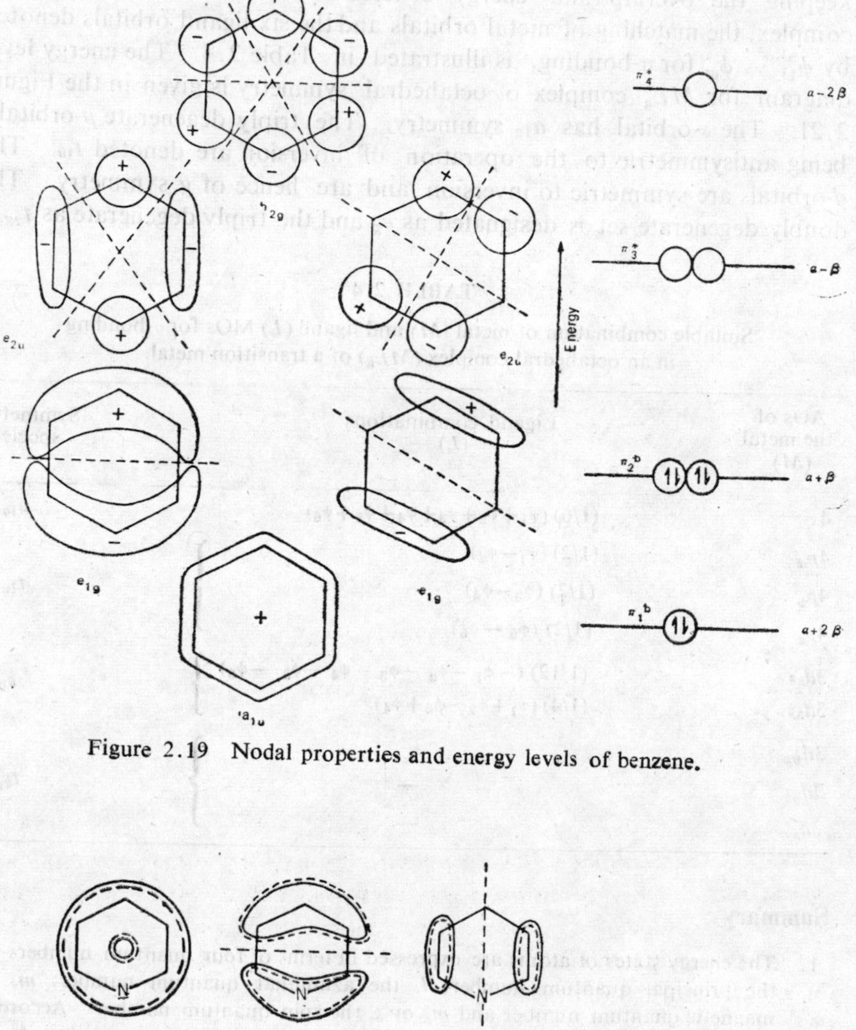

Figure 2.19 Nodal properties and energy levels of benzene.

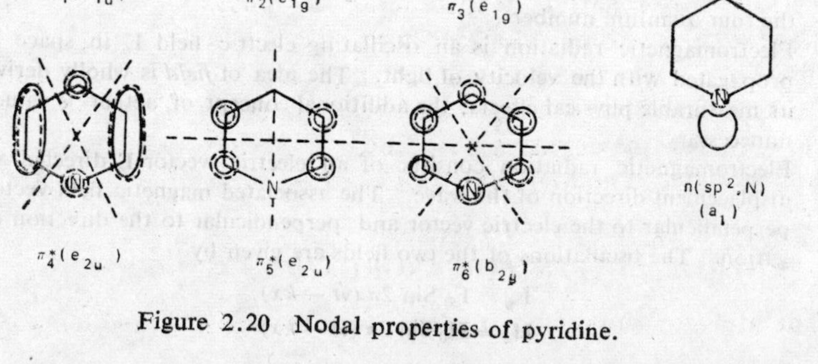

Figure 2.20 Nodal properties of pyridine.

the valence electrons of the metal ion with the AOs or MOs of the ligands, keeping the overlap and energy criteria in mind. For an octahedral complex, the matching of metal orbitals and the six ligand orbitals denoted by $\phi_1, \ldots, \phi_6$, for σ-bouding, is illustrated in Table 2.4. The energy level diagram for ML_6 complex of octahedral symmetry is given in the Figure 2.21. The *s*-orbital has a_{1g} symmetry. The triply degenerate *p*-orbitals, being antisymmetric to the operation of inversion are denoted t_{1u}. The *d*-orbitals are symmetric to inversion and are hence of *g* symmetry. The doubly degenerate set is designated as e_g and the triply degenerate as t_{2g}.

TABLE 2.4

Suitable combination of metal (*M*) and ligand (*L*) MOs for σ-bonding in an octahedral complex (ML_6) of a transition metal.

AOs of the metal (*M*)	Ligand combinations (*L*)	Symmetry species
4*s*	$(1/6)(\phi_1+\phi_2+\phi_3+\phi_4+\phi_5+\phi_6)$	a_{1g}
$4p_x$	$(1/2)(\phi_1-\phi_2)$	t_{1u}
$4p_y$	$(1/2)(\phi_3-\phi_4)$	
$4p_z$	$(1/2)(\phi_5-\phi_6)$	
$3d_{z^2}$	$(1/12)(-\phi_1-\phi_2-\phi_3-\phi_4-\phi_5-\phi_6)$	e_g
$3d_{x^2-y^2}$	$(1/4)(\phi_1+\phi_2-\phi_3+\phi_4)$	
$3d_{yx}$	—	t_{2g}
$3d_{yz}$	—	
$3d_{zx}$	—	

Summary

1. The energy states of atoms are expressed in terms of four quantum numbers : *n*, the principal quantum number; *l*, the azimuthal quantum number; *m*, the magnetic quantum number and m_s or *s*, the spin quantum number. According to Pauli's exclusion principle, no two electrons can have the same values for all the four quantum numbers.
2. Electromagnetic radiation is an oscillating electric field **E** in space which is propagated with the velocity of light. The idea of *field* is wholly derived from its measurable physical effects; the additional concept of 'aether' is valueless and unnecessary.
3. Electromagnetic radiation consists of an electric vector **E** directed along the displacement direction of the wave. The associated magnetic field vector **H**, lies perpendicular to the electric vector and perpendicular to the direction of propagation. The oscillations of the two fields are given by

$$E_y = E_0 \sin 2\pi(\nu t - kx)$$
$$H_z = H_0 \sin 2\pi(\nu t - kx)$$

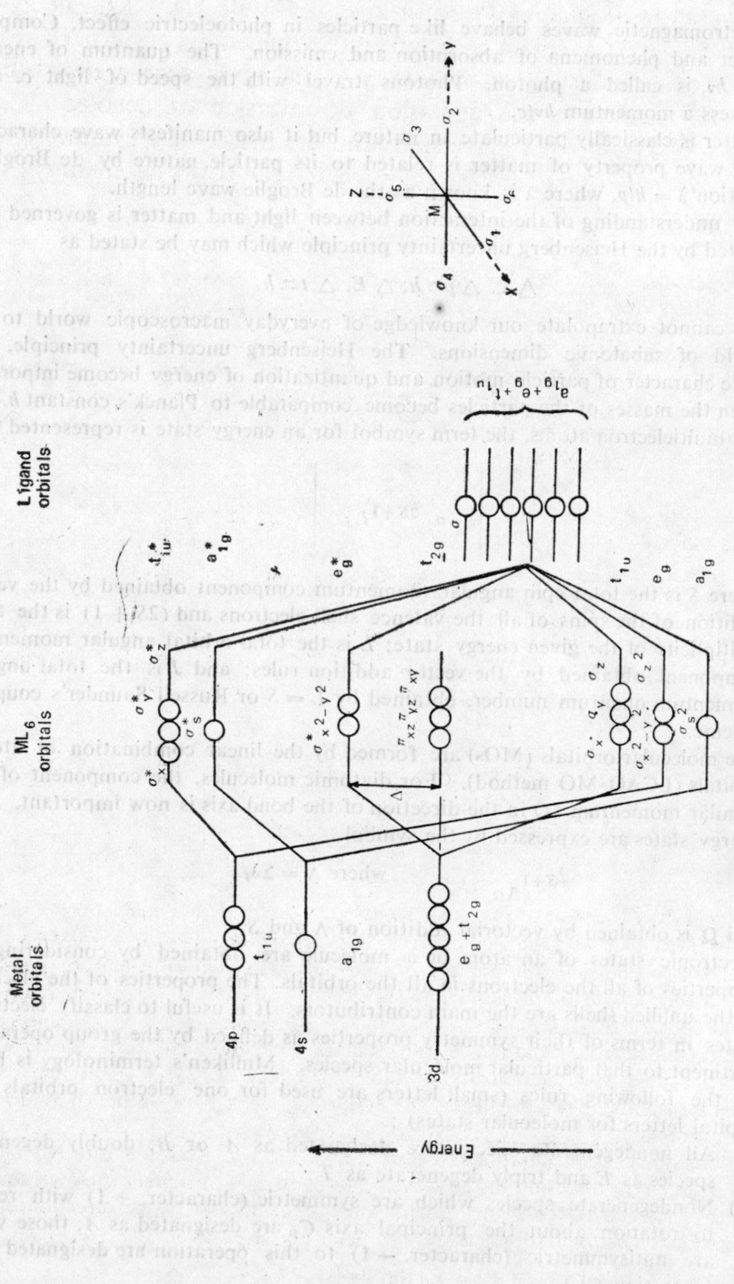

Figure 2.21 Relative molecular orbital energies of an octahedral complex ML_6 (σ-bonding with the ligands).

and $E_0/H_0 = \sqrt{\mu/\epsilon}$ μ = magnetic permeability

ϵ = dielectric constant

$k = 1/\lambda$

4. Electromagnetic waves behave like particles in photoelectric effect, Compton effect and phenomena of absorption and emission. The quantum of energy, $E = h\nu$, is called a photon. Photons travel with the speed of light c, and possess a momentum $h\nu/c$.
5. Matter is classically particulate in nature, but it also manifests wave character. The wave property of matter is related to its particle nature by de Broglie's relation $\lambda = h/p$, where λ is known as the de Broglie wave length.
6. Our understanding of the interaction between light and matter is governed and limited by the Heisenberg uncertainty principle which may be stated as

$$\Delta x . \Delta p \simeq h; \Delta E . \Delta t \simeq h$$

7. We cannot extrapolate our knowledge of everyday macroscopic world to the world of subatomic dimensions. The Heisenberg uncertainty principle, the wave character of particle motion and quantization of energy become important when the masses of the particles become comparable to Planck's constant h.
8. For multielectron atoms, the term symbol for an energy state is represented as

$$n \; {}^{2S+1}L_J$$

where S is the total spin angular momentum component obtained by the vector addition of the spins of all the valence shell electrons and $(2S + 1)$ is the total multiplicity of the given energy state; L is the total orbital angular momentum component obtained by the vector addition rules; and J is the total angular momentum quantum number, obtained by $L = S$ or Russell Saunder's coupling scheme.

9. The molecular orbitals (MOs) are formed by the linear combination of atomic orbitals (LCAO–MO method). For diatomic molecules, the component of the angular momentum (λ) in the direction of the bond axis is now important. The energy states are expressed by the symbol

$${}^{2S+1}\Lambda_{\Omega} \qquad \text{where } \Lambda = \Sigma\lambda_i$$

and Ω is obtained by vectorial addition of Λ and S.

10. Electronic states of an atom or a molecule are obtained by considering the properties of all the electrons in all the orbitals. The properties of the electrons in the unfilled shells are the main contributors. It is useful to classify electronic states in terms of their symmetry properties as defined by the group operations pertinent to that particular molecular species. Mulliken's terminology is based on the following rules (small letters are used for one electron orbitals and capital letters for molecular states) :
   (i) All nondegenerate species are designated as A or B; doubly degenerate species as E and triply degenerate as T.
   (ii) Nondegenerate species which are symmetric (character, + 1) with respect to rotation about the principal axis C_p are designated as A; those which are antisymmetric (character, − 1) to this operation are designated as B (operation C_p^1).
   (iii) If the species is symmetric with respect to a C_2 operation perpendicular to

the principal axis then subscript 1 (symmetric) or subscript 2 (antisymmetric) is added to A and B (operation C^1_{2v}).

(iv) Single primes and double primes are added to A and B when the species is symmetric or antisymmetric respectively to reflection in a plane perpendicular to the principal axis of symmetry (operation σ_h).

(v) For centrosymmetric molecules with a centre of inversion i, subscripts g and u are added if the species is symmetric or antisymmetric respectively to the operation of inversion through this centre (operation i).

11. In saturated and unsaturated organic molecules, singlet and triplet energy states are conveniently designated as $^1(\pi, \pi^*)$, $^3(\pi, \pi^*)$, $^1(n, \pi^*)$, $^3(n, \pi^*)$ etc. according as they are generated by $(\pi \rightarrow \pi^*)$ transition or $(n \rightarrow \pi^*)$ transition. For most common terminology, S_0, S_1, S_2, ... and T_1, T_2, T_3, etc. are used for singlet and triplet states.

12. For inorganic complex compounds, group theoretical nomenclatures are used.

THREE

Mechanism of Absorption and Emission of Radiation of Photochemical Interest

3.1 ELECTRIC DIPOLE TRANSITIONS

When electromagnetic radiation falls on an atom or a molecule, the electric field of the radiation tends to disturb the charge cloud around the atom or the molecule. The situation is analogous to the case when a particle composed of positive and negative charges is brought near an electric field. When the field is applied from the upper side (Figure 3.1),

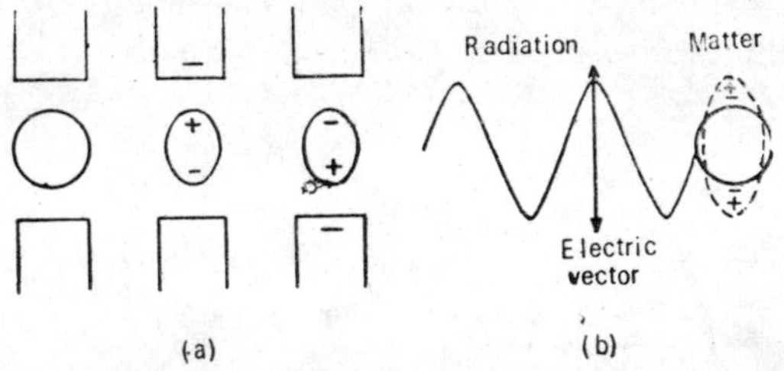

Figure 3.1 Creation of an oscillating dipole by interaction between the charge cloud of an atom, (a) an oscillating electric field; (b) electromagnetic radiation wave.

the positive charge density is attracted towards it and the negative charge density repelled, generating a dipole moment in the particle. If now the field is applied from the lower side, the direction of the dipole is reversed.

If the field oscillates between the upper and the lower positions, the induced dipole will also oscillate with the frequency of oscillation of the field. The oscillating electric field of the electromagnetic radiation acts in a similar fashion to create an oscillating dipole in the atom or the molecule with which it is interacting. The dipole is generated in the direction of the electric vector of the incident radiation.

From electrodynamics, we know that when a positive and a negative charge oscillate with respect to each other, it becomes a source of electromagnetic radiation. The radiation emanating from the source is propagated in all directions like a sound wave from a ringing bell. A similar situation applies to the present case. The disturbed molecule becomes a source of electromagnetic radiation of the same frequency as the frequency of the incident radiation. This is the mechanism of *scattering* of radiation by a particle of molecular dimensions. The secondary radiation thus scattered uniformly in all directions interferes with the primary incident radiation. As a result, radiation waves are cancelled out by destructive interference in all directions except that of *reflection* or *refraction*.

As long as the frequency of incident radiation ν_i is not close to that of the natural frequency ν_n of the molecule as given by its energy states, the oscillations are due to forced distortion of the molecule by the electromagnetic wave. But when $\nu_i \approx \nu_n$, and a resonance condition is established between the two interacting partners (the photon and the molecule), the oscillations classically become 'free'. Under this condition a quantum of radiation or a photon is *absorbed* by the atom or the molecule, promoting it to a higher energy state. An oscillating dipole moment μ (defined by the product of charge e times the distance of separation r between the centres of gravity of positive and negative charges) is created and designated as the *transition moment*. The mechanism can be illustrated as follows for the case of a hydrogen atom (Figure 3.2). The electric vector of the plane polarized incident radiation distorts the normal hydrogen atom in its spherically symmetric s state to a state which creates a dipole in the direction of the vector field. Obviously it is a p state with a nodal plane

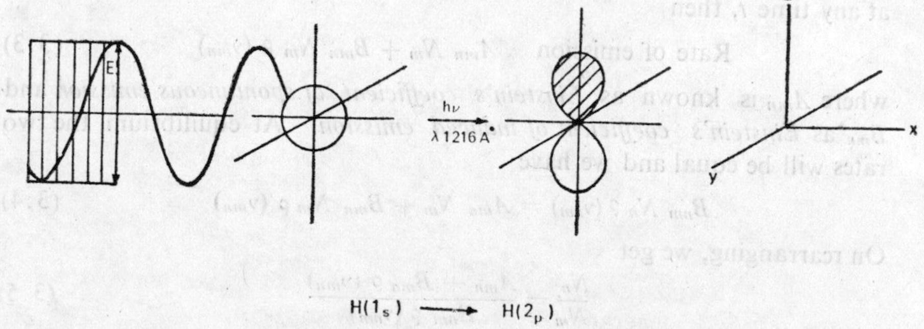

Figure 3.2 Mechanism of absorption of radiation by H atom.

perpendicular to the electric vector. Under the resonance condition, promotion to the upper state occurs with absorption of appropriate radiation. This is known as an *electric dipole transition.*

3.2 EINSTEIN'S TREATMENT OF ABSORPTION AND EMISSION PHENOMENA

According to the classical electromagnetic theory, a system of accelerated charged particles emits radiant energy. When exposed to thermal radiation at temperature T, it also absorbs radiant energy. The rates of absorption and emission are given by the classical laws. These opposing processes are expected to lead to a state of equilibrium. Einstein in 1916 treated the corresponding problem for quantized systems such as atoms and molecules in the following way.

Let us consider two nondegenerate stationary states m and n of a system, with energy values E_m and E_n where $E_m > E_n$. According to Bohr's theory, the frequency of the absorbed radiation is given by

$$\nu_{mn} = \frac{E_m - E_n}{h} \tag{3.1}$$

Let us assume that the system is in the lower state n and exposed to a radiation of density $\rho(\nu)$ defined as the energy of radiation per unit volume between the frequencies ν and $\nu + d\nu$. The probability per unit time that it will absorb the radiation and will thereby be raised to the upper state m, is proportional to the number of particles N_n in the state n and the density of radiation of frequency ν_{mn}. Hence

$$\text{Rate of absorption} = B_{nm}\, N_n\, \rho(\nu_{nm}) \tag{3.2}$$

where B_{nm} is the proportionality constant and is known as *Einstein's coefficient of absorption.*

The probability of return from m to n consists of two parts, one which is spontaneous and hence independent of radiation density and the other proportional to it. If N_m be the number of particles in the upper state m at any time t, then

$$\text{Rate of emission} = A_{mn}\, N_m + B_{mn}\, N_m\, \rho(\nu_{mn}) \tag{3.3}$$

where A_{mn} is known as *Einstein's coefficient of spontaneous emission* and B_{mn} as *Einstein's coefficient of induced emission.* At equilibrium, the two rates will be equal and we have

$$B_{nm}\, N_n\, \rho(\nu_{nm}) = A_{mn}\, N_m + B_{mn}\, N_m\, \rho(\nu_{mn}) \tag{3.4}$$

On rearranging, we get

$$\frac{N_n}{N_m} = \frac{A_{mn} + B_{mn}\, \rho(\nu_{mn})}{B_{nm}\, \rho(\nu_{nm})} \tag{3.5}$$

The ratio of the number of particles in the states n and m with energy E_n

and E_m, respectively, can also be obtained from the Boltzmann distribution law:

$$\frac{N_n}{N_m} = e^{-(E_n - E_m)/kT} = e^{h\nu/kT} \qquad (3.6)$$

Solving for $\rho(\nu_{nm})$ and substituting for N_n/N_m, from (3.5) and (3.6), we have

$$\rho(\nu_{nm}) = \frac{A_{mn}}{B_{nm}\, e^{h\nu/kT} - B_{mn}} \qquad (3.7)$$

On the other hand, from Planck's derivation of energy density for a black body radiation at temperature T (Sec. 1.4), we know that

$$\rho(\nu_{nm}) = \frac{8\pi\, h\nu^3}{c^3}\left(\frac{1}{e^{h\nu/kT} - 1}\right) \qquad (3.8)$$

Comparing (3.7) and (3.8), we find

$$B_{n \to m} = B_{m \to n} \qquad (3.9)$$

$$A_{m \to n} = \frac{8\pi\, h\nu^3}{c^3}\, B_{m \to n} \qquad (3.10)$$

3.2.1 Probability of Induced Emission and Its Application to Lasers

From the condition of equilibrium between the rates of absorption and emission, we have from equations (3.5) and (3.6)

$$\frac{N_n}{N_m} = \frac{A_{mn} + B_{mn}\,\rho(\nu_{mn})}{B_{nm}\,\rho(\nu_{nm})}$$

$$= \frac{A_{mn}}{B_{mn}\,\rho(\nu_{mn})} + 1 \quad (\text{assuming } B_{mn} = B_{nm})$$

$$= e^{h\nu/kT}$$

Hence,

$$\frac{A_{mn}}{B_{mn}\,\rho(\nu_{mn})} = e^{h\nu/kT} - 1 \qquad (3.11)$$

when $h\nu \gg kT$, 1 is negligible in comparison to $e^{h\nu/kT}$, and $A_{mn} \gg B_{mn}\,\rho(\nu)$. Therefore, at high frequency regions such as visible and near ultraviolet, spontaneous emission has a large probability but that of induced emission is very small. On the other hand, when $h\nu \ll kT$, $e^{h\nu/kT}$ can be subjected to a series expansion. Considering only the first two terms,

$$\frac{A_{mn}}{B_{mn}\,\rho(\nu_{mn})} = 1 + \frac{h\nu}{kT} + \ldots - 1 \simeq \frac{h\nu}{kT} \qquad (3.12)$$

Since $h\nu/kT$ is small, the ratio of the two transition probabilities is small and $A_{mn} \ll B_{mn}\,\rho(\nu_{mn})$. This condition is obtained in the microwave region and is utilized in the construction of *masers* (microwave amplification by stimulated emission of radiation).

The importance of induced or stimulated emission lies in the fact that it has phase coherence. The spontaneous emission is a random process obeying the unimolecular rate law and the emitted radiation is incoherent (Figure 3.3a). The divergence of radiation in all directions causes loss of light intensity on propagation. But if a beam of radiation is generated with all the photon waves vibrating in the same phase (Figure 3.3b), the energy of radiation is localized in a small angle of divergence and the intensity is preserved even on propagation to long distances. This property of stimulated emission has found immense use in maser and laser technology.

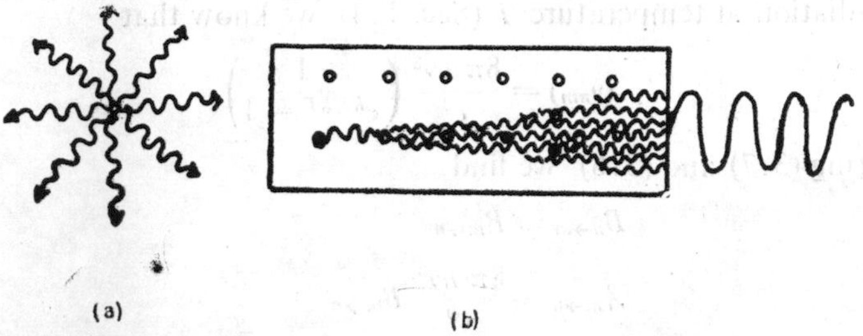

Figure 3.3 (a) Radiation from a normal source—incoherent beam, (b) Radiation from a laser source—coherent beam.

To obtain the *laser* action or light amplification by stimulated emission of radiation, the probability of induced emission in the visible region must be increased. The probability of induced emission is given as $B_{mn}N_m \rho(\nu_{mn})$. When B_{mn} is small, the rate of such emission is expected to be small since the population N_m of the upper state m is in general small. One way to increase the rate of phase-coherent emission, even when the emission probability B_{mn} is low, is to somehow increase the number of particles N_m in the higher energy state. For any given value of B_{mn}, which is equal to B_{nm}, whether absorption will take place or emission stimulated, depends on the difference $(N_n - N_m)$, where N_n and N_m are the populations of the energy states n and m, respectively. In the absence of a radiation field, the numbers are governed by the Boltzmann distribution law. For electronic energy states at ordinary temperatures, practically only the ground state is populated. When a stream of photons is allowed to impinge on such a system, absorption is preferred because N_n is large. The population of state m starts building up and after a time, the rate of absorption becomes equal to the rate of emission. A *photostationary equilibrium* is established. As long as the irradiation source is maintained the populations of the two energy levels remain constant.

On the other hand, if by some means the population of the excited state is increased, the impinging photon is more likely to encounter an excited particle than an unexcited one, thereby stimulating emission of a photon rather than its absorption. To each photon impinging on the system, an extra photon is added to the beam. Under the circumstances the emitted intensity will be larger than the incident intensity. This is known as the *laser action.* Therefore, the important aspect of laser system is the inherent possibility of bringing about *population inversion.* The radiation which promotes the molecules to upper energy state is known as pump radiation and the radiation which stimulates emission is known as *laser radiation* (Figure 3.4).

The photon thus induced to be emitted has the same phase relationship as the inducing photon. Further amplification of this coherent emission is brought about in a resonant optical cavity containing two highly reflecting mirrors, one of which allows the amplified beam to come out, either through a pin-hole or by a little transmission (Section 10.4).

3.3 TIME-DEPENDENT SCHRÖDINGER EQUATION

The phenomena of absorption and emission can be handled mathematically by the time-dependent Schrödinger equation only. It is represented in one dimension as

$$-\frac{h^2}{8\pi^2 m}\frac{\partial^2\Psi(x,t)}{\partial x^2}+V(x)\,\Psi_{(x,t)}=\frac{-h}{2\pi i}\frac{\partial\Psi(x,t)}{\partial t} \qquad (3.13)$$

or,

$$H\Psi_{(x,t)}=-\frac{h}{2\pi i}\frac{\partial\Psi(x,t)}{\partial t} \qquad (3.14)$$

where Ψ is a function of position coordinate x, and time t, and $(-h/2\pi i)\,\partial/\partial t)$ is the quantum mechanical operator for the total energy E in a nonconservative system. If we denote $\psi(x)$ as a function of x only and $\phi(t)$ as a function of t only, then

$$\Psi(x,t)=\psi(x)\,\phi(t)$$

Substituting

$$-\frac{h^2}{8\pi^2 m}\frac{\partial^2\psi(x)}{\partial x^2}\phi(t)+V(x)\,\psi(x)\,\phi(t)=\frac{-h}{2\pi i}\,\psi(x)\frac{\partial\phi(t)}{\partial t}$$

On separation of variables, we have

$$-\frac{1}{\psi(x)}\frac{h^2}{8\pi^2 m}\frac{\partial^2\psi(x)}{\partial x^2}+V(x)\,\psi(x)=\frac{-h}{2\pi i}\frac{1}{\phi(t)}\frac{\partial\phi(t)}{\partial t} \qquad (3.15)$$

Each side must be equal to a constant which is identified with energy E, and we have,

$$-\frac{h^2}{8\pi^2 m}\frac{d^2\psi(x)}{dx^2}+V(x)\,\psi(x)=E\,\psi(x) \qquad (3.16)$$

$$-\frac{h}{2\pi i}\frac{d\phi(t)}{dt}=E\,\phi(t) \qquad (3.17)$$

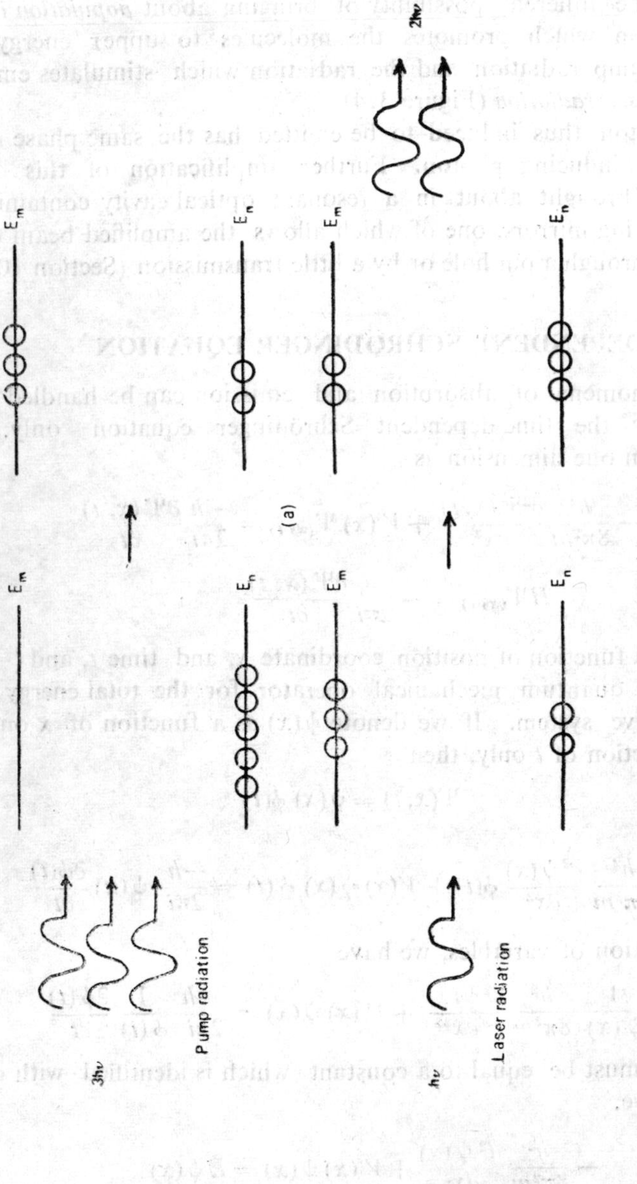

Figure 3.4 Population inversion and laser action.

The time-dependent part can be easily solved

$$\frac{d\phi(t)}{\phi(t)} = -\frac{2\pi i}{h} E\, dt$$

$$\phi(t) = \exp \frac{-2\pi i\, Et}{h} \tag{3.18}$$

The total wave function for one dimensional Schrödinger equation for the nth energy state is written as,

$$\Psi_n(x,t) = \psi_n(x)\, e^{-2\pi i E_n t/h}$$

and the probability distribution function is given by,

$$\int \Psi_n^*(x,t)\, \Psi_n(x,t)\, d\tau = \int \psi_n^*(x)\, \psi_n(x)\, e^{2\pi i E_n t/h}\, e^{-2\pi i E_n t/h}\, d\tau \tag{3.19}$$

$$= \int \psi_n^*(x)\, \psi_n(x)\, d\tau \tag{3.20}$$

The time-dependent part disappears if we consider the probability distribution in the same energy state. This defines the concept of *stationary states.* These are the allowed states that are obtained as solutions of the *time-independent* Schrödinger equation.

3.4 TIME-DEPENDENT PERTURBATION THEORY

On absorption or emission of radiation, a system goes from one stationary state to another under the influence of incident electromagnetic radiation. It is therefore, necessary to investigate how a system is disturbed or perturbed so that a transition is induced.

Let us consider two stationary states n and m of an unperturbed system represented by the wave function Ψ°_n and Ψ°_m such that $E_m > E_n$. Let us assume that at $t = 0$, the system is in the state n. At this time, the system comes under the perturbing influence of the radiation of a range of frequencies in the neighbourhood of ν_{mn} of a definite field strength **E**.

The description of the unperturbed system is given by the wave equation including time

$$\mathcal{H}^\circ \Psi^\circ = -\frac{h}{2\pi i} \frac{\partial \Psi^\circ}{\partial t} \tag{3.21}$$

where $\mathcal{H}^\circ$ is the Hamiltonian for the unperturbed system. The normalized general solution of this equation is

$$\Psi = \sum_n a_n \Psi^\circ_n \tag{3.22}$$

In this expression a_n are constants or weighting coefficients which are not functions of x but may be functions of time t. Furthermore, $\Sigma a_n^* a_n = 1$, as required for normalized systems. Ψ°_n are wave functions of stationary states corresponding to the energy values $E_0, E_1, \ldots, E_n$. When the system comes under the perturbing influence of the field of the electromagnetic radiation, the potential energy of the system changes due to the interaction. This change can be represented by the additional term $\mathscr{H}'$ in the Hamiltonian of the unperturbed system.

The unperturbed Hamiltonian $\mathscr{H}^\circ$ is independent of t, whereas the perturbing term $\mathscr{H}'$ may be a function of time as well as the coordinates of the system. For example, $\mathscr{H}'$ might be zero except during the period $t_1 < t < t_2$, the perturbation then being effective only during this period.

The time-dependent perturbation of the Schrödinger equation is given as

$$(\mathscr{H}^\circ + \mathscr{H}')\,\Psi = -\frac{h}{2\pi i}\frac{\partial \Psi}{\partial t} \tag{3.23}$$

The wave function satisfying this equation is a function of time and the coordinates of the system. For a given value of t, say t', $\Psi(t')$ is a function of coordinates only. The general solution can be represented as

$$\Psi(x_1 \ldots, z_n, t') = \sum_n a_n \Psi^\circ_n\,(x_1, \ldots, z_n, t')$$

The coefficients a_n are constant for time t' alone. A similar set of equations can be written for any other time t'', but a_n will have different values. In general,

$$\Psi(x_1 \ldots z_n, t) = \sum_n a_n(t)\,\Psi^\circ_n(x_1 \ldots z_n, t) \tag{3.24}$$

The quantities $a_n(t)$ are functions of t alone such as to cause Ψ to satisfy the Schrödinger equation for the perturbed system. Substituting (3.24) in equation (3.23), we get

$$(\mathscr{H}^\circ + \mathscr{H}')\sum_n a_n(t)\Psi^\circ_n = -\frac{h}{2\pi i}\frac{\partial}{\partial t}\sum_n a_n(t)\Psi^\circ_n \tag{3.25}$$

$$\sum_n a_n(t)\,\mathscr{H}^\circ\Psi^\circ_n + \sum a_n(t)\,\mathscr{H}'\Psi^\circ_n$$

$$= -\frac{h}{2\pi i}\frac{\partial \sum a_n(t)}{\partial t}\Psi^\circ_n - \frac{h}{2\pi i}\sum a_n(t)\frac{\partial \Psi^\circ_n}{\partial t}$$

$$= -\frac{h}{2\pi i}\frac{\partial \sum a_n(t)}{\partial t}\Psi^\circ_n + \sum a_n(t)\,\mathscr{H}^\circ\Psi^\circ_n \tag{3.26}$$

On simplifying and rearranging (3.26), we have

$$-\frac{h}{2\pi i}\frac{\partial \sum a_n(t)}{\partial t}\Psi^\circ_n = \sum_n a_n(t)\,\mathscr{H}'\Psi^\circ_n \tag{3.27}$$

Multiplying throughout by $\Psi^{\circ *}_m$ and integrating over the entire configuration space, equation (3.27) becomes

$$-\frac{h}{2\pi i}\frac{\partial \sum a_n(t)}{\partial t}\int \Psi^{\circ *}_m \Psi^\circ_n\,d\tau = \sum a_n(t)\int \Psi^{\circ *}_m \mathscr{H}'\Psi^\circ_n\,d\tau \tag{3.28}$$

All terms on the left vanish except that for $n = m$ due to orthogonality property of wave functions and from the conditions of normalization,

$$\frac{d\,a_m(t)}{dt} = -\frac{2\pi i}{h} \sum_n a_n(t) \int \Psi_m^{\circ *} \mathcal{H}' \Psi_n^{\circ}\, d\tau \qquad (3.29)$$

This is a set of simultaneous differential equations in the function $a_m(t)$. If we consider the simple case consisting of only two states n and m then on expanding

$$\frac{d\,a_m(t)}{dt} = -\frac{2\pi i}{h} a_n(t) \int \Psi_m^{\circ *} \mathcal{H}' \Psi_n^{\circ}\, d\tau - \frac{2\pi i}{h} a_m(t) \int \Psi_m^{\circ *} \mathcal{H}' \Psi_m^{\circ}\, d\tau \quad (3.30)$$

Since initially the system is in the lower energy state n, we have at $t = 0$, $a_n = 1$ and $a_m = 0$. Therefore, the final term will not contribute initially, and we have

$$\frac{d\,a_m}{dt}(t) = -\frac{2\pi i}{h} \int \Psi_m^{\circ *} \mathcal{H}' \Psi_n^{\circ}\, d\tau \qquad (3.31)$$

This equation gives the rate at which a system changes from one stationary state to another under the perturbing influence of the radiation. The oscillating electric field of the radiation disturbs the potential energy of the molecule and causes it to escape from its initial stationary state n. The rate with which the coefficient a_m increases corresponds to the rate with which the description of the system changes from Ψ_n° to Ψ_m°.

To evaluate the equation, we must find out the nature of perturbation imposed on a system of molecules when the electromagnetic radiation falls on it. When the oscillating electric field of the electromagnetic radiation interacts with the molecules composed of positive and negative charges, the electric vector of the incident radiation induces a dipole moment er in the molecule along its direction (Sec. 3.1). The perturbation energy along the x-axis as given by the perturbing Hamiltonian $\mathcal{H}'$ is

$$\mathcal{H}' = \mathbf{E}_x \sum e_j x_j \qquad (3.32)$$

where $\mathbf{E}_x$ is the electric field strength in x-direction and e_j represents the charge and x_j, the x-component of the displacement for the jth particle of the system. The expression $\sum e_j x_j$ is known as *the component of the electric dipole moment of the system along the* **x**-*axis* and is represented by μ_x. When the time variation of the field is introduced (Section 3.2). $\mathcal{H}'$ becomes

$$\mathcal{H}' = \mathbf{E}_x^0(\nu)\,(e^{2\pi i\nu t} + e^{-2\pi i\nu t}) \sum e_j x_j \qquad (3.33)$$

Substituting the value of $\mathcal{H}'$ in the expression (3.31),

$$\frac{d\,a_m(t)}{dt} = -\frac{2\pi i}{h} \int \Psi_m^{\circ *} \left[\mathbf{E}_x^0(\nu)\,(e^{2\pi i\nu t} + e^{-2\pi i\nu t}) \sum e_j x_j \right] \Psi_n^{\circ}\, d\tau \quad (3.34)$$

Separating the space and the time dependent parts (equation 3.19), $\Psi^{\circ}_m = \psi^{\circ}_m e^{-\frac{2\pi i}{h}E_m t}$, expression (3.34) becomes

$$\frac{d}{dt}a_m(t) = -\frac{2\pi i}{h}\mathbf{E}^0_x(\nu)\int \psi^{\circ *}_m e^{\frac{2\pi i E_m t}{h}}\left\{e^{2\pi i\nu t}+e^{-2\pi i\nu t}\right\} \Sigma e_j x_j \psi^{\circ}_n e^{-\frac{2\pi i E_n t}{h}} d\tau$$

$$= -\frac{2\pi i}{h}\mathbf{E}^0_x(\nu)\int \psi^{\circ *}_m \Sigma e_j x_j \psi^{\circ}_n d\tau\left[e^{\frac{2\pi i}{h}(E_m - E_n + h\nu)t} + e^{\frac{2\pi i}{h}(E_m - E_n - h\nu)t}\right]$$

$$= -\frac{2\pi i}{h}\mathbf{E}^0_x|\mu_{xnm}|\left[e^{\frac{2\pi i}{h}(E_m - E_n + h\nu)t} + e^{\frac{2\pi i}{h}(E_m - E_n - h\nu)t}\right] \quad (3.35)$$

where $|\mu_{xnm}| = \int \psi^{\circ *}_m \Sigma e_j x_j \psi^{\circ}_n d\tau$ is known as the x-component of the *transition moment integral* corresponding to the transition $n \rightarrow m$. On integration of the expression (3.35) and obtaining the integration constant from the condition that when $t = 0$, $a_m(0) = 0$, we have

$$a_m(t) = |\mu_{xmn}|\mathbf{E}^0_x(\nu)\left[\frac{1 - e^{\frac{2\pi i}{h}(E_m - E_n + h\nu)t}}{E_m - E_n + h\nu} + \frac{1 - e^{\frac{2\pi i}{h}(E_m - E_n - h\nu)t}}{E_m - E_n - h\nu}\right] \quad (3.36)$$

Of the two terms in this equation only one is important at any one time and that also only when frequency happens to lie close to $\nu = (E_m - E_n)/h$. For a single frequency the term $|\mu_{xnm}|\mathbf{E}^0_x(\nu)$ is always small and the expression is small unless the denominator is also small. If $E_m > E_n$, corresponding to light absorption, the second denominator tends to zero for the condition $h\nu \simeq E_m - E_n$. This is just the Bohr's frequency relationship. When $E_m < E_n$, corresponding to light emission (induced) the first denominator approximates to zero. (Here E_m and E_n signify energies of the final and initial states, respectively.) So we find that due to the presence of the socalled resonance denominator $(E_m - E_n - h\nu)$ the coefficient $a_m(t)$ of the final energy state will have a large value only under the condition of Bohr's frequency rule.

Therefore, neglecting the first term, *the probability of absorption per unit time* is given by the product $a^*_m(t)a_m(t)$ and for a single frequency we obtain after minor rearrangement

$$a^*_m(t)\,a_m(t) = 4\,|\mu_{xnm}|^2\,\mathbf{E}^{0^2}_x(\nu)\frac{\sin^2\left\{\frac{\pi}{h}(E_m - E_n - h\nu)t\right\}}{(E_m - E_n - h\nu)^2} \quad (3.37)$$

where the imaginary exponential is replaced by the sine function. This expression, however, includes only the term due to a single frequency. Since under experimental conditions, a range of frequencies is involved, we can write $\mathbf{E}^0_x(\nu_{nm})$ for $\mathbf{E}^0_x(\nu)$ and integrate the expression over the limits $-\infty$ to $+\infty$,

$$a_m^*(t)\,a_m(t) = 4|\mu_{xnm}|^2 \int_{-\infty}^{+\infty} \mathbf{E}_x^{0^2}(\nu_{nm}) \frac{\sin^2 \frac{\pi}{h}(E_m - E_n - h\nu)\,t}{(E_m - E_n - h\nu)^2}\,d\nu \qquad (3.38)$$

This is justified as the effects of light of different frequencies are additive and the integrand will be large over only a small region where the denominator is small, i.e. in the region of ν near ν_{nm}. Taking the constant $\mathbf{E}_x^0(\nu_{nm})$ out of the integral,

$$a_m^*(t)\,a_m(t) = 4|\mu_{xmn}|^2 \mathbf{E}_x^{0^2}(\nu_{nm}) \int_{-\infty}^{+\infty} \frac{\sin^2 \frac{\pi}{h}(E_m - E_n - h\nu)\,t}{(E_m - E_n - h\nu)^2}\,d\nu \qquad (3.39)$$

Let $x = \frac{\pi}{h}(E_m - E_n - h\nu)\,t$ and $dx = \pi t\,d\nu$ so that on substitution,

$$a_m^*(t)\,a_m(t) = 4|\mu_{xmn}|^2 \mathbf{E}_x^{0^2}(\nu_{nm}) \frac{\pi_t}{h^2} \int_{-\infty}^{+\infty} \frac{\sin^2 x}{x^2}\,dx$$

Using the relationship, $\int_{-\infty}^{+\infty} \frac{\sin^2 x}{x^2}\,dx = \pi$, we obtain

$$a_m^*(t)\,a_m(t) = \frac{4\pi^2}{h^2} |\mu_{xnm}|^2 \mathbf{E}_x^{0^2}(\nu_{nm})\,t \qquad (3.40)$$

It is seen that as a result of integration over a range of frequencies, the probability of transition to the state m in time t becomes proportional to t.

The rate of change of probability or the *transition probability per unit time* is given by,

$$\frac{d}{dt} a_m^* a_m = \frac{4\pi^2}{h^2} \mu_{xnm}^2 \mathbf{E}_x^{0^2}(\nu_{nm}) \qquad (3.41)$$

Now we can introduce the energy density $\rho(\nu_{nm})$ to transform this result into the Einstein coefficient of absorption, viz. the probability that the molecule (or atom) will absorb a quantum in unit time under unit radiation density. The probability of absorption in the Einstein expression is given by $B_{nm}\,\rho(\nu_{nm})$. Under the influence of the radiation polarized in x-directions, the relationship between the field strength $\mathbf{E}_x$ in x-direction and the radiation density is deduced as follows:

$$\rho(\nu_{nm}) = \frac{1}{4\pi} \bar{\mathbf{E}}^2(\nu_{nm}) \qquad (3.42)$$

where $\bar{\mathbf{E}}^2(\nu_{nm})$ is the average value of the square of the field strength.

For isotropic radiation ($\bar{\mathbf{E}}_x^2 = \bar{\mathbf{E}}_y^2 = \bar{\mathbf{E}}_z^2$)

$$\tfrac{1}{3}\bar{\mathbf{E}}^2(\nu_{nm}) = \bar{\mathbf{E}}_x^2(\nu_{nm}) \qquad (3.43)$$

Hence, $$\rho(\nu_{nm}) = \frac{3}{4\pi}\bar{\mathbf{E}}_x^2(\nu_{nm})$$

$$= \frac{3}{4\pi}(2\bar{\mathbf{E}}_x^0(\nu_{nm})\cos 2\pi\nu t)^2 \tag{3.44}$$

The average value of $\cos^2 2\pi\nu t = \frac{1}{2}$, giving

$$\rho(\nu_{nm}) = \frac{3}{2\pi}\mathbf{E}_x^{0^2}(\nu_{nm}) \tag{3.45}$$

and $$\mathbf{E}_x^{0^2}(\nu_{nm}) = \frac{2\pi}{3}\rho(\nu_{nm}) \tag{3.46}$$

Substituting for $\bar{\mathbf{E}}_x^{0^2}(\nu_{nm})$ in (3.41):

$$\text{Probability of transition} = \frac{4\pi^2}{h^2}|\mu_{xnm}|^2\frac{2\pi}{3}\rho(\nu_{nm})$$

$$= \frac{8}{3}\frac{\pi^3}{h^2}|\mu_{xnm}|^2\rho(\nu_{nm}) \tag{3.47}$$

Similar expressions are obtained for the *component of the transition moment integral in y and z directions.* Therefore, for an isotropic radiation of frequency ν_{nm} the transition probability is given as

$$\frac{d}{dt}(a_m^* a_m) = \frac{8\pi^3}{3h^2}\left[\mu_{xnm}^2 + \mu_{ynm}^2 + \mu_{znm}^2\right]\rho(\nu_{nm}) \tag{3.48}$$

$$= B_{nm}\rho(\nu_{nm}) \quad \text{(from Einstein's equation)}$$

Comparing we have,

$$B_{nm} = \frac{8\pi^3}{3h^2}\left[\mu_{xnm}^2 + \mu_{ynm}^2 + \mu_{znm}^2\right] \tag{3.49}$$

By completely analogous treatment in which values of $a_n(0) = 0$ and $a_m(0) = 1$, are used, the Einstein coefficient for induced emission B_{mn} is found to be given by the equation:

$$B_{mn} = \frac{8\pi^3}{3h^2}\left[\mu_{xmn}^2 + \mu_{ymn}^2 + \mu_{zmn}^2\right] \tag{3.50}$$

$B_{nm} = B_{mn}$ as required and

$$A_{mn} = \frac{8\pi h\nu_{mn}^3}{c^3}B_{mn} \tag{3.10}$$

$$= \frac{64\pi^4\nu_{mn}^3}{3hc^3}\left[\mu_{xmn}^2 + \mu_{ymn}^2 + \mu_{zmn}^2\right] \tag{3.51}$$

$$= \frac{64\pi^4\nu_{mn}^3}{3hc^3}|\mathbf{M}_{mn}|^2 \tag{3.52}$$

The square of the transition moment integral $|\mathbf{M}_{mn}|^2$, a vector quantity is the sum of the square of the components in x, y and z directions:

$$| \mathbf{M}_{nm} |^2 = [\mu^2_{xmn} + \mu^2_{ymn} + \mu^2_{zmn}] \quad (3.53)$$

3.5 CORRELATION WITH EXPERIMENTAL QUANTITIES

Experimental determination of intensity of absorption is based on Lambert-Beer law (Section 1.2) which is repeated here for convenience. When the radiation of frequency (ν) is incident on an absorbing system composed of atoms or molecules, the fractional decrease in intensity on passage through the solution of molar concentration C and layer thickness dl can be expressed as:

$$-\frac{dI}{I} = \alpha_\nu C dl \quad (1.1)$$

α_ν is the proportionality constant and is a function of incident frequency ν. On integration over the path length l the expression becomes

$$\ln \frac{I_0}{I} = \alpha_\nu \, Cl \quad (1.2)$$

where I_0 = incident intensity at $l = 0$, and I = transmitted intensity, at $l = l$ (Figure 1.1). Hence α_ν expresses the probability of absorption per unit time per unit concentration per unit length of the optical path. When converted into decadic logarithm ϵ_ν (= 2.303 α_ν)

$$\epsilon_\nu = \frac{1}{Cl} \log \frac{I_0}{I} \text{ mol}^{-1} \text{ cm}^{-1} \text{ s}^{-1} \quad (3.54)$$

$$= \frac{1}{Cl} \text{optical density}$$

For a range of frequencies,

$$\int_{\nu_{nm}} \epsilon_\nu \, d\nu = \frac{1}{Cl} \int_{\nu_{nm}} \log \frac{I_0}{I} \, d\nu = A \quad (3.55)$$

= molar integrated intensity over the absorption band.

In terms of wavenumbers,

$$A = c \int_{\bar{\nu}_{nm}} \epsilon_{\bar{\nu}} \, d\bar{\nu} = c \bar{A} \text{ and } \bar{A} = \frac{A}{c} \, l \text{ mol}^{-1} \text{ cm}^{-2} \quad (3.56)$$

where c is the velocity of light.

Looking at the problem at the molecular level, the absorption of each quantum of radiation of frequency ν_{nm} removes an amount of energy $h\nu_{nm}$ from the radiation. If there are N' molecules per cm³, and if the probability of absorption per unit time per unit radiation density is given by B_{nm}, then the decrease in intensity per unit time, when a beam of radiation of density $\rho(\nu_{nm})$ passes through a solution of path length dl, can be expressed as

$$-dI = B_{nm} \, N' \, h\nu \, \rho(\nu_{nm}) \, dl \quad (3.57)$$

Since I, the intensity of radiation is the energy flux per unit area per unit time, it is related to the radiation density $\rho(\nu_{nm})$ by the factor c, the velocity of light:

$$I = \rho(\nu_{nm})\, c$$

and $$-dI = B_{nm}\, N'\, h\nu_{nm} \left(\frac{I}{c}\right) dl$$

Expressing in terms of concentration C mol litre^{-1}, $C = \frac{N'}{N} \times 1000$,

$$-\frac{dI}{I} = B_{nm} \frac{Nh\nu_{nm}}{1000\, c} C dl \tag{3.58}$$

$$\ln \frac{I_0}{I} = B_{nm} \frac{Nh\nu_{nm}}{1000\, c} Cl = \alpha_\nu\, Cl \tag{3.59}$$

$$\log \frac{I_0}{I} = B_{nm} \frac{Nh\nu_{nm}}{2303\, c} Cl = \epsilon_\nu\, Cl \tag{3.60}$$

Comparing with Beer's law expression (3.54)

$$\epsilon_\nu = B_{nm} \frac{Nh\nu_{nm}}{2303\, c}$$

Therefore $$B_{nm} = \frac{2303\, c}{Nh\nu_{nm}} \epsilon_\nu$$

For a range of frequencies

$$\int B_{nm}\, d\nu \equiv B_{nm} = \frac{2303\, c}{Nh} \int \frac{\epsilon_\nu\, d\nu}{\nu}$$

$$= \frac{2303\, c}{Nh\nu_{max}} \int \epsilon_\nu\, d\nu \tag{3.61}$$

if the average frequency ν is considered as the band maximum ν_{max}. Substituting for B_{nm} from equation (3.49) and rearranging, we have, for integrated absorption intensity A,

$$A = \int \epsilon_\nu\, d\nu = \frac{8\pi^3}{3h^2} \frac{Nh\nu}{2303\, c} |\mathbf{M}|^2 \tag{3.62}$$

Thus the integrated absorption intensity is found to depend on the square of the transition moment integral defined as

$$|\mathbf{M}_{nm}| = \int_{-\infty}^{+\infty} \psi_m\, \hat{\mu}_{nm}\, \psi_n\, d\tau \tag{3.63}$$

where μ_{nm} is the operator for transition moment.

Experimentally, the intensity of absorption is measured in a spectrophotometer for a range of suitable wavelengths. The molar extinction value at each wavelength is plotted as a function of wavenumber

(Figure 3.5a). The total area under the curve gives the integrated absorption intensity, $\int \epsilon_{\bar{\nu}}\, d\bar{\nu}$ and Einstein's coefficient of absorption can be calculated after expressing (3.61) in terms of wave number.

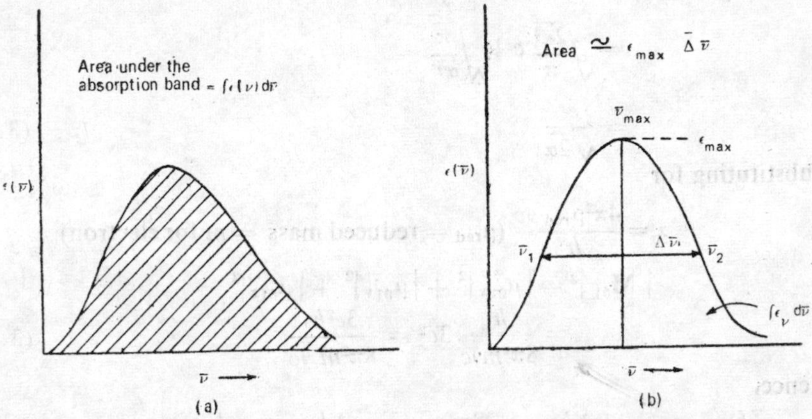

Figure 3.5 Absorption curve, (a) Plot of molar extinction coefficient $\epsilon_{\bar{\nu}}$ vs $\bar{\nu}$; (b) $\int \epsilon_{\bar{\nu}}\, d\bar{\nu} \simeq \epsilon_{max}\, \Delta\bar{\nu}$, where $\Delta\bar{\nu}$ is the half band width.

3.6 INTENSITY OF ELECTRONIC TRANSITIONS

3.6.1 Theoretical Absorption Intensity

The intensity of absorption for an electronic transition is the probability of absorption between two given energy states. It can be theoretically computed by using the expression for the transition moment integral. The transition moment integral $\mathbf{M}_{01}$ between the ground and the first excited state is expressed as

$$|M_{01}| = \int \psi_0\, \hat{\mu}\, \psi_1\, d\tau \tag{3.64}$$

where ψ_0 and ψ_1 are the wave functions for the two combining states and $\hat{\mu}$ is the dipole moment operator ($= er$). If the assumption is made that the optical electron, i.e. the electron which is excited to the higher energy state, is attracted to the centre of the molecule with a Hooke's law type of force, the harmonic oscillator wave functions can be used to evaluate $|\mathbf{M}_{01}|$.

Substituting the values for $\psi_{v=0}$ and $\psi_{v=1}$ respectively, in the harmonic oscillator problem,

$$\psi_0 = (\alpha/\pi)^{1/4}\, e^{-\alpha r^2/2} \tag{3.65}$$

$$\psi_1 = (\alpha/\pi)^{1/4}\, \sqrt{2\alpha}\; r\, e^{-\alpha r^2/2} \tag{3.66}$$

the component of the transition moment integral in the x-direction is,

$$|\mu_{01x}|\,x = \int (\alpha/\pi)^{1/4} e^{-\alpha x^2/2} (ex) (\alpha/\pi)^{1/4} \sqrt{2\alpha}\, x\, e^{-\alpha x^2/2}\, dx$$

$$= (\alpha/\pi)^{1/2} \sqrt{2\alpha} \int x^2 e^{-\alpha x^2}\, dx$$

$$= \sqrt{\frac{2\alpha^2}{\pi}}\, e\, \frac{1}{2} \sqrt{\frac{\pi}{\alpha^3}}$$

$$= \frac{e}{\sqrt{2\alpha}} \tag{3.67}$$

Substituting for

$$\alpha = \frac{4\pi^2 \mu_{red}\, \bar{\nu} c}{h} \quad (\mu_{red} = \text{reduced mass} = m \text{ for electron})$$

$$|\mathbf{M}_{01}|^2 = |\mu_{01x}|^2 + |\mu_{01y}|^2 + |\mu_{01z}|^2$$

$$= \frac{h}{8\pi^2 m \bar{\nu} c}\, 3e^2 = \frac{3e^2 h}{8\pi^2 m\, \bar{\nu} c} \tag{3.68}$$

Hence,

$$\int \epsilon_{\bar{\nu}}\, d\nu = B_{nm} \frac{Nh\bar{\nu}_{01}}{2303c} = \frac{8\pi^3}{3h^2} |\mathbf{M}_{01}|^3 \frac{Nh\bar{\nu}_{01}}{2303c} = \frac{N\pi e^2}{2303mc^2} \tag{3.69}$$

$$= \frac{6.02\times 10^{23} \times 3.14 \times (4.8 \times 10^{-10}\ \text{esu})^2}{9.1\times 10^{-28} g \times 2303 \times (3 \times 10^{10}\ \text{cm } s^{-1})^2}$$

$$= 2.309 \times 10^8\ l\ \text{mol}^{-1}\ \text{cm}^{-1}$$

This numerical value for the integrated intensity of absorption is expected when all the N molecules are transferred to the higher energy state, each absorbing a quantum of radiation. This value is considered as a reference standard for expression of actual absorption intensities.

3.6.2 Strength of an Electronic Transition—Oscillator Strength

The strength of an electronic transition is generally expressed in terms of a quantity called *oscillator strength f*. It is defined as the ratio of the experimental transition probability to that of the ideal case of a harmonic oscillator, that is

$$f = \frac{\left[\int \epsilon_{\bar{\nu}}\, d\bar{\nu}\right]_{\text{expt.}}}{\left[\int \epsilon_{\bar{\nu}}\, d\bar{\nu}\right]_{\text{ideal}}} = \frac{\int \epsilon_{\bar{\nu}}\, d\bar{\nu}}{2.31\times 10^8}$$

$$f = 4.33 \times 10^{-9} \int \epsilon_{\bar{\nu}}\, d\bar{\nu} \tag{3.70}$$

For a one-electron transition the ideal value of unity is obtained when all the molecules are transferred to the higher energy state, i.e. when the transition probability is unity. The oscillator strength can also be expressed as

$$f = \frac{|\mathbf{M}_{nm}|^2}{|\mathbf{M}_{01}|^2_{\text{ideal}}} = \frac{|\mathbf{M}_{nm}|^2}{3e^2 h / 8\pi^2 m \bar{\nu}_{nmc}} = \frac{8\pi^2 m \bar{\nu}_{nmc}}{3he^2} |\mathbf{M}_{nm}|^2 \quad (3.71)$$

$$= 4.704 \times 10^{29} \bar{\nu}_{nm} |\mathbf{M}_{nm}|^2 \quad (3.72)$$

$|\mathbf{M}_{nm}|^2$ is sometimes written as $|\mathbf{R}_{nm}|^2$, where $|\mathbf{R}_{nm}|$ is known as the *matrix element of the electric dipole moment* and has the same meaning as the transition moment integral. There are a few other related quantities which are used for expressing the strength of an electronic transition.

Dipole strength of transition

$$D = |\mathbf{M}_{nm}|^2 \text{ or } |\mathbf{R}_{nm}|^2 = e^2 Q^2 \quad (3.73)$$

where, $$Q = \int \psi_n \mathbf{r} \psi_m \, d\tau = \text{dipole length} \quad (3.74)$$

The concept of oscillator strength was developed from the classical theory of dispersion. It is related to the molar refraction R of a substance by

$$R = \frac{M}{d} \frac{n^2 - 1}{n^2 + 2} = \frac{Ne^2}{3\pi m} \sum \frac{f_{mn}}{\nu_{nm}^2 - \nu^2} \quad (3.75)$$

where M is molecular weight of the substance, d the density, n the refractive index, ν the frequency of the radiation field and f_{mn} the oscillator strength for transitions between the states n and m which absorbs a frequency corresponding to ν_{nm}. The summation is over all the bands or energy states to which the transition can take place from the ground state n, including the continuum.

3.7 THE RULES GOVERNING THE TRANSITION BETWEEN TWO ENERGY STATES

3.7.1 The Basis of Selection Rules

As already discussed, the interaction of electromagnetic radiation with matter leads to absorption only if a dipole moment is created as a result of such interaction. During the process of emission the dipole is destroyed. This may be stated symbolically as $d\mu/dt$ is positive and $d^2\mu/dt^2 \neq 0$.

The strength or intensity of absorption is related to the dipole strength of transition D or square of the transition moment integral $|\mathbf{M}_{nm}|^2$, and is expressed in terms of oscillator strength f or integrated molar extinction $\int \epsilon_{\bar{\nu}} \, d\bar{\nu}$. A transition with $f = 1$, is known as totally *allowed* transition. But the transitions between all the electronic, vibrational or rotational states are not equally permitted. Some are *forbidden* which can become allowed under certain conditions and then appear as weak absorption bands. The rules which govern such transitions are known as *selection rules*. For atomic energy levels, these selection rules have been empirically obtained from a comparison between the number of lines theoretically expected in the spectrum of a given atom and those experimentally obtained (vide Section 2.6). The rules so derived state: $\Delta m = 0, \pm 1$ and $\Delta l = \pm 1$ where m and l are magnetic and orbital annglar momentum quantum numbers, respectively. There is no such restriction imposed for n. These selection rules can be justified from the mechanics of interaction between matter and radiation.

When a plane polarized radiation with the electric vector directed along z-axis is imposed on a hydrogen atom in its ground state ($l = 0$), the radiation distorts the molecule in z direction such that an oscillating dipole is necessarily created as shown diagrammatically in Figure 3.2. If the radiation is of a frequency in resonance with the transition $1s \rightarrow 2p_z$, the radiation is absorbed. Thus, a dipolar radiation can bring about changes only from $s \rightarrow p$, or $p \rightarrow d$ or $d \rightarrow f$. Each transition involves the creation of a node at right angles to the direction of polarization of the electric vector, and the change in angular momentum quantum number $\Delta l = 1$ in each case. When the atom reverts to its original energy state $p \rightarrow s$, $\Delta l = -1$. For an $s \rightarrow d$ transition, a quadrupole radiation is required since two nodal planes have to be created by the incident radiation field. Such transitions have very low probabilities, since the intensity of electric quadrupole component in the radiation field is only 5×10^{-6} of that of the dipole component. Also $1s \rightarrow 2s$, or $2p \rightarrow 3p$ transitions are forbidden, as no transition moment is created or destroyed conforming to the selection rule $\Delta l \neq 0$.

The same conclusions can be drawn from the quantum mechanical condition that the square of the transition moment integral $|\mathbf{M}_{nm}|^2$ must be nonzero for a transition to be induced by electromagnetic radiation. That is

$$|\mathbf{M}_{nm}|^2 = \int_{-\infty}^{+\infty} \psi_n^* \, \hat{\mu}_{nm} \, \psi_m \, d\tau \neq 0 \tag{3.76}$$

where $\hat{\mu}_{nm}$ is the dipole moment operator. To understand the selection rules we have to consider the symmery properties of the three functions ψ_n^*, $\hat{\mu}_{nm}$ and ψ_m within the limits of integration. A qualitative picture of the problem can be obtained if we consider one electron wave functions to be located with the nucleus as the centre and work out the symmetry of each function, in terms of odd and even character. For a one-dimensional case in the x direction, an odd function changes sign on changing the coordinate from x to $-x$. With this criterion, the dipole moment vector is always odd since it changes sign at the origin. The one electron wave functions are either odd (−) or even (+).

Let us consider two types of possible transitions, $1s \rightarrow 2s$ and $1s \rightarrow 2p$, by pictorial representation of s and p one-electron orbital functions (Figure 3.6). For convenience, we divide the configuration space into

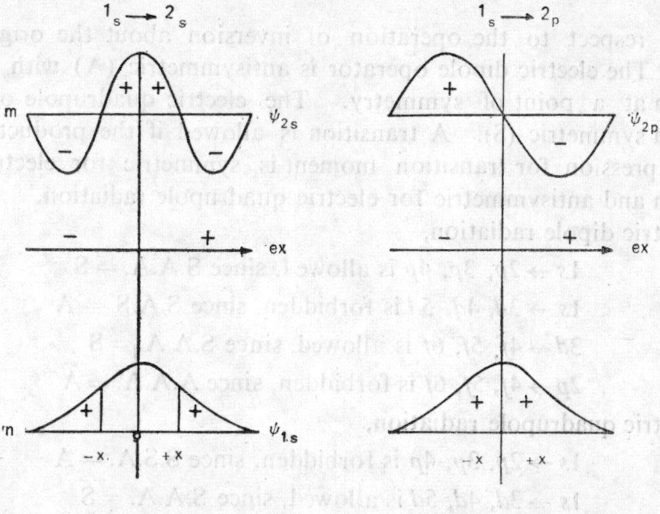

(A) (B)

Figure 3.6 Graphical integration of transition moment for $1s \rightarrow 2s$ and $1s \rightarrow 2p$ transition to emphasize the selection rule: $|M| \neq 0$ for an allowed transition.

$$\text{(A) } 1s \rightarrow 2s: \ |M_{nmx}| = \int_{-\alpha}^{+\alpha} \psi_{1s}^{*}(ex)\,\psi_{2s}\,dx = \int_{-\alpha}^{-x} (+)(-)(-) + \int_{-\alpha}^{0} (+)(-)(+)$$

$$+ \int_{0}^{+x} (+)(+)(+) + \int_{+x}^{-\alpha} (+)(+)(-) = (+) + (-) + (+) + (-) = 0; \text{(B) } 1s \rightarrow 2p: |M_{nmx}|$$

$$= \int_{-\alpha}^{+\alpha} \psi_{1s}^{*}(ex)\,\psi_{2p}\,dx = \int_{-\alpha}^{0} (+)(-)(+) + \int_{0}^{+\alpha} (+)(+)(-) = (-) + (-) \neq 0$$

two parts : $-\infty$ to 0 and 0 to $+\infty$. The expression for transition moment can be written as

$$|\mu_{mnx}| = \int_{-\infty}^{0} \psi_n\,(ex)\,\psi_m\,dx + \int_{0}^{+\infty} \psi_n\,(ex)\,\psi_m\,dx \qquad (3.77)$$

With reference to Figure 3.6, the integral can be further subdivided for point by point multiplication of odd and even functions. It is observed that a nonzero value of transition moment is obtained only when an even atomic wave function *s*, combines with an odd function *p*. Besides establishing the selection rule $\Delta l = 1$, it also says that a transition is allowed between a *g* state and an *u* state only. The transition $g \rightarrow g$ is forbidden. These two statements are symbolically written as $g \rightarrow u$ (allowed), $g \not\rightarrow g$ (forbidden) and are applicable for systems with a centre of symmetry.

These observations can be further generalized. The atomic wave functions *s, p, d, f*, etc. are alternatively symmetric (S) and antisymmetric

(A) with respect to the operation of inversion about the origin of the system. The electric dipole operator is antisymmetric (A) with respect to inversion at a point of symmetry. The electric quadrupole operator is inversion symmetric (S). A transition is allowed if the product function in the expression for transition moment is symmetric for electric dipole radiation and antisymmetric for electric quadrupole radiation.

For electric dipole radiation,

$1s \rightarrow 2p,\ 3p,\ 4p$ is allowed, since S.A.A. = S

$1s \rightarrow 3d,\ 4d,\ 5d$ is forbidden, since S.A.S. = A

$3d \rightarrow 4f,\ 5f,\ 6f$ is allowed, since S.A.A. = S

$2p \rightarrow 4f,\ 5f,\ 6f$ is forbidden, since A.A.A. = A

For electric quadrupole radiation,

$1s \rightarrow 2p,\ 3p,\ 4p$ is forbidden, since S.S.A. = A

$1s \rightarrow 3d,\ 4d,\ 5d$ is allowed, since S.A.A. = S

3.7.2 Selection Rules for Molecular Transitions

For molecules, the rules governing the transition between two given energy states are

(i) $\Delta\Lambda = 0, \pm 1$: allowed changes in the component of the total orbital angular momentum in the direction of the molecular axis.

(ii) $\Delta S = 0$: the spin conservation rule.

(iii) Symmetry properties of the energy states must be conserved.

If in the total wave function (ψ), one electron orbital (ϕ) is factored out from the vibrational function (χ) and the spin function (S), we get

$$\psi_i = \phi_i\, \chi_i\, S_i \tag{3.78}$$

$$|\mathbf{M}_{nm}| = \int \psi_n\, (e\mathbf{r})\, \psi_m\, d\tau = \int \phi_n \chi_n S_n\, (e\mathbf{r})\, \phi_m \chi_m S_m\, d\tau$$

$$= \int \phi_n\, e\mathbf{r}\, \phi_m\, d\tau_e \int \chi_n \chi_m\, d\tau_v \int S_n S_m\, d\tau_s \tag{3.79}$$

$d\tau_e$, $d\tau_v$ and $d\tau_s$ are the configuration space for electronic, vibrational and spin functions, assumed to be independent of each other. The factorization is possible because only the electronic functions are involved in the creation of transition dipole (Born-Oppenheimer approximation). The nuclear motion or the electron spins are not affected. The first integral is the *electronic transition moment* which has already been discussed. The second expression gives the overlap requirement of the molecular vibrational eigenfunctions and is a quantum mechanical expression of the *Franck-Condon principle* (Section 4.4.1). The third integral designates the *spin conservation rule*. The spin conservation rule is the most stringent of requirements and dictates that *during an electronic transition, total spin angular momentum must be conserved.* Since the two spin functions α and β are orthogonal, the integral has a nonzero value only if both the functions are the same. As a consequence allowed transitions are only between *singlet-singlet*, or *triplet-triplet* states. The *singlet-triplet*, transitions are forbidden, unless some perturbing influences bring about *spin-orbital* coupling and thereby change the 'pure singlet' character of the energy states.

The observed oscillator strength for any given transition is seldom found to be unity. A certain degree of 'forbiddenness' is introduced for each of the factors which cause deviation from the selection rules. The oscillator strength of a given transition can be conveniently expressed as the product of factors which reduce the value from that of a completely allowed transition:

$$f = f_m f_o f_s f_p \ldots$$

where, f_m = forbiddenness due to spin multiplicity rule, 10^{-5} to 10^{-6}

f_o = forbiddenness due to overlap criterion, 10^{-2} to 10^{-4}

f_s = forbiddenness due to symmetry consideration (orbitally forbidden), 10^{-1} to 10^{-3}

f_p = forbiddenness due to parity consideration (odd-even functions), 10^{-1} to 10^{-2}

Only when $f_m = f_o = f_s = f_p = 1$, that we get $f = 1$ and the transition is totally allowed.

Although oscillator strength is proportional to the integrated intensity of absorption $\int \epsilon_{\bar{\nu}}\, d\bar{\nu}$, there is often a fairly good correlation between f and ϵ_{max}, the molar absorption coefficient at the band maximum. This correlation is valid if we assume a Lorenzian shape for the absorption band and replace the integral by $\epsilon_{max}\, \Delta\bar{\nu}$, where $\Delta\bar{\nu}$ is the half-band-width of the absorption band (Figure 3.5b). The *half-band-width* is defined as the width of the absorption band (in cm^{-1}) where the value of $\epsilon = \frac{1}{2}\epsilon_{max}$. Hence

$$f = 4.31 \times 10^{-9}\, \epsilon_{max}\, \Delta\bar{\nu} \tag{3.80}$$

For very intense transitions as observed for the dye molecules, $\Delta\bar{\nu}$ is $\simeq 2000$ cm^{-1} which predicts

$$\epsilon_{max} \simeq \frac{f \times 10^9}{4.31 \times 2000} \simeq 10^5 \tag{3.81}$$

Thus for $f = 1$, $\epsilon_{max} \simeq 10^5$; $f = 0.1$, $\epsilon_{max} \simeq 10^4$; $f = 0.01$, $\epsilon_{max} \simeq 10^3$; $f = 0.001$, $\epsilon_{max} \simeq 10^2$, and so on.

3.7.3 Modification of Selection Rules

(i) Spin-orbital interaction. The rule governing the transition between the states of like multiplicities is most stringently obeyed. Transitions between ideal singlet and triplet states are strictly forbidden. But such transitions

do occur under the influence of intramolecular and intermolecular perturbations which can mix pure singlet and pure triplet states. These perturbations are functions of the magnetic field near the nucleus and are therefore a function of atomic mass (*heavy atom effect*). The Hamiltonian operator which causes the mixing of states of unlike multiplicities is expressed as

$$\mathcal{H}_{so} = K\xi(L\cdot S) \tag{3.82}$$

where ξ is a function which depends on the field of the nucleus, $(L\cdot S)$ is the scalar product of orbital and spin angular momentum vectors respectively ($L\cdot S = \alpha \cos\theta$, where θ is the angle between the two vectors) and K is a constant for the molecule. The wave function obtained on such spin-orbit coupling interactions which cause mixing of the pure triplet Ψ°_T and pure singlet Ψ°_S is expressed as

$$\Psi_{SO} = \Psi^{\circ}_T + \lambda\Psi^{\circ}_S \tag{3.83}$$

where λ indicates the degree of mixing and is given by

$$\lambda = \frac{\int \Psi^{\circ}_S \; \mathcal{H}_{so} \; \Psi^{\circ}_T \, d\tau}{|E_S - E_T|} \simeq \frac{V_{so}}{|E_S - E_T|} \tag{3.84}$$

E_S and E_T are the energies of the singlet and triplet states respectively, and V_{so} is the interaction energy which *flips* the electronic spin. Thus, smaller is the energy gap between singlet and triplet states, the larger is the mixing coefficient λ. V_{so} will be large also if the molecule is paramagnetic.

Therefore, under spin-orbital coupling interaction the transition moment $|\mathbf{M}|$ for transition from singlet ground state to a mixed excited state is given as

$$\begin{aligned} |\mathbf{M}| &= \int \Psi_{SO}\hat{\mu}\Psi_1 \, d\tau \\ &= \int (\Psi^{\circ}_T + \lambda\Psi^{\circ}_S)\hat{\mu}\Psi_1 \, d\tau \\ &= \underset{\text{(forbidden)}}{\int \Psi^{\circ}_T\hat{\mu}\Psi_1 \, d\tau} + \lambda \underset{\text{(allowed)}}{\int \Psi^{\circ}_S\hat{\mu}\Psi_1 \, d\tau} \end{aligned} \tag{3.85}$$

For the singlet ground state, the first term is zero but the second term contributes. The transition intensity is proportional to λ. From expression (3.82) and (3.84)

$$|\mathbf{M}| \propto \frac{\xi \int \Psi^{\circ}_S \, \mathrm{L\cdot S} \, \Psi^{\circ}_T \, d\tau}{|E_S - E_T|} \tag{3.86}$$

and is seen to be directly related to ξ and inversely to the energy separation between the singlet and triplet states. ξ is a function of the potential field near a nucleus and has a high value for an orbital which can penetrate close to the nucleus of a heavy atom such as iodine. The values for several atoms have been calculated from atomic spectral data and are presented in the Table 3.1.

TABLE 3.1

Values of ξ for atoms obtained from spectral data

Atom	ξ cm^{-1}
Carbon	28
Nitrogen	70
Oxygen	152
Chlorine	587
Bromine	2460
Iodine	5060
Lead	7294

A linear correlation with the atomic number Z, for $^1S \rightarrow {}^3P_1$ transition in Gr II atoms is illustrated in Figure 3.7. The heavy atom Hg has considerable intensity for intercombination transitions. The S→T transition is said to borrow intensity from S→S transition.

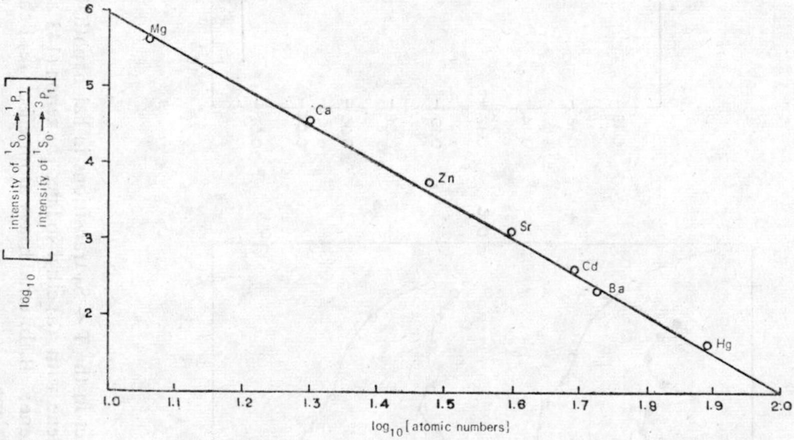

Figure 3.7 Spin-orbit coupling. Linear relation between atomic number Z and ratio of intensities $^1S_0 \rightarrow {}^1P_1$ and $^1S_0 \rightarrow {}^3P_1$ transitions in Gr II atoms. (After R.H. Hocstrasser, *Behaviour of Electrons in Atoms*, New York: W.A. Benjamin, 1964)

The effect is observed when the heavy atom is substituted in the molecule (intramolecular effect) as also when the molecule collides with a heavy atom containing perturber (intermolecular or external effect). The dramatic enhancement of S—T transition due to intramolecular and intermolecular heavy atom perturbations are respectively shown in Figure 3.8 for chloronaphthalenes in ethyl iodide and other perturbants. Molecular oxygen has a perturbing effect on S—T absorption spectra of organic molecules in solution. Under a pressure of 100 atm of O_2, well defined

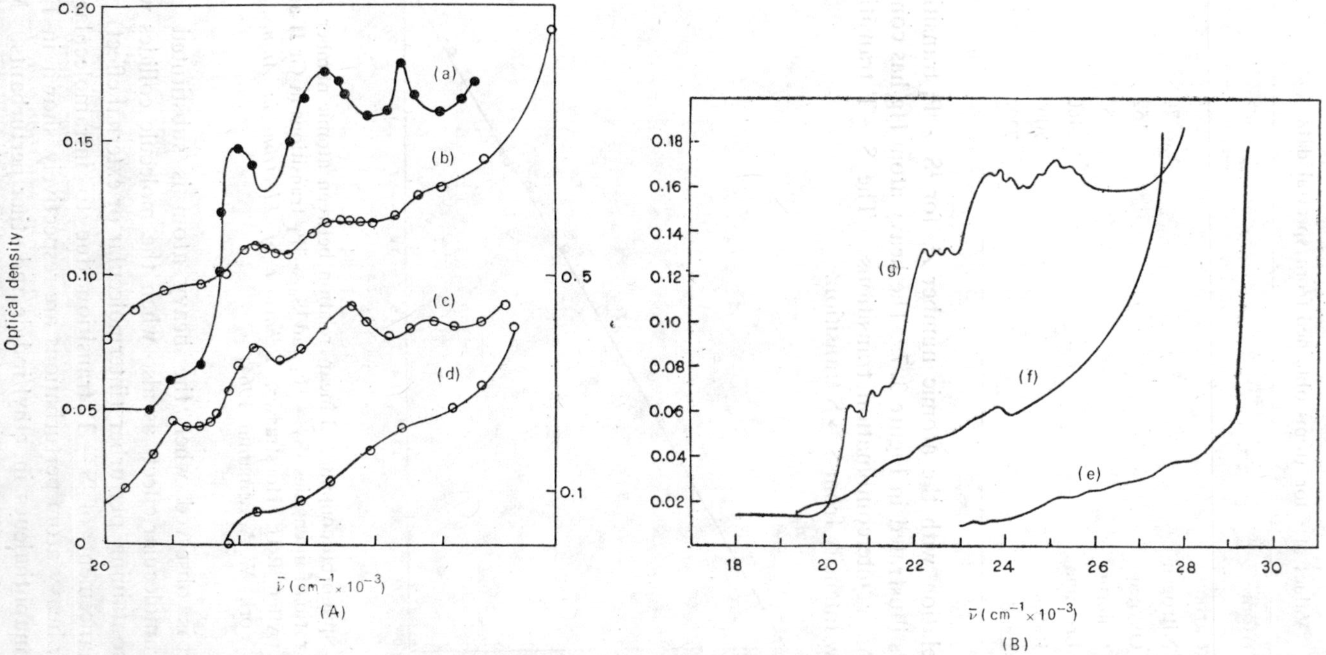

Figure 3.8 Heavy atom effect in the T ← S_0 transitions in halonaphthalenes. A. External heavy atom effect: l-chloronaphthalene with (a) ethyl iodide, (b) xenon (143 atm), (c) oxygen (30 atm) and (d) pure l-chloronaphthalene; B. Internal heavy atom effect: (e) l-chloronaphthalene, (f) 2-iodonaphthalene, (g) l-iodonaphthalene.

S – T absorption spectra are obtained for aromatics, heterocyclics, olefines and acetylenes. The transition disappears on removal of oxygen. It is suggested that a charge-transfer complex ($A^+O_2^-$) (Section 3.10.1) is formed which relaxes the spin multiplicity restrictions.

(*ii*) *Symmetry forbidden transitions–vibronic interactions.* It can be shown from group theoretical considerations (Section 3.8) that a $n \rightarrow \pi^*$ transition in formaldehyde should not occur. It is overlap forbidden since non-bonding orbitals on O-atom are directed perpendicular to the plane of π^* antibonding orbital in C=O group (Figure 3.9). In pyridine, the lone pair on nitrogen occupies an sp^2 hybrid orbital and because of the s-character, such a transition gains in intensity (Figure 3.9). When such transitions do occur though with very low intensity, they are said to be induced by unsymmetrical molecular vibrations.

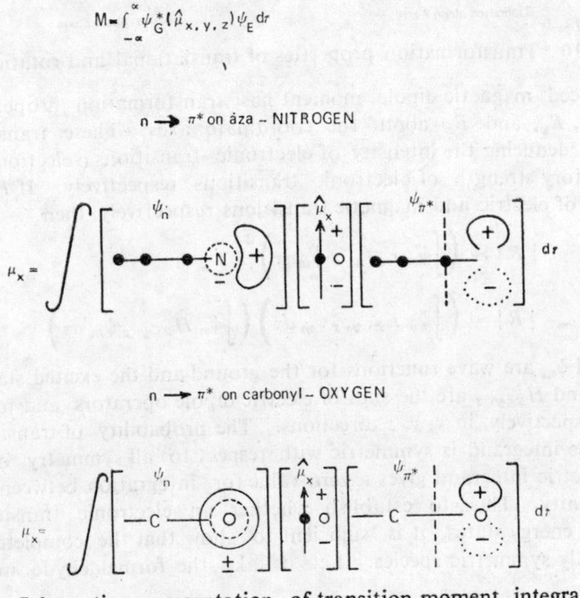

Figure 3.9 Schematic representation of transition moment integral in formaldehyde and pyridine. (Ref. 3.4).

A different kind of symmetry restriction is the cause of the small intensity of 260 nm absorption in benzene (${}^1B_{2u} \leftarrow {}^1A_{1g}$). The electric dipole moment vector of the electromagnetic radiation cannot generate ${}^1B_{2u}$ form of excited state from the totally symmetric hexagonal ground state designated as ${}^1A_{1g}$ (Figure 2.19). The restriction of the symmetry selection rule can be broken down by *vibration-electronic* (*vibronic*) *interactions.* The molecular vibration which can change the molecular symmetry for 260 nm transition in benzene is designated as E_{2g}.

3.7.4 Group Theoretical Approach to the Selection Rules

If we draw an arrow to the coordinate axes to which the symmetry of a given molecule is referred, then the transformation properties of these translation vectors under the symmetry operation of the group are the same as the electric dipole moment vector induced in the molecule by absorption of light (Figure 3.10).

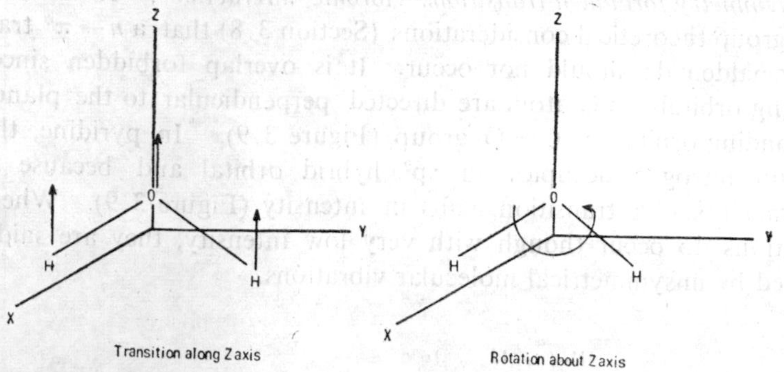

Figure 3.10 Transformation properties of translational and rotational vectors.

The induced magnetic dipole moment has transformation properties similar to rotations R_x, R_y, and R_z about the coordinate axes. These transformations are important in deducing the intensity of electronic transitions (selection rules) and the optical rotatory strength of electronic transitions respectively. If P and R are the probabilities of electric and magnetic transitions respectively, then

$$|P| \sim \left(\int \psi_n \hat{E}_{x,y,z} \psi_m d\tau\right)^2 \quad (3.87)$$

$$|R| \sim \left(\int \psi_n \hat{E}_{x,y,z} \psi_m d\tau\right)\left(\int \psi_n \hat{H}_{x,y,z} \psi_m d\tau\right) \quad (3.88)$$

where ψ_n and ψ_m are wave functions for the ground and the excited states respectively and $E_{x,y,z}$ and $H_{x,y,z}$ are the sum of electric dipole operators and magnetic dipole operators, respectively, in x, y, z directions. The probability of transition is nonzero only when the integrand is symmetric with respect to all symmetry operations since an antisymmetric integrand gives a zero value for integration between minus infinity and plus infinity. Thus, to establish whether an electronic transition is allowed between two energy states, it is sufficient to show that the complete integrand belongs to totally symmetric species. Let us take the formaldehyde molecule for an example.

The state symmetries of formaldehyde can easily be deduced from group theoretical considerations. The wave functions of π, n and π^* orbitals are pictorially represented in Figure 3.11 with coordinate axes as used for the molecule.

Like water, it has the symmetry of point group C_{2v}. When the MOs of formaldehyde molecule as given in the figure are subjected to the symmetry operations of this group, π-orbital is observed to transform as b_1 and n orbital as b_2 as shown below:

$$E(\pi) = +1(\pi)$$
$$C_2(\pi) = -1(\pi)$$
$$\sigma_v^{xz}(\pi) = +1(\pi)$$
$$\sigma_v^{yz}(\pi) = -1(\pi)$$

hence π transforms as b_1

$$E(n) = +1(n)$$
$$C(n) = -1(n)$$
$$\sigma_v^{xz}(n) = -1(n)$$
$$\sigma_v^{yz}(n) = +1(n)$$

hence n transforms as b_2

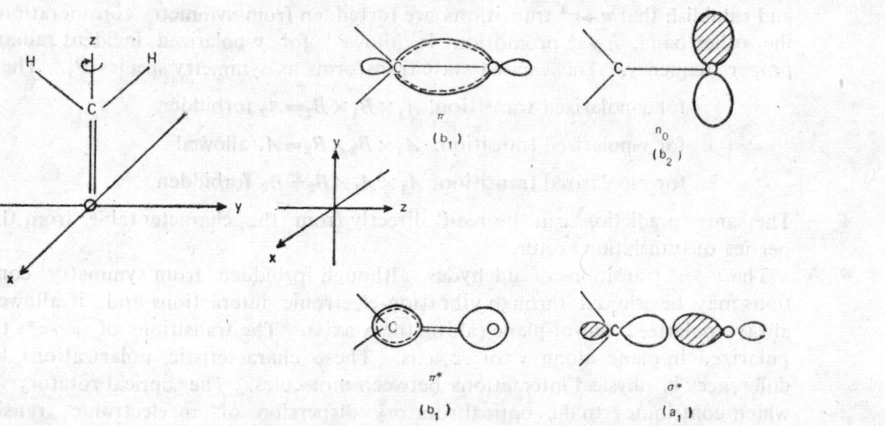

Figure 3.11 Symmetry and charge distribution in molecular orbitals of formaldehyde: n (b_2), π (b_1) and π^* (b_1) and σ^* (a_1).

Similarly, π^* and σ^* transform as the symmetry species b_1 and a_1 respectively.

The electronic configuration of formaldehyde under different types of electronic transitions can be represented in terms of electron occupancy (Table 3.2). To obtain the molecular state symmetry the symmetry of singly occupied orbitals need only be considered for the direct product (Section 2.9).

TABLE 3.2

The energy state, electronic configuration, molecular state symmetry and the nature of transition in formaldehyde molecule

Energy state	Electronic configuration	Molecular state symmetry	Selection rule for polarization of transition dipole
S_o	$\sigma^2\pi^2n^2$	A_1	
$S_{n\pi}*$	$\sigma^2\pi^2n^1\pi^{*1}$	$b_2\times b_1=A_2$	overlap forbidden
$S_{\pi\pi}*$	$\sigma^2\pi^1n^2\pi^{*1}$	$b_1\times b_1=A_1$	z-polarized transition
S_n	$\sigma^2\pi^2n^1\sigma^{*1}$	$b_2\times a_1=B_2$	y-polarized transition

Once the state symmetries have been established it only remains to be shown that the direct product of the species of the ground state symmetry, the coordinate translational symmetries and the excited state symmetry belong to the totally symmetric species A_1. Let us take the $n\rightarrow\pi^*$ transition in formaldehyde. The ground state total wave function has the symmetry A_1. The coordinate vectors x, y and z transform as B_1, B_2 and A_1 respectively (refer Character Table for C_{2v}; Section 2.9, Table 2.2), The excited state transforms as symmetry species A_2. The direct products are:

for x-polarized transition, $A_1\times B_1\times A_2=B_2$ forbidden

for y-polarized transition, $A_1\times B_2\times A_2=B_1$ forbidden

for z-polarized transition, $A_1\times A_1\times A_2=A_2$ forbidden and establish that $n \rightarrow \pi^*$ transitions are forbidden from symmetry considerations. On the other hand, $n \rightarrow \sigma^*$ promotion is allowed for y-polarized incident radiation of proper frequency. This excited state transforms as symmetry species B_2. Therefore,

for x-polarized transition, $A_1 \times B_1 \times B_2 = A_2$ forbidden

for y-polarized transition, $A_1 \times B_2 \times B_2 = A_1$ allowed

for z-polarized transition, $A_1 \times A_1 \times B_2 = B_2$ forbidden

The same predictions can be read directly from the character table from the properties of translation vector.

The $n \rightarrow \pi^*$ transitions of aldehydes, although forbidden from symmetry considerations may be allowed through vibration-electronic interactions and, if allowed, are always polarized out-of-plane (along the x axis). The transitions of ($\pi \rightarrow \pi^*$) type are polarized in-plane along y or z axis. These characteristic polarizations lead to differences in physical interactions between molecules. The optical rotatory strength which contributes to the optical rotatory dispersion of an electronic transition is given by a scalar product of two different vector integrals (equation 3.88). The first integral is the transition moment integral and the second integral involves the magnetic moment operators which have the symmetry properties of rotation R_x, R_y and R_z. The translations T_x, T_y and T_z and the rotations R_x, R_y and R_z transform in a complementary manner. If the molecule has a plane of symmetry or a centre of inversion, the product of the two integrals may be zero. This explains the fact that optical activity can be observed only for asymmetric molecules, with no plane or centre of symmetry (inversion).

3.8 DIRECTIONAL NATURE OF LIGHT ABSORPTION

Since the electric vector of the incident radiation is responsible for the creation of a dipole moment during the act of absorption, a difference in the strength of absorption is expected for different orientations of an isotropic molecule. The absorption spectra as obtained with ordinary light do not give this information. For the determination of the direction of polarization of the oscillator, absorption spectra need to be obtained with polarized radiation in rigid systems of properly oriented molecules such as crystals. The directional nature of light absorption can be explained from the considerations of symmetry properties of the energy levels involved in the transition. It is best understood by discussing few specific examples.

Example 1

Let us consider the simplest molecule H_2. In the ground state the two electrons occupy the lowest energy level of symmetry Σ_g^+. Since the molecule is anisotropic two possibilities arise—if we use a plane polarized light, the electric vector of the light is either parallel to the molecular axis or it is perpendicular to it (Figure 3.12). In each case, the dipole direction will be along the direction of the electric vector (e v) and a new node will be formed at right angles to the e v. The symmetry of the two lowest excited states of H_2-molecule corresponds to Σ_u and π_u respectively. When the incident radiation is polarized parallel to the molecular axis, the transition dipole created is also in the parallel direction with the nodal plane perpendicular to it. If the frequency of radiation ν_1 is correctly chosen, an electron will be promoted from Σ_g to Σ_u. On the other hand, when the plane of polarization directs the electric

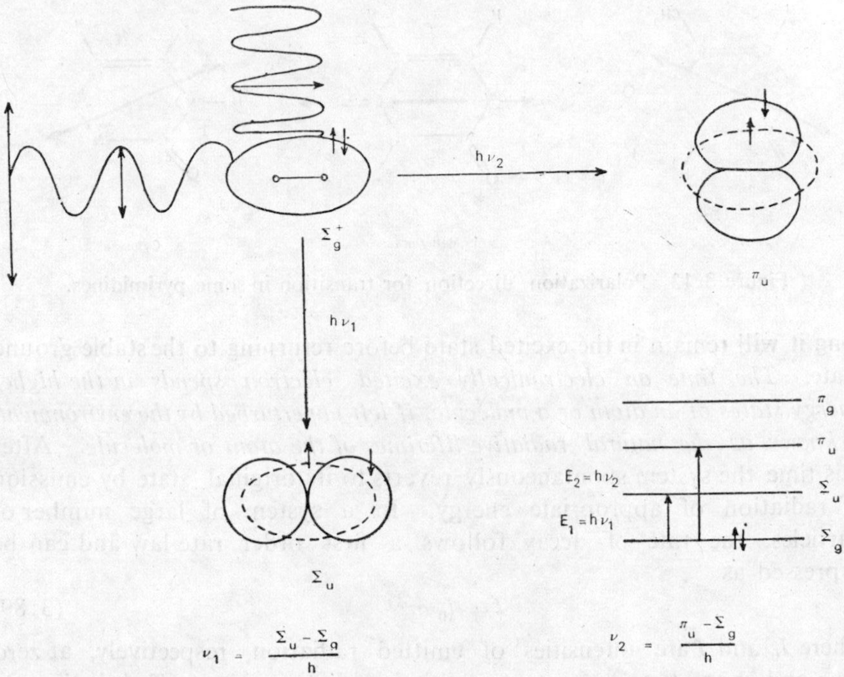

Figure 3.12 Directional nature of light absorption in H_2 molecule.

vector perpendicular to the molecular axis, the charge density can only change in a perpendicular direction with the nodal plane along the molecular axis. The transition is to π_u state if the choice of frequency ν_2 is correct. It is assumed that the molecule is fixed in its orientation.

Example 2

Anthracene absorbs at two wavelengths, at 360 nm and at 260 nm. The flat molecule is anisotropic and it has long axis along x coordinates and a short axis along y coordinate. The absorption at 360 nm is short axis polarized (L_a type). A substituent at 9, 10 or 1, 4, 5, and 8 positions may help or retard the creation of dipole in this direction. Therefore, the intensity or position of this absorption region may be influenced by such substitutions. The absorption at 260 nm is long axis polarized (L_b type) and is perturbed by substitution at 2, 3, 6 and 7 positions. Such substituent effects are sometimes used to identify the polarization directions of a given electronic transition.

The polarization directions for transition in some pyrimidine molecules are illustrated in Figure 3.13.

3.9 LIFETIMES OF EXCITED ELECTRONIC STATES OF ATOMS AND MOLECULES

When once an electron is raised to the higher energy state by absorption of a quantum of radiation, the question naturally arises as to how

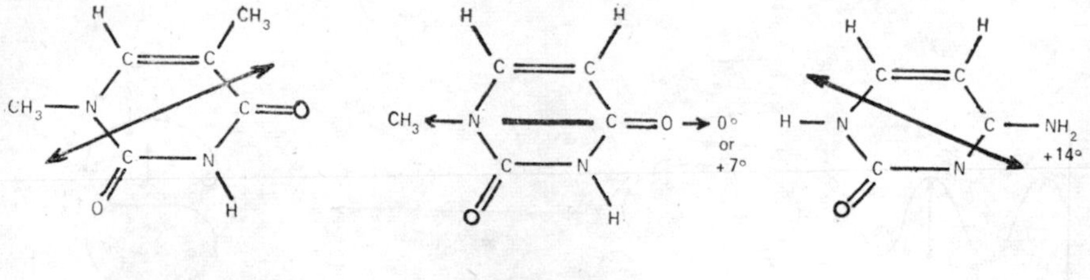

Figure 3.13 Polarization direction for transition in some pyrimidines.

long it will remain in the excited state before returning to the stable ground state. *The time an electronically excited electron spends in the higher energy states of an atom or a molecule, if left unperturbed by the environment, is known as the natural radiative lifetime of the atom or molecule.* After this time the system spontaneously reverts to its original state by emission of radiation of appropriate energy. In a system of large number of particles, the rate of decay follows a first order rate law and can be expressed as

$$I = I_0 \, e^{-kt} \tag{3.89}$$

where I_0 and I are intensities of emitted radiation, respectively, at zero time and at any time t after the exciting radiation is cut off; k is the rate constant for the emission process and has the dimension of reciprocal time.

$$\text{If } k = 1/\tau \text{ then } I = I_0 \, e^{-t/\tau} \tag{3.90}$$

and for a time $t = \tau$

$$I = I_0 \, e^{-1} = I_0 \, \frac{1}{e} \tag{3.91}$$

Therefore the lifetime is defined as *the time taken for the radiation intensity to decay to 1/e th of its original value* (Figure 3.14). Thus the natural radiative lifetime τ_N is inversely related to the rate constant for the spon-

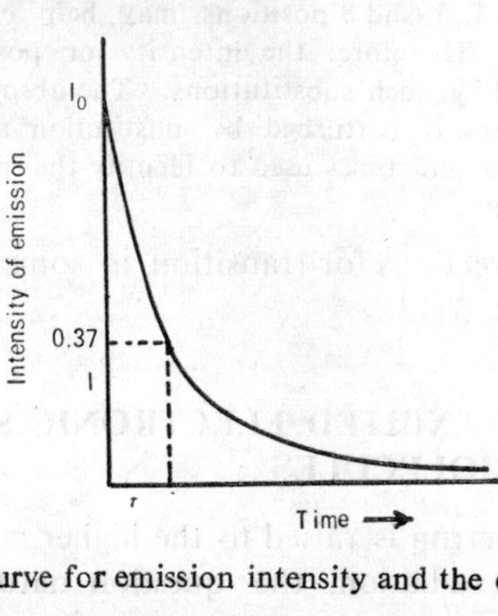

Figure 3.14 Decay curve for emission intensity and the definition of lifetime τ.

taneous emission process. The latter is identical with Einstein's spontaneous emission probability A_{mn} from all the upper energy states of the system to the ground energy state. From equation 3.10

$$k = A_{mn} = \frac{8\pi h\nu^3}{c^3} B_{mn} \tag{3.92}$$

On substitution of the value for B_{mn}, from (3.61)

$$k = 8\pi \bar{\nu}^2_{max}\, c \frac{2303}{N} \int \epsilon_{\bar{\nu}}\, d\bar{\nu} \tag{3.93}$$

or

$$\frac{1}{\tau_N} = \frac{2303}{N}\, 8\pi \bar{\nu}^2_{max}\, c \int \epsilon_{\bar{\nu}}\, d\bar{\nu} \tag{3.94}$$

and

$$\tau_N = \frac{3.47 \times 10^8}{\bar{\nu}^2_{max}} \frac{1}{\int \epsilon_{\bar{\nu}}\, d\bar{\nu}} = \frac{1.5}{\bar{\nu}^2_{max} f} \tag{3.95}$$

A more general expression for molecules dissolved in a medium of refractive index n, is

$$\tau_N = \frac{3.47 \times 10^8}{n^2 \bar{\nu}^2_{max}} \frac{g_m}{g_n} \frac{1}{\int \epsilon_{\bar{\nu}}\, d\bar{\nu}} \tag{3.96}$$

where g_m/g_n is a multiplicity correction factor, useful for intercombination transition between a singlet and a triplet state. Here g_m and g_n are the spin statistical factors for the upper and the lower energy states, respectively. For singlet ← singlet transition, $g_m = 1$, and $g_n = 1$; for triplet ← singlet transition $g_m = 3$, $g_n = 1$. The expression (3.95) is truly applicable to atomic transitions only but it gives fairly good results for molecular cases also.

A more refined expression is given by Förster,

$$\frac{1}{\tau_N} = 2.88 \times 10^{-9}\, n^2 \int \frac{(2\bar{\nu}_0 - \bar{\nu})^3}{\bar{\nu}} \epsilon_{\bar{\nu}}\, d\bar{\nu} \tag{3.97}$$

where $\bar{\nu}_0$ is the wave number of the approximate mirror image reflection plane of absorption and fluorescence spectra plotted as function of $\bar{\nu}$ (Section 5.7). A spin statistical factor should be included for intercombinational transitions. The theoretical basis for the empirical mirror symmetry relation between fluorescence and absorption spectra for large molecules depends on the approximate parallel nature of the potential energy surfaces of ground and excited states of molecules such as dyes where bond lengths are little affected because of delocalization of the electronic energy. Such relationship may not exist for small molecules.

The natural radiative lifetime gives an upper limit to the lifetime of an excited molecule and can be calculated from the integrated absorption intensity. The quantity to be plotted is $\epsilon_{\bar{\nu}}$ vs $\bar{\nu}$ if (3.96) is used and

$(2\bar{\nu}_0 - \bar{\nu})^3 \epsilon_{\bar{\nu}}/\bar{\nu}$ vs $\bar{\nu}$, if (3.97) is used. More accurate experimental methods are given in Section 10.2. For fully allowed transitions, $f = 1$ and from 3.96,

$$\tau_N = \frac{1.5}{\bar{\nu}_{max}^2 f} \simeq 10^{-9}\, s \qquad (3.98)$$

The value of τ_N is greater for a forbidden transition whose f-number is less than unity. Therefore, weak absorption bands imply long lifetimes and strong absorption bands imply short lifetimes. In Table 3.3 are recorded some experimental values and compared with those calculated from (3.96). The factor g_m/g_n is equal to 1 and 3 for singlet and triplet transitions, respectively.

TABLE 3.3

Experimental and calculated lifetimes for singlet-singlet (*S–S*) and singlet-triplet (*S–T*) transitions of ($\pi \rightarrow \pi^*$) and ($n \rightarrow \pi^*$) type

Compound	(Exptl) in s	(Calc) in s	Transition
Anthracene	20.0×10^9	16.0×10^9	$S-S$ $\pi \rightarrow \pi^*$
Perylene	5.1 ,,	5.6 ,,	$S-S$ $\pi \rightarrow \pi^*$
9, 10 diphenyl anthracene	8.9 ,,	8.8 ,,	$S-S$ $\pi \rightarrow \pi^*$
9, 10 dichloro anthracene	11.00 ,,	15.4 ,,	$S-S$ $\pi \rightarrow \pi^*$
Fluorescein	4.7 ,,	5.0 ,,	$S-S$ $\pi \rightarrow \pi^*$
Rhodamine B	6.0 ,,	6.0 ,,	$S-S$ $\pi \rightarrow \pi^*$
Anthracene	90×10^3	0.1×10^3	$S-T$ $\pi \rightarrow \pi^*$
Biacetyl	1.5 ,,	0.7 ,,	$S-T$ $n \rightarrow \pi^*$
Acetophenone	8.0 ,,	3.0 ,,	$S-T$ $n \rightarrow \pi^*$
2-Iodonaphthalene	2.0 ,,	3.0 ,,	$S-T$ $\pi \rightarrow \pi^*$

A correlation table for f, ϵ and τ_N is given in Figure 3.15.

The natural radiative lifetime is independent of temperature, but is susceptible to environmental perturbations. Under environmental perturbation, such as collisions with the solvent molecules or any other molecules present in the system, the system may lose its electronic excitation energy by nonradiative processes. Any process which tends to compete with spontaneous emission process reduces the life of an excited state. In an actual system the average lifetime τ is less than the natural radiative lifetime τ_N as obtained from integrated absorption intensity. In many polyatomic molecules, nonradiative intramolecular dissipation of energy may occur even in the absence of any outside perturbation, lowering the inherent lifetime.

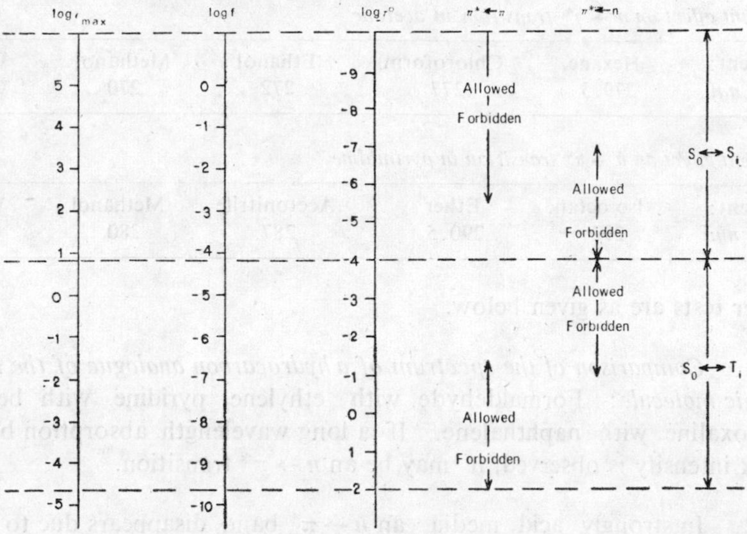

Figure 3.15 Correlation between oscillator strength f, molar extinction coefficient ϵ and lifetime τ for different transition types.

3.10 TYPES OF ELECTRONIC TRANSITIONS IN ORGANIC MOLECULES

As already discussed (Section 2.10), the types of electronic orbitals that may be present in organic molecules are σ, π, n, π^* and σ^*. Energy-wise, they are arranged as given in the Figure 2.17. Photochemists designate them by an arrow from the lower energy state to the upper energy state as $n \rightarrow \pi^*$, $\pi \rightarrow \pi^*$, $n \rightarrow \sigma^*$, $\sigma \rightarrow \sigma^*$, etc. They appear in the appropriate regions of the spectrum. $n \rightarrow \pi^*$ and $\pi \rightarrow \pi^*$ are normally observed in the near UV and visible regions and are most characteristic transitions. Kasha has developed tests described below for distinguishing between these two types of transtitions, of which solvent perturbation technique is the most convenient.

Kasha's tests for identification of $n \rightarrow \pi^*$ and $\pi \rightarrow \pi^*$ transitions. Solvent perturbation technique is a useful way to identify transitions as $n \rightarrow \pi^*$ or $\pi \rightarrow \pi^*$ for complex molecules. While comparing bands of different orbital promotion types in hydroxylic solvents such as water and ethanol with those in hydrocarbon or nonpolar solvents, if the band shifts towards the high frequency or shorter wavelength side (blue-shift) then the transition is probably $n \rightarrow \pi^*$. If there is a small red shift, the transition is likely to be $\pi \rightarrow \pi^*$. The effect of solvents on the $n \rightarrow \pi^*$ transition in acetone and pyrimidine is shown in the Table 3.3.

TABLE 3.3

Solvent effect on $n \rightarrow \pi^*$ transition in acetone

Solvent:	Hexane	Chloroform	Ethanol	Methanol	Water
λ_{max} *nm*	279.5	277	272	270	266.5

Solvent effect on $n \rightarrow \pi^*$ transition in pyrimidine

Solvent:	Iso octane	Ether	Acetonitrile	Methanol	Water
λ_{max} *nm*	292	290.5	287	280	271

Other tests are as given below.

1. *Comparison of the spectrum of a hydrocarbon analogue of the hetero-atomic molecule*: Formaldehyde with ethylene, pyridine with benzene, quinoxaline with naphthalene. If a long wavelength absorption band of weak intensity is observed, it may be an $n \rightarrow \pi^*$ transition.

2. In strongly acid media, an $n \rightarrow \pi^*$ band disappears due to protonation of the lone pair. The protonation may increase the excitation energy to an extent that the band may shift far out into the UV region and not be observed.

3. The intensity of the absorption band of an $n \rightarrow \pi^*$ transition, even when symmetry allowed, is much smaller than that of a $\pi \rightarrow \pi^*$ transition. This is because of the poor overlap of the wave functions of ground and excited states in $n \rightarrow \pi^*$ transitions.

4. There are significant differences in the polarization directions of absorption and emission spectra for two types of transition. In the case of cyclic compounds, like pyridine, the transition vectors lie in the plane of the molecule for $\pi \rightarrow \pi^*$ transition but perpendicular to it for $n \rightarrow \pi^*$ transition.

TABLE 3.4

Different nomenclatures for several important transitions in polyatomic molecules

	Ethylene λ_{max}=165 nm $\epsilon_{max} \simeq 1.5 \times 10^4$	Benzene (perimeter model) λ_{max}=256 nm $\epsilon_{max} \simeq 1.6 \times 10^2$	Formaldehyde λ_{max}=304 nm $\epsilon_{max} \simeq 18$
Group theory	${}^1B_{1u} \leftarrow {}^1A$	${}^1B_{2u} \leftarrow {}^1A_{1g}$	${}^1A_2 \leftarrow {}^1A_1$
Mulliken state	$V \leftarrow N$	$V \leftarrow N$.	$Q \leftarrow N$
Platt	${}^1B \leftarrow {}^1A$	${}^1L_b \leftarrow {}^1A$	${}^1U \leftarrow {}^1A$
Molecular orbital	$\sigma^2\pi^2 \rightarrow \sigma^2\pi\,\pi^*$	$(a_{1u})^2\,(e_{1g})^3\,(e_{2u})^1 \leftarrow (a_{1u})^2\,(e_{1g})^4$ degenerate	$\sigma^2\pi^2 p_y^2 \rightarrow \sigma^2\pi^2 p_y \pi^*$
Kasha	$\pi \rightarrow \pi^*$	$\pi \rightarrow \pi^*$	$n \rightarrow \pi^*$

In *spectroscopic convention*, the upper energy state is written first followed by the lower energy state between which the transition is taking place. For absorption, the arrow points from left to right, e.g. for the lowest energy transition in benzene at 260 nm, in absorption: $^1B_{2u} \leftarrow {}^1A_{1g}$ and in emission: $^1B_{2u} \rightarrow {}^1A_{1g}$. Other nomenclatures are given in the Table 3.4.

3.10.1 Charge Transfer Transitions

When two molecules which are likely to form loose complexes are mixed together, very often a new featureless band appears (Figure 3.16),

$$D + A \rightleftharpoons (DA)_{complex} \overset{h\nu}{\rightleftharpoons} (D^+ - A^-)_{complex}$$

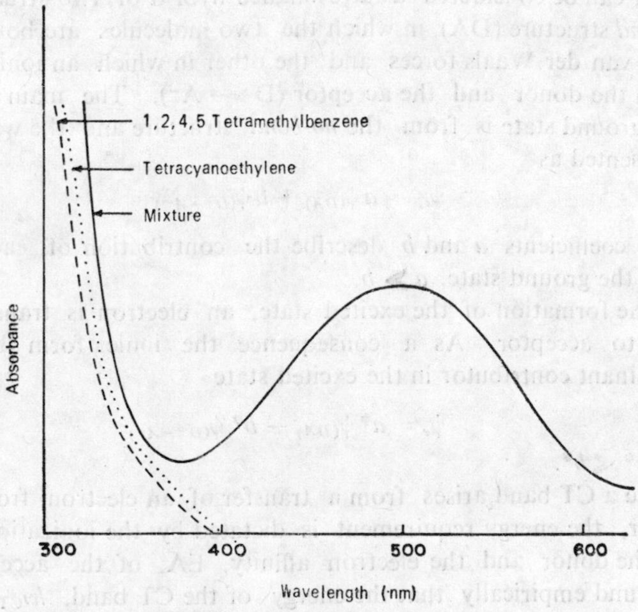

Figure 3.16 Charge transfer band for the complex between 1, 2, 4, 5-tetramethylbenzene and tetracyanoethylene. [Adapted from Britain, Gerrge and Wells, *Introduction to Molecular Spectroscopy-Theory and Experiment*. Academic Press, 1972.]

This band arises due to the transition of an electron from the molecule with high charge density (donor) to the molecule with low charge density (acceptor). The transition occurs from the highest filled molecular orbital of the donor to the lowest empty molecular orbital of the acceptor. Promotion of an electron to higher unoccupied orbitals of the acceptor can occur to give additional charge transfer absorption bands. Schematic representation of CT (charge transfer) transitions is given in Figure 3.17.

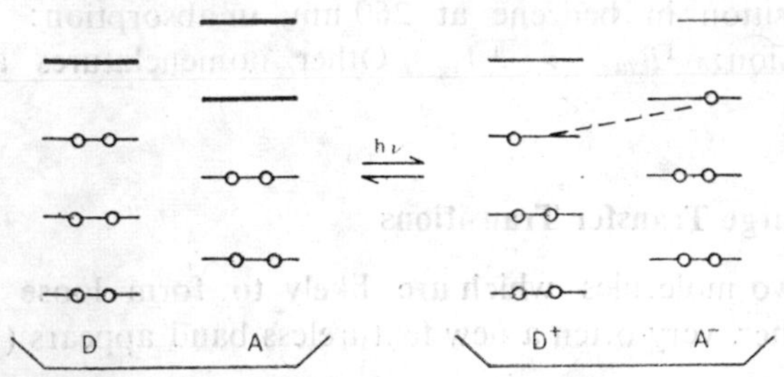

Figure 3.17 Schematic representation of CT transition.

The molecular complex formed by the interaction of donor and acceptor can be considered as a resonance hybrid of two structures: one a *no-bond* structure (DA) in which the two molecules are bound together only by van der Waals forces and the other in which an ionic *bond* exists between the donor and the acceptor ($D^+ - A^-$). The main contribution to the ground state is from the *no-bond* structure and the wave function is represented as

$$\psi_G = a\,\psi_{(DA)} + b\,\psi_{(D^+-A^-)} \tag{3.99}$$

The coefficients a and b describe the contribution of each structure and for the ground state, $a \gg b$.

In the formation of the excited state, an electron is transferred from donor to acceptor. As a consequence the ionic form becomes the predominant contributor in the excited state

$$\psi_e = a^*\,\psi_{(DA)} - b^*\,\psi_{(D^+-A^-)} \tag{3.100}$$

where $a^* \ll b^*$.

Since a CT band arises from a transfer of an electron from donor to acceptor, the energy requirement is dictated by the ionization potential, IP, of the donor and the electron affinity, EA, of the acceptor. It has been found empirically that the energy of the CT band, $h\nu_{CT}$, for a series of donors and the same acceptor can be expressed as

$$h\nu_{CT} = IP - EA + \Delta \tag{3.101}$$

where Δ is a constant for the related series of donors.

The intensity of an $S_1 \leftarrow S_0$ charge transfer band is given approximately by

$$f = 1.093 \times 10^{-5} s^2 d^2 \bar{\nu} \tag{3.102}$$

where S is the overlap integral given by $S = \int \psi_n \psi_m \, d\tau$, ψ_n and ψ_m being donor and acceptor orbitals respectively, d is the distance between the donor and the acceptor centres in Å units, and $\bar{\nu}$ is the transition energy in wavenumbers (cm^{-1}). The *f*-numbers can vary over a large range and are sensitive to *S* and *d* values.

There are two types of donors, π-donors and *n*-donors. In the former, the electron available for donation is located in the π-MO of the molecule, e.g. aromatic hydrocarbons, alkenes, alkynes. They are said to form *π-complexes.* In the second type of donors, a nonbonding electron is transferred. The examples of *n*-donors are alcohols, amines, ethers, etc.

There are many types of acceptor molecules. Some of the inorganic acceptors are halogens, metal halides, Ag^+, etc. Of the organic acceptors such compounds as tetracyanoethylene, trinitrobenzene and others which contain highly electronegative substituents are more important. In transition metal complexes, intramolecular CT transitions may be observed in those cases in which an electron is transferred from the ligand to the metal ion. Such CT bands shift to the red when ligands with lower electron affinity are substituted as in the case of $[Co(NH_3)_5X]^{3+}$ where *X* is a halide ligand of F, Cl, Br, and I. The *intramolecular* CT transitions occur in derivatives of benzene with a substituent containing *n* electrons such as carbonyl or amino group. Charge transfer transitions are important in biological systems.

3.10.2 Characteristics of π → π*, n → π* and CT Transition

Aromatic carbonyl compounds provide examples in which absorption bands can be identified as due to π → π*, *n* → π* and CT transitions. A rough guideline for identification of a given band due to any of these transitions and properties of the corresponding energy states have been given in Table 3.5

TABLE 3.5

Characteristics of the excited states of aromatic carbonyl compounds

State	(*n*, π*)	(π, π*)	CT
Transitions	*n*→π*	π→π*	CT
Intensity (log ε)	2	3–4	3–4
Energy range (cm^{-1})	30,000	40,000	30,000
Solvent shift (cm^{-1})	−800	+600	+2300
T–S splitting (cm^{-1})	3,000	8,000	3,000

(Table 3.5 Contd.)

State	(n, π^*)	(π, π^*)	CT
Triplet lifetime at 77° K (s)	<0.02	>0.1	>0.1
Electron distribution on C=0	$\delta-$ $\delta+$ >C=O	>C=0	$\delta+$ $\delta-$ >C=O
Polarization of transition moment	perpendicular to the molecular plane	in the molecular plane	perpendicular to the molecular plane
Reactivity or H-abstraction from iso-propanol	1	0.1	0.0

3.10.3 Charge Transfer to Solvent Transitions

The absorption spectra of anions are very sensitive to the composition of solvents in which they are embedded. In general, they are solvated, i.e. they are surrounded by a solvent shell. The molecules composing the solvent shell constantly exchange position with those in the bulk of the solvent. In these transitions, an electron is ejected not into the orbitals of a single molecule, but to a potential well defined by the group of molecules in the solvation shell. Such transitions are known as charge-transfer-to-solvent (CTTS) transitions.

A classic example is the absorption spectra of alkali iodide ions which are found to depend on the nature of the solvent but not on the nature of the alkali. The spectra consist of two absorption maxima at energies 529.3 kJ and 618.4 kJ, which closely correspond to the energy difference of 90.92 kJ, between I ($5^2P_{3/2} - 5^2P_{1/2}$) of I atoms in the gaseous state. The steps in the transition process can be represented as follows:

$$X^-_{aq} \overset{h\nu}{\rightleftharpoons} (X^\cdot + e^-)_{aq} \rightleftharpoons X^\cdot_{aq} + e^-_{aq}$$

In aqueous solutions, it has been shown that the solvated electron e^-_{aq} circulates in the solvation shell until it is captured by I_2 to given an I_2^- radical ion which is finally stabilized to I_3^-. The solvated electrons have a characteristic absorption band near 700 nm which has been detected in flash photolytic studies of aqueous KI. The orbital of the excited electron may be considered to be spherically symmetric like that of a hydrogen atom, with its centre coinciding with the centre of the cavity containing the ion.

Similar CTTS spectra have been observed for $Fe(CN)_6^{4-}$, aqueous solutions of phenate ions and many other inorganic ions. Some examples are given in Table 3.6.

TABLE 3.6
CTTS processes in aqueous solution

Ion	λ (nm)	Process
I^-	253.7	$I\cdot + e^-_{aq}$
X^- ($X = Cl^-, Br^-$)	184.9	$X\cdot + e^-_{aq}$
OH^-	184.9	$\cdot OH + e^-_{aq}$
SO_4^{2-}	184.9	$\cdot SO_4^- + e^-_{aq}$
CO_3^{2-}	184.9	$\cdot CO_3^- + e^-_{aq}$
$Fe(CN)_6^{4-}$	253.7	$Fe(CN)_6^{3-} + e^-_{aq}$
Fe^{2+}_{aq}	253.7	$Fe^{3+} + e^-_{aq}$

3.11 TWO-PHOTON ABSORPTION SPECTROSCOPY

The study of short lived excited states is limited by the low concentrations in which they are created on excitation with normal light sources. The use of high intensity sources such as flash lamps with suitable flashing rates and laser sources have been helpful in this respect. Triplet-triplet absorption, absorption by excited singlet state to higher singlet state and absorption by exciplexes (Section 6.6.1) can be effectively observed by *sequential* biphotonic processes.

The *simultaneous* absorption of two visible photons in a single transition can also be made to occur using high light intensities available from lasers. *Two-photon absorption spectroscopy* is able to identify and assign atomic and molecular states which are not accessible to the usual one-photon absorption. The electric dipole (ED) allowed one-photon absorption has high transition probability between states of opposite parity $g \rightarrow u$, $u \rightarrow g$ ($\Delta l = \pm 1$), but two-photon transition can occur between states of same parity $g \rightarrow g$, $u \rightarrow u$, ($\Delta l = \pm 2$). The absorption results from two ED transitions. The quantum picture for two-photon simultaneous process is that if two photons λ and μ of frequencies $\bar{v}_\lambda$ and $\bar{v}_\mu$ each of which individually is not in resonance with any of the energy levels of the absorbing molecules but are so when combined, can induce free oscillations in the molecule. The two photons are destroyed simultaneously and the molecule is raised to an excited state. If the two photons strike at right angles to each other, they can overcome the symmetry barrier for the selection rules for angular momentum for single photon transitions. On the other hand, transitions between states of opposite parity become disallowed.

Thus for naphthalene, the transition (A) is forbidden for single photon spectroscopy both for x and y polarized process, the transition moment integral $\mathbf{M} = 0$. But it becomes allowed for two photon process (B) as shown in Figure 3.18. The nonzero value of transition moment integral indicates allowedness of transition. $\mathbf{M}_{gg}$ is transition moment integral for $g \rightarrow g$ transition.

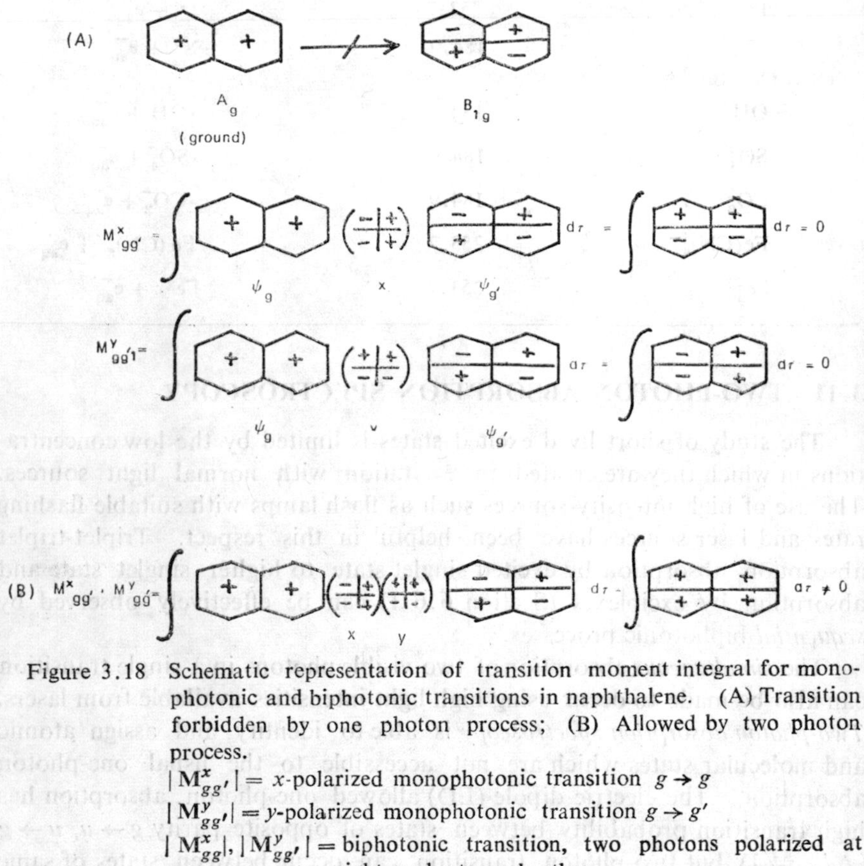

Figure 3.18 Schematic representation of transition moment integral for monophotonic and biphotonic transitions in naphthalene. (A) Transition forbidden by one photon process; (B) Allowed by two photon process.
$|\mathbf{M}^x_{gg'}| =$ x-polarized monophotonic transition $g \rightarrow g$
$|\mathbf{M}^y_{gg'}| =$ y-polarized monophotonic transition $g \rightarrow g'$
$|\mathbf{M}^x_{gg}|, |\mathbf{M}^y_{gg'}| =$ biphotonic transition, two photons polarized at right angles to each other.
[Adapted from W.H. McClain—Acc. Chem. Res. 7 (1974) 129]

Summary

1. Electric dipole radiation is the most important component involved in normal excitation of atoms and molecules. The electric dipole operator has the form $\Sigma e_j x_j$ where e is the electronic charge in esu and x_j is the displacement vector for the jth electron in the oscillating electromagnetic field.
2. The dipole moment induced by the light wave is a transitory dipole moment and is not related to the permament dipole moment of the molecule. It is known as *transition moment.*
3. The transition moment is measured by the *transition moment integral* which for a transition between the energy states $m \rightarrow n$ is expressed as

$$|\mathbf{M}_{mn}| = \int \psi_m \hat{\mu} \psi_n \, d\tau$$

where $\hat{\mu}$ is the transition moment operator. It is derived from the Schrödinger equation including time by application of time dependent perturbation imposed by oscillating electric field.

4. The symmetries of the initial and the final wave functions and of the electromagnetic radiation operator determine the allowedness or forbiddenness of an electronic transition. The transition moment integrand must be totally symmetric for an allowed transition such that $|M_{mn}|^2 \neq 0$.
5. A transition may be forbidden due to operation of the law of conservation of spin momenta $\Delta S = 0$.
6. Spin selection rule can be modified by spin-orbit coupling interactions.
7. Symmetry forbidden transition can be made partially *allowed* by vibronic interactions.
8. A totally allowed transition has oscillator strength $f = 1$ and molar extinction coefficient $\sim 10^5$. Different factors may reduce the f values to different extents. Oscillator strengths f are related to integrated absorption intensities by the expression

$$f = 4.31 \times 10^{-9} \int \epsilon_{\bar{\nu}} \, d\bar{\nu}$$

9. The probability of absorption given by the Einstein coefficient of induced absorption B_{mn} can be expressed in terms of $|M|^2$,

$$B_{nm} = \frac{8\pi^3}{3h^2} |M|^2$$

and Einstein's coefficient for spontaneous emission.

$$A_{mn} = \frac{64\pi^4 \nu^3_{mn}}{3hc^3} |M|^2$$

10. Excited state lifetimes are related to the Einstein coefficients of spontaneous emission A_{mn} and can be approximately calculated from the expression

$$\frac{1}{\tau_N} = 2.88 \times 10^{-9} \, n^2 \frac{g_{m-2}}{g_n} \nu_{max} \int \epsilon_{\bar{\nu}} \, d\bar{\nu}$$

11. $\pi \rightarrow \pi^*$, $n \rightarrow \pi^*$ and charge transfer (CT) transitions can be identified by solvent perturbation technique and by the shape and intensity of the corresponding absorption bands.
12. Some solvated anions like I^-, $Fe(CN)_6^{3-}$, phenate, etc. eject an electron into the solvation shell on electronic excitation. Such a solvated electron e^-_{aq}, thus generated, absorbs at $\sim$700 nm as observed by flash photolysis.
13. The two-photon absorption spectroscopy can overcome the symmetry barrier imposed by the selection rule for angular momenta in the one-photon process. Thus, the technique is able to identify and assign molecular and atomic states which are not accessable to one-photon spectroscopy.

FOUR

Physical Properties of the Electronically Excited Molecules

4.1 NATURE OF CHANGES ON ELECTRONIC EXCITATION

A molecule in the electronically excited state can be a completely different chemical species with its own wave function and nuclear geometry. Since the charge densities are different it shows a different chemistry from the normal ground state molecule, more so because it has excess energy but weaker bonds. Certain other physical properties like dipole moment, *pK* values, redox potentials also differ in comparison to the ground state values. Excited states, in general, have less deep minima in their potential energy surfaces, indicative of the weakening of attractive interactions. Usually the equilibrium internuclear distances increase. Some of the states may be completely repulsional, leading to direct dissociation on transition to them.

A transition from a bonding to an antibonding orbital can cause a triple bond to become double, a double bond to become single and a single bond to disrupt. Bond distances may increase by about 15% or more. Bond angles may change by as much as 80° because of changes in the degree of *s-p* hybridization. For polyatomic molecules with the same electronic configuration but differing in spin multiplicity, the differences in electron-electron repulsion terms in the singlet and triplet states may give rise to different geometry and bonding properties. These changes in the geometries are reflected in the respective absorption bands in the form of changes in the vibrational envelop and intensities.

4.2 ELECTRONIC, VIBRATIONAL AND ROTATIONAL ENERGIES

Each electronic energy state is associated with its vibrational and rotational energy levels (Figure 4.1). The quantized vibrational energies are given by

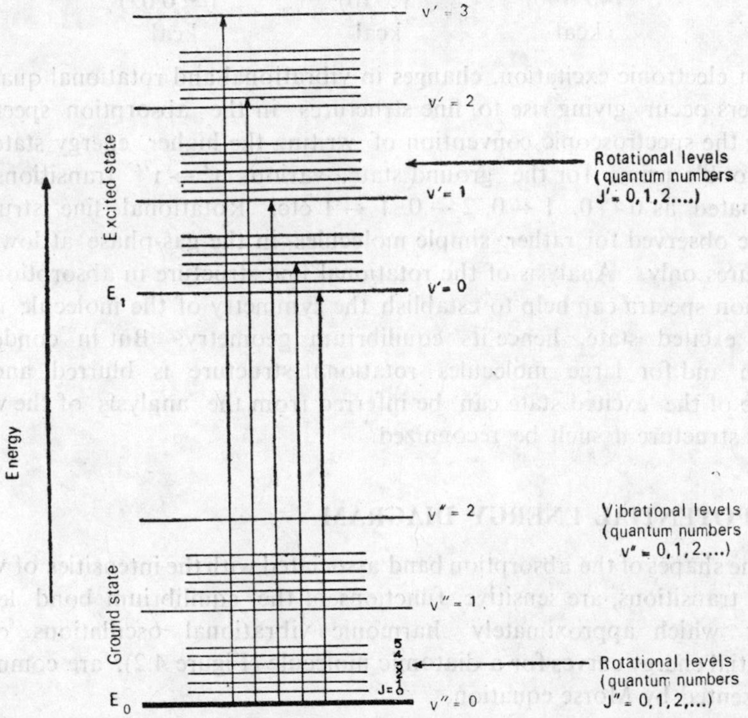

Figure 4.1 Electronic energy levels of molecules with associated vibrational and rotational levels.

$$E_v = (v + \tfrac{1}{2})\, h\nu \tag{4.1}$$

as for a simple harmonic oscillator where v, the vibrational quantum number has values 0, 1, 2, 3, etc. The potential function $V(r)$ for simple harmonic motion as derived from Hooke's law is given by

$$V(r) = \tfrac{1}{2}\, k(r - r_e)^2 \tag{4.2}$$

where r defines the displacement from the equilibrium position r_e, and k is known as the force constant. The rotational energy, E_J, is expressed as

$$E_J = J\,(J+1)\,\frac{h^2}{8\pi^2 I} \tag{4.3}$$

as for a rigid rotator where J is the angular momentum quantum number and can have values 0, 1, 2, etc. I is the moment of inertia.

The total energy E, of the system is the sum

$E =$	E_e	+	E_v	+	E_J
	160–600 kJ/mole		20–40 kJ/mole		$\simeq$ 0.08 kJ/mole
	(40–140) kcal		(5–10) kcal		($\simeq$ 0·02) kcal

On electronic excitation, changes in vibrational and rotational quantum numbers occur giving rise to fine structures in the absorption spectrum. Using the spectroscopic convention of writing the higher energy state first and double prime for the ground state, various $v' \rightarrow v''$ transitions are designated as $0 \leftarrow 0$, $1 \leftarrow 0$, $2 \leftarrow 0$, $1 \leftarrow 1$ etc. Rotational fine structure can be observed for rather simple molecules in the gas-phase at low temperatures only. Analysis of the rotational fine structure in absorption and emission spectra can help to establish the symmetry of the molecule in the given excited state, hence its equilibrium geometry. But in condensed system and for large molecules rotational structure is blurred and the nature of the excited state can be inferred from the analysis of the vibrational structure if such be recognized.

4.3 POTENTIAL ENERGY DIAGRAM

The shapes of the absorption band associated with the intensities of vibrational transitions, are sensitive functions of the equilibrium bond length, about which approximately harmonic vibrational oscillations occur. Potential energy curves for a diatomic molecule (Figure 4.2), are commonly represented by Morse equation,

$$V(r) = D_e \left\{1 - e^{-a(r - r_e)^2}\right\} \tag{4.4}$$

where D_e, the dissociation energy, is equal to the depth of the potential well measured from the zero vibrational level; a is a constant and $(r - r_e)$ is the extent of displacement from the equilibrium internuclear distance r_e. It is obvious that in real molecules the vibrational oscillations are not harmonic except near the lowest vibrational levels ($v = 0$). For higher values of v, the motion becomes anharmonic and the energy levels are not equidistant as predicted for simple harmonic motion (equation 4.1). This is as it should be because with increasing vibrational energy, the restoring force is weakened and at a certain value, the molecule dissociates into its component atoms. The dissociation energy D_e is given as $D_e = D_o - \frac{1}{2}h\nu$ where $\frac{1}{2}h\nu$ is known as the zeropoint energy and exists as a consequence of the Heisenberg uncertainty principle ($\Delta x\ \Delta p \approx \hbar$).

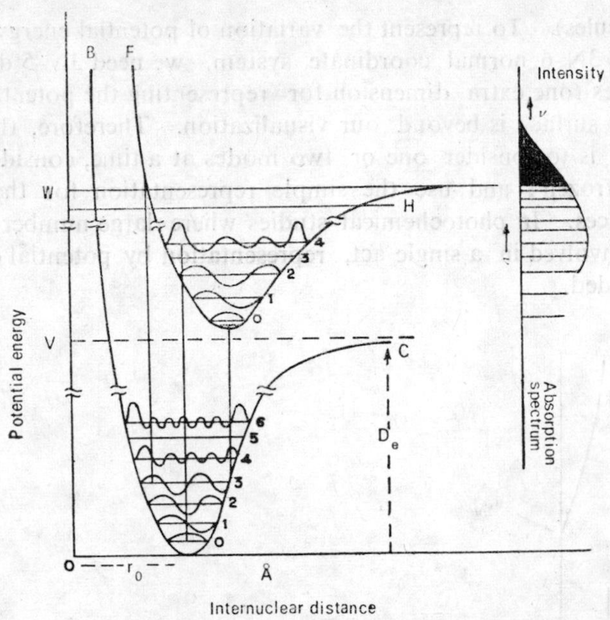

Figure 4.2 Electronic absorption and Franck-Condon principle.

The vibrational energy levels are usually represented as horizontal lines across the potential energy curves. The distribution function as obtained from the solution of the Schrödinger equation for various values of v are represented diagramatically in Figure 4.2 for each energy state E_v. The total energy E_v is a constant for any given vibrational level but the relative contributions of kinetic and potential energies change as the bond oscillates about the equilibrium position. The energy is totally kinetic at the centre and totally potential at the turning points. Therefore, in course of oscillation the molecule classically spends most of its time near the turning point. This situation is true for higher vibrational levels but for the zero vibrational level, the situation is reversed.

The most probable separation between the atoms is the equilibrium internuclear distance which is identified with the bond length. Each excited state has its own characteristic potential energy curve and its own equilibrium internuclear distance. the depth of the minima being an indication of the bond strength. The potential ene gy PE* curves for I_2 molecule for the ground and some excited energy states are given in Figure 4.3.

The representation as a two-dimensional potential energy diagram is simple for diatomic molecules. But for polyatomic molecules, vibrational motion is more complex. If the vibrations are assumed to be simple harmonic, the net vibrational motion of N-atomic molecule can be resolved into $3N$–6 components termed normal modes of vibrations ($3N$–5 for

*Henceforward potential energy will be abbreviated as PE,

linear molecules). To represent the variation of potential energy as a function of the 3N–6 normal coordinate system, we need $3N$–5 dimensional hypersurfaces (one extra dimension for representing the potential energy). But such a surface is beyond our visualization. Therefore, the best that can be done is to consider one or two modes at a time, considering other modes as 'frozen', and use the simple representation for the potential energy surfaces. In photochemical studies where large number of excited states are involved in a single act, representation by potential energy surfaces is avoided.

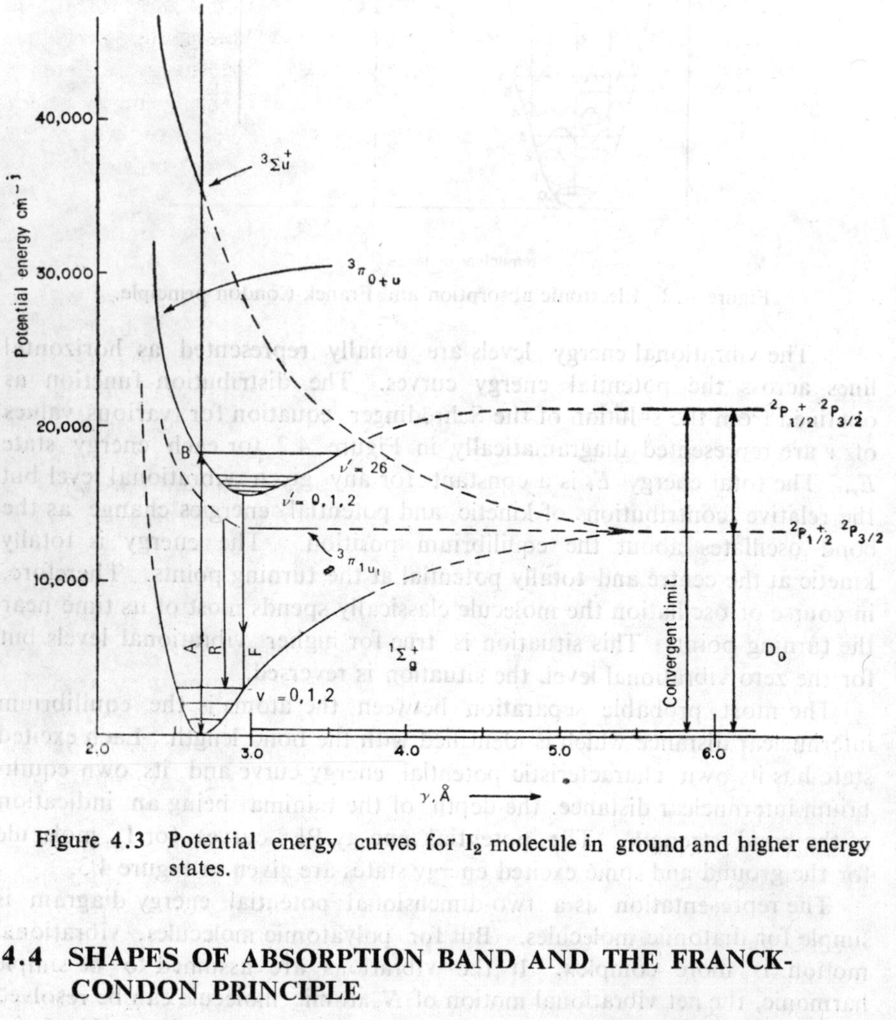

Figure 4.3 Potential energy curves for I_2 molecule in ground and higher energy states.

4.4 SHAPES OF ABSORPTION BAND AND THE FRANCK-CONDON PRINCIPLE

At normal temperatures most of the molecules reside in the zero vibrational level of the ground state potential function. This information is easily obtained from the Maxwell-Boltzmann distribution law:

$$\frac{N}{N_0}=e^{-\Delta E/RT} \tag{4.5}$$

where N_0, N are the populations of lower and upper states and ΔE is the energy difference between the two. It is evident that the population is controlled by the ratio $\Delta E/RT$ where RT is the thermal energy. For electronic levels ΔE is 160–600 kJ mol^{-1} and the electronically excited state is not expected to be populated at the room temperature. If a large amount of thermal energy is supplied, only the vibrational and rotational motions increase to the extent that the molecule may dissociate (*pyrolysis*) before being excited to the higher electronic state. The energy difference between vibrational levels are about 20 kJ mol^{-1} and thermal energy under laboratory condition is approximately 4 kJ mol^{-1}. Therefore, again the zero vibrational level is mostly populated with only few molecules occupying the higher vibrational states.

The wave function for the zero vibrational level has a maximum in the centre, indicating the region of maximum probability (Figure 4.2). Therefore, the most probable transition during the act of light absorption is that, which originates from the centre of $v=0$ vibrational level. The time taken for an electronic transition is of the order of 10^{-15} s, the reciprocal of the radiation frequency. On the other hand, the time period for vibration is about 10^{-13} s which is nearly 100 times as slow. As a consequence, the internuclear distances do not change during the act of light absorption. Hence, the transition process can be represented by a vertical line which is parallel to the potential energy axis and originates from the lower potential curve to the upper curve. This fact forms the basis of formulation of the Franck-Condon principle which is stated as follows: *Electronic transitions are so fast (10^{-15} s) in comparison to the nuclear motion (10^{-13} s) that immediately after the transition, the nuclei have nearly the same relative position and momentum as they did just before the transition.* This principle essentially implies that it is difficult to convert electronic energy rapidly into vibrational kinetic energy and the most probable transitions between different electronic and vibrational levels are those for which the momentum and the position of the nuclei do not change very much.

This most probable transition appears as the most intense absorption line at a frequency given by Bohr relationship $\Delta E=h\nu$. Few transitions are also possible from the other positions of the $v=0$ level adding to the width of the absorption band. Since they are less probable, these absorption lines are of lower intensities. Furthermore, those molecules which are able to reside in higher vibrational levels as a consequence of the Boltzmann distribution add to the vibrational transitions of lower probabilities giving the familar contour of an absorption band. If a transition occurs to a position corresponding to energy greater than the excited state dissociation energy, the molecule dissociates directly on excitation (*photolysis*) and discrete vibrational bands are not observed. On photodissociation, an excited and a ground state molecule may be formed, the excess energy being converted into the kinetic energy of the partners. Since kinetic energy is not quantized, a continuum is more likely to appear for simple diatomic molecules. For polyatomic molecules of photochemical interest, the continuum may be in the far UV region.

For the internuclear geometry as represented by the potential energy (PE) surfaces of the ground and excited electronic states in Figure 4.4 b, the vibrational excitation to $v' = 4$ level is coupled with the electronic transition, giving an intensity distribution pattern as shown at the upper left side of the figure. If a different equilibrium bond length is indicated in the excited state PE function, the shape of the absorption curve will change accordingly. Thus, the shape of the absorption contour as defined by the vibrational intensities, is governed by the shape and disposition of the potential energy function with respect to the ground state function (Figure 4.4).

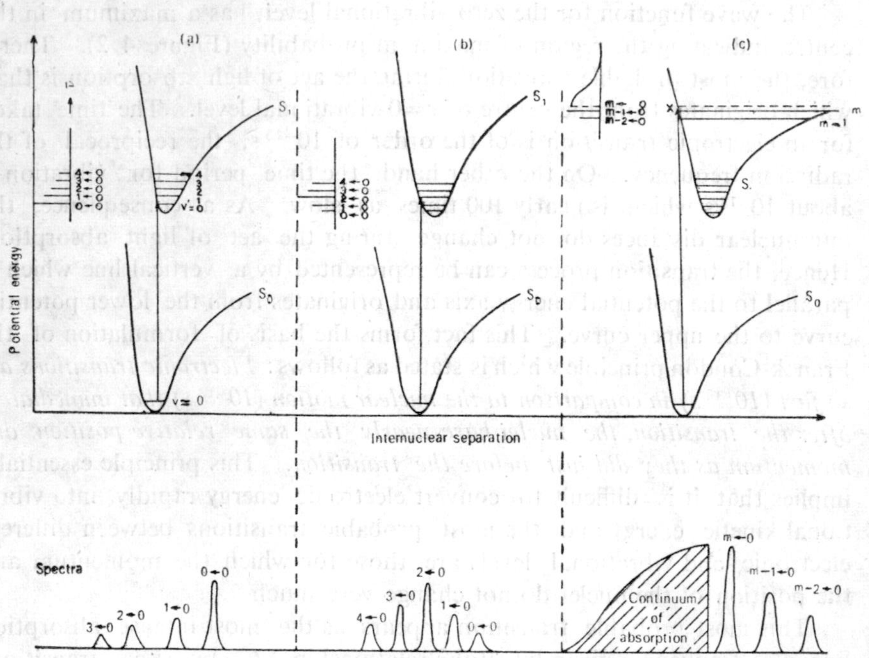

Figure 4.4 Internuclear configurations and shapes of absorption bands.

4.4.1 Quantum-Mechanical Formulation of the Franck-Condon Principle

For transitions in which electronic transitions are coupled with vibrational transitions as well, the transition moment integral must include the nuclear coordinates also (rotational coordinates are neglected).

$$| M |_{(e, v)} = \int \psi^*_{(e, v)_2} \hat{\mu} \psi_{(e, v)_1} d\tau_e d\tau_v$$

$$= \int \psi_2^* \hat{\mu} \psi_1 d\tau_e \int x_2^{v'} x_1^{v''} d\tau_v \quad (4.6)$$

Since nuclear and electronic coordinates are separable (Born-Oppenheimer approximation) and the electric dipole transition involves the electron coordinates only, the above factorization is justified. The intensity of transition is observed to depend on the integral $\int x_2^{v'} x_1^{v''} d\tau_v$ also. This integral is known as the *Franck-Condon integral* and demands a considerable overlap of the wave functions of the two combining states. The transition to that energy state will be most probable for which the terminal maximum of the upper energy wave function, overlies vertically above the maximum for $v = 0$ wave function of the ground state. The width of the vibrational progression is determined mainly by the range of values covered by the nuclei in $v = 0$ level which in ultimate analysis is governed by the uncertainty principle ($\Delta E \times \Delta t \approx \hbar$). The Franck-Condon overlap criterion is important for an understanding of radiationless or nonradiative transitions as well, as we will see in Section 5.2. Hence, it is of photochemical interest. Figure 4.5 illustrates good and poor overlap of wave functions.

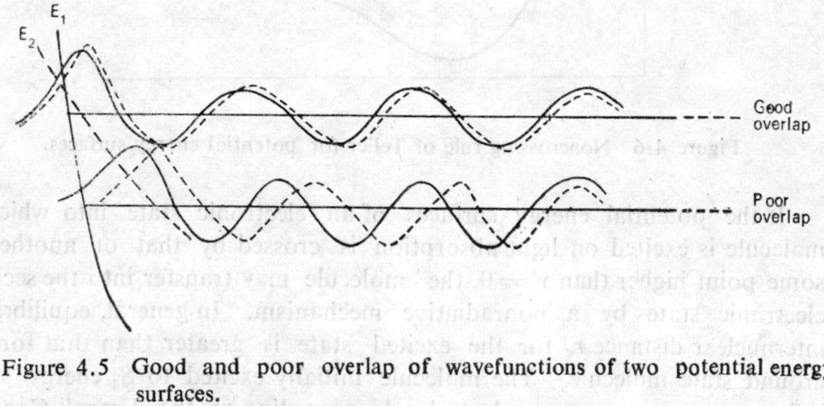

Figure 4.5 Good and poor overlap of wavefunctions of two potential energy surfaces.

4.4.2 Crossing of Potential Energy Surfaces

The beginning of the continuum in the absorption spectrum of simple gas-phase molecules implies the dissociation of the bond in question (Figure 4.3). In some molecules a blurring of the spectrum appears before the dissociative continuum suggesting dissociation at a lower energy requirement. This is known as *predissociation.* An understanding of the origin of predissociation spectra can help us to appreciate the consequences of crossing of potential energy surfaces which play a big role in predicting the physical and chemical behaviour of vibronically excited molecules. We might do well here to define the *noncrossing rule* of Teller: *when the potential energy curves of a molecule are drawn, no two levels with completely identical characters with respect to all symmetry operations can cross.* Figure 4.6 illustrates the PE surfaces which cross and which avoid crossing.

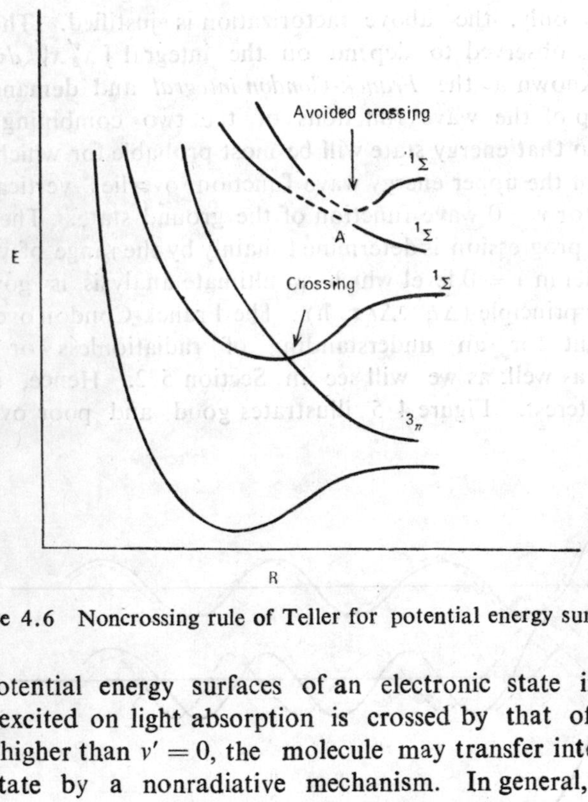

Figure 4.6 Noncrossing rule of Teller for potential energy surfaces.

If the potential energy surfaces of an electronic state into which a molecule is excited on light absorption is crossed by that of another at some point higher than $v' = 0$, the molecule may transfer into the second electronic state by a nonradiative mechanism. In general, equilibrium internuclear distance r_e for the excited state is greater than that for the ground state molecule. The molecule initially excited to S_1 energy state is formed as a compressed molecule according to the Franck-Condon principle. Since a few vibrational quantum states are also excited, the molecule immediately starts vibrating along the potential well of the state S_1. At the crossing point (isoenergetic point), if there is good overlap of vibrational wave functions, i.e. Franck-Condon overlap integral is large, there is a possibility that the molecule, instead of moving along the PE surface of S_1, goes over to that of S_2. Such transfers have time periods greater than vibrational time periods $\approx 10^{-13}$ s but less than the rotational time periods $\approx 10^{-11}$ s. Once the molecule is transferred to the second excited state, it may dissociate or behave as dictated by the PE surface of ths second state. If a simple molecule dissociates into fragments from electronically excited state at least one of the partners is likely to be electronically excited. When the absorbed energy exceeds that of the dissociation limit, the excess energy appears as the kinetic energy of the partners and may increase the chemical reactivity and subsequent course of the chemical change. The transfer of the molecule from one excited state to the other at the point of intersection is governed by the selection rule for radiationless transition which will be discussed in Section 5.2.

These ideas can only be applied qualitatively to polyatomic molecules because of the lack of structure in their spectra.

4.5 EMISSION SPECTRA

The emission of radiation is also governed by the Franck-Condon principle, since what is probable for absorption must be probable for emission. In the case of atoms or simple molecules at very low pressures where collisional perturbations are absent, the excited species may return to the ground state directly by emitting the same frequency of radiation as it has absorbed. This is known as *resonance radiation.* The mercury resonance radiation at 253.7 nm in due to such transitions.

For polyatomic molecules and molecules in the condensed state, the excess vibrational energy gained in a vibration-coupled electronic transition, is quickly lost to the surrounding in a time period $\simeq 10^{-13}$ s. The molecule comes to stay at the zero vibrational level of the first excited state for about 10^{-8} s for an allowed transition and more for a partially forbidden transition. If it is permitted to return to the ground state by radiative transition *fluorescence emission* is observed. On the return transition, it again follows the Franck-Condon principle and arrives at a higher vibrational level of the ground state. Because of the nature of wave function in $v' = 0$ state and the Boltzmann distribution of energy in the excited state, there is a distribution of frequencies around the most probable transition. The most probable transition again depends on the internuclear equilibrium geometries of the molecules in the two states. The vibrational transitions, associated with reverse transition of electron are written as $0 \rightarrow 0$, $0 \rightarrow 1$, $0 \rightarrow 2$, $0 \rightarrow 3, \ldots$, etc. (Figure 4.7). It is clear that whereas the spacings between the vibrational band in the absorption spectrum correspond to vibrational energy differences $\Delta E_{v'}$ in the excited state, those in the emission reflect the energy differences in the ground state $\Delta E_{v''}$. If the internuclear geometries are not very different, as in many polyatomic molecules, and $E_{v'} \approx E_{v''}$ the fluorescence spectrum appears as a *mirror image* of the corresponding absorption spectrum with a region of overlap. A frequency ν_0 will lie in the plane of the mirror image symmetry defined as

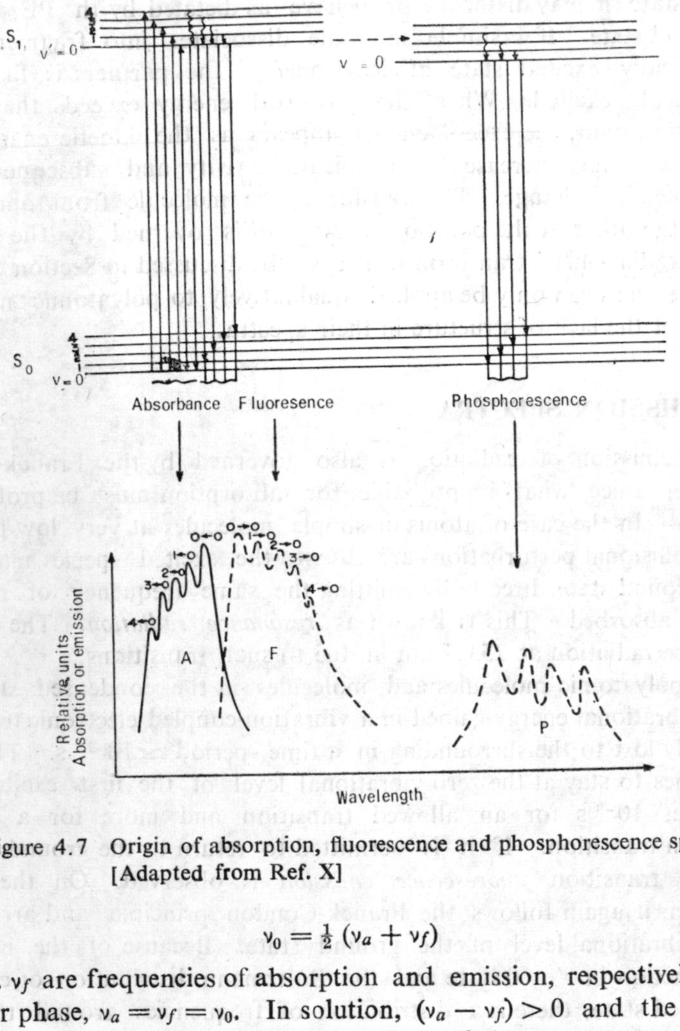

Figure 4.7 Origin of absorption, fluorescence and phosphorescence spectra. [Adapted from Ref. X]

$$\nu_0 = \frac{1}{2}(\nu_a + \nu_f)$$

ν_a and ν_f are frequencies of absorption and emission, respectively. In the vapour phase, $\nu_a = \nu_f = \nu_0$. In solution, $(\nu_a - \nu_f) > 0$ and the extent of overlap varies depending on the PE surface of the upper state relative to the ground state. Since the size of the emitted quantum $h\nu_f$ is smaller than the size of absorbed quantum, $h\nu_a$, $(h\nu_f < h\nu_a)$ emission occurs at a longer wavelength. This is known as the *Stokes shift*. If the system crosses over to the triplet state subsequent to promotion to a singlet excited state, emission may be observed at a still longer wavelength. This is known as *phosphorescence emission* and does not show a mirror image relationship with $S_1 \leftarrow S_0$ absorption band. Since in condensed systems, emission occurs from the thermally equilibrated upper state, the intensity distribution in fluorescence and phosphorescence spectra is independent of the exciting wavelength.

4.6 ENVIRONMENTAL EFFECT ON ABSORPTION AND EMISSION SPECTRA

Because of the variation of electron density in different electronic states, an interaction with the environment affects differently the various electronic states of a molecule. A shift in the spectrum may result due to such differences in the two combining states. It is important to realize that the electronic origin or O — O band is expected to be affected differently from the band maxima if they do not nearly coincide.

There are two types of solute-solvent interactions: (1) universal interaction, and (2) specific interaction.

(1) The *universal interaction* is due to the collective influence of the solvent as a dielectric medium and depends on the dielectric constant D and the refractive index n of the solvent. Reasonably large environmental perturbations may be caused by van der Waals dipolar or ionic fields in solution, in liquids, in solids or in gases at high pressures. In gases, liquids and solutions the perturbation must be averaged over the molecular fluctuations.

The van der Waals interactions include (i) London dispersion force interactions, (ii) induced dipole interactions, and (iii) dipole-dipole interactions. These are attractive interactions. The dispersion force interactions depend on the molecular polarizability which increases with the electron density in the molecule. In large molecules, the local polarizability, such as carbon-halogen bond polarizability in halogen-substituted compounds, is more important than the overall molecular polarizability in explaining the spectral shift. Dispersion force interactions are present even for nonpolar solutes in nonpolar solvents.

The repulsive interactions are primarily derived from exchange forces (nonbonded repulsion) as the electrons of one molecule approach the filled orbitals of the neighbour. In general, both the attractive and repulsive interactions have larger coefficients in the excited state, the former, because of the greater polarizability and the latter, because of the increase in the orbital size. The attractive interaction B tends to stabilize the excited energy state with respect to the ground state, whereas the repulsive interaction A has the opposite effect. The resultant shift in energy ΔE is the sum of these two opposing effects,

$$\Delta E = \sum \left(\frac{\Delta A}{r^{12}} - \frac{\Delta B}{r^{6}} \right) \tag{4.7}$$

where ΔA and ΔB are differences between the repulsive and attractive coefficients in the two electronic states; r the distance between the interacting molecules with their respective exponents and the summation is over all the neighbours. The spectrum shifts towards low frequency region when ΔE is negative (*red shift*), and towards high frequency region when ΔE is positive (*blue shift*) (Figure 4.8).

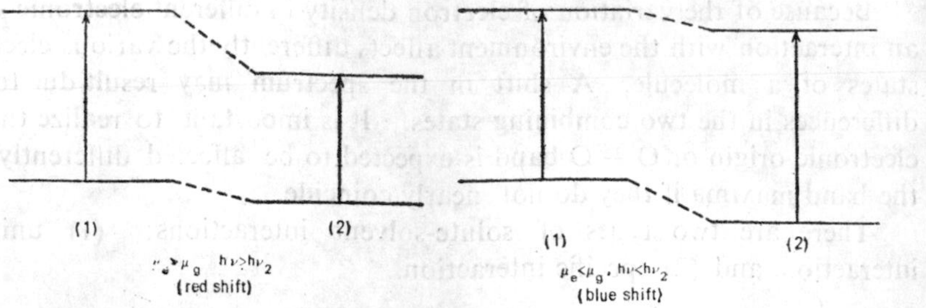

Figure 4.8 Solvent red shift and blue shift in absorption spectra.

If the solute molecule has a dipole moment, it is expected to differ in various electronic energy states because of the differences in charge distribution. If the solvent is nonpolar, then the rough description of the interaction is dipole-induced dipole type. In polar solvents, dipole-dipole interactions also become important. The London forces are always present. For the calculation of dipole-dipole interaction energy, point dipole approximations are made which are poor description for large extended molecules.

(2) Many solute-solvent interactions are somewhat more complex and *specific* than the one discussed so far. If there are low lying unfilled orbitals in the solvent, it is likely to have a strong affinity for electrons. If the solvent accepts electrons from the solute molecules, a *charge-transfer-to-solvent* (CTTS) complex is said to be formed. The charge-transfer complexes bridge the gap between weak van der Waals complexes and molecules bonded by strong valence forces. The donor capacity of the solute in the ground and excited states are also expected to be different. If it is greater in the ground state then the ground state energy level is stabilized and blue shift occurs and vice versa. Still another specific type of solute-solvent interaction is the formation of *hydrogen bonded complexes* and *exciplexes*. These involve short range forces.

Now, the use of solvent perturbation technique for the identification of ($n \rightarrow \pi^*$) and ($\pi \rightarrow \pi^*$) transitions can be rationalized. In general, the electron charge distribution of a (π, π^*) excited state is more extended than that of the ground state and the excited state is, therefore, more polarizable. Change to a polar from a nonpolar solvent increases the solvent interaction with both states but the corresponding decrease in energy is greater for the excited states resulting in the red shift. On the other hand, in $n \rightarrow \pi^*$ transition, nonbonding lone pair on the heteroatom is hydrogen bonded in the ground state. This results in greater decrease in ground state energy for more polar and hydrogen bonding solvents.

The excited state is not much depressed as the promotion of the n-electron into the π^*-orbital reduces the hydrogen bonding forces in the excited state. The result is a blue shift as illustrated in Figure 4.8. A charge-transfer (CT) state has a much greater permanent dipole moment than the ground state and, therefore, change to a polar solvent results in a considerably larger red shift.

In addition to the spectral shifts, the vibrational structures of the fluorescence and absorption spectra are influenced by interaction with the solvent environment. The line spectrum at low temperature in the vapour phase is usually broadened and sometimes its vibrational pattern is completely lost in solvents which can specifically interact with it. Some broadening also results due to thermal vibrations.

4.7 EXCITED STATE DIPOLE MOMENT

The effect of solvent on the absorption and emission spectra may give useful information regarding such physical properties as dipole moment changes and polarizability of the molecule in the two combining states. Theories have been worked out in terms of Onsager theory of dielectrics and Franck-Condon principle. According to these, the energy shift between the O—O band in absorption and emission is related to the dielectric constant D, and the refractive index n of the solvent and the difference in dipole moments of the solute molecule in the two combining energy states. The short range interactions such as H-bonding is neglected.

$$\Delta E = h\nu_a - h\nu_f = 2\left\{\frac{D-1}{2D+1} - \frac{n^2-1}{2n^2+1}\right\}\left\{\frac{\mu_g - \mu_e}{a^3}\right\}^2 \qquad (4.8)$$

where μ_g and μ_e are dipole moments of the ground and the excited molecules respectively, and a is the radius of the cavity in which the solute molecule resides. The latter is defined by Onsager's reaction field assuming the solvent as continuous dielectric.

The shift in O – O transition due to solvent interaction in the two states of different polarity can be explained with the help of Franck-Condon principle (Figure 4.9) for absorption and emission processes.

In the ground state, a solute molecule having a dipole moment μ_g residing in a spherical cavity of radius a in a medium of static dielectric constant D polarizes the medium. The reaction field R_0 produced on such interaction is

$$R_0 = \frac{2\,\mu_g}{a^3}\left\{\frac{D-1}{2D+1}\right\} \qquad (4.9)$$

If S_0 is the unperturbed vapour phase energy level of the molecule in the ground state, in solution it is depressed by an amount proportional to R_0.

On excitation, the molecule is promoted to the Franck-Condon excited state (FC)*. In the excited state, the dipole moment may not only have a

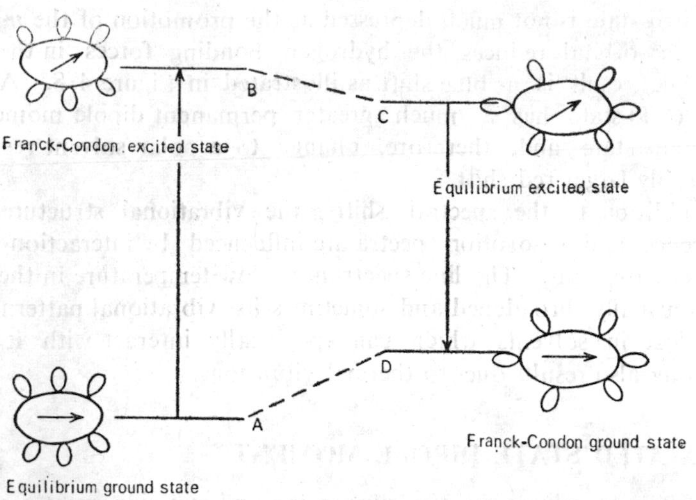

Figure 4.9 Effect of refractive index and dielectric constant of solvent on O—O transitions of a polar molecule.

different magnitude, μ_e but different direction also. Since the electronic transition is much faster than the dielectric relaxation time, the ground state solvent orientation will not have the time to change in tune with the new situation. Only the electronic polarizability is affected immediately on excitation. The reaction field in (FC)* state is

$$R^*_{FC} = \frac{2\,\mu_g}{a^3}\,\frac{(n^2-1)}{(2\,n^2+1)} \tag{4.10}$$

where n is the refractive index and n^2 is equal to high frequency dielectric constant at the frequency of absorption. The energy of absorption is, therefore, given as

$$h\,\nu_a = (h\,\nu_a)_0 + (R^*_{FC} - R_0)\,(\mu_e - \mu_g) \tag{4.11}$$

where $(h\,\nu_a)_0$ is the ene.gy in the gas phase.

If the average lifetime τ_f of excited state is greater than the dielectric relaxation time τ_D then excited state dipole moment μ_e polarizes the solvent and the solvent molecules adjust to the new situation. A reaction field

$$R^* = \frac{2\,\mu_e}{a^3}\left\{\frac{D-1}{2\,D+1}\right\} \tag{4.12}$$

is produced. The FC excited state (FC)* is stabilized by this amount to give equilibrium excited state S_1. The O—O fluorescence transition occurs from this energy state. Again a strained ground state (FC) is produced in accordance with the Franck-Condon principle in which solvent orientation does not have time to change. The Onsager's field is

$$R_{FC} = \frac{2\,\mu_e}{a^3}\left\{\frac{n^2-1}{2n^2+1}\right\} \tag{4.13}$$

The energy of the O—O transition in emission is given as

$$h\nu_f = (h\nu_f)_0 + (R_{FC} - R^*)(\mu_e - \mu_g) \qquad (4.14)$$

Finally, the FC ground state relaxes to the equilibrium ground state.

The difference between O—O absorption and O—O fluorescence energies is obtained from the two relationships, (4.11) and (4.14) assuming $(h\nu_a)_0 \approx (h\nu_f)_0$, and expressing in wavenumbers,

$$hc\,\Delta\bar{\nu} = hc\,(\bar{\nu}_a - \bar{\nu}_f)$$
$$= (R^*_{FC} - R_0 - R_{FC} + R^*)(\mu_e - \mu_g) \qquad (4.15)$$

On substituting the values for the reaction fields from (4.12), 4.13), (4.15) and (4.16) the final expression for $\Delta\bar{\nu}$, the difference in the O—O energy in absorption and emission, is obtained as

$$\Delta\bar{\nu} = \frac{2(\mu_e - \mu_g)^2}{hc\,a^3}\left\{\frac{D-1}{2D+1} - \frac{n^2-1}{2n^2+1}\right\} \qquad (4.16)$$

$$= \frac{2(\mu_e - \mu_g)^2}{hc\,a^3}\,\Delta f \qquad (4.17)$$

where $\Delta f = \left\{\frac{D-1}{2D+1} - \frac{n^2-1}{2n^2+1}\right\}$

For a nonpolar solvent $D \simeq n^2$ and the interaction energy is zero.

When the quantity $\Delta\bar{\nu}$, is plotted as function of Δf of the solvent the slope is given by $2(\mu_e - \mu_g)^2/hc\,a^3$ from which $(\mu_e - \mu_g)$, the difference in the moments for the two states involved in the transition, can be calculated

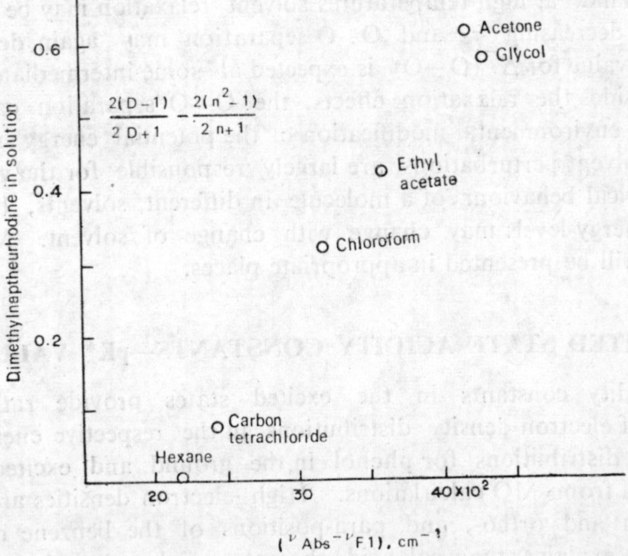

Figure 4.10 Plot of $(\bar{\nu}_a - \bar{\nu}_f)$ vs $\Delta f = [2(D-1)/(2D+1) - 2(n^2-1)/(2n^2+1)]$

if a reasonable assumption about the radius of the Onsager's cavity a be made. If the excited state dipole is greater than the ground state. which is normally the case, a positive slope is obtained (Figure 4.10). A negative slople is obtained in the reverse case. Thus, it is a very useful method for the determination of excited state dipole moments and the results agree with more rigorous calculations (Table 4.1).

TABLE 4.1

Dipole moment of some compounds in the ground and the excited states

Compound	a (in Å)	$\Delta\mu$ Debye	μ_g Debye	μ_e Debye
p-dimethylamino-p'-nitrostilbene	8	24	7.6	32
p-dimethylamino-p'-cyanostilbene	8	23	6.1	29
Dimethylaminonaphtheurhodin	7	10	2	12
p-amino-p'-nitrophenyl†			6 0	22.2
p-amino-p'-cyanodiphenyl†			6.5	23.4
p-amino-p'-nitrostilbene†			6.0	13 0

†Measured by fluorescence depolarization studies. (Czebella)

At high viscosities or low temperatures, dielectric relaxation time τ_d may be larger than the mean radiative lifetime τ_f of the molecule. This may decrease the O—O separation between absorption and emission. On the other hand, at high temperatures solvent relaxation may be promoted thermally decreasing τ_d and O—O separation may again decrease. A maximum value for $\Delta\nu$ (O—O) is expected at some intermediate temperatures. Besides the relaxation effects, the O—O separation can also be affected by environmental modification of the potential energy surfaces.

Such solvent perturbations are largely responsible for the variation of photochemical behaviour of a molecule in different solvents, because the order of energy levels may change with change of solvent. Many such examples will be presented in appropriate places.

4.8 EXCITED STATE ACIDITY CONSTANTS—pK* VALUES

The acidity constants in the excited states provide rather direct measures of electron density distributions in the respective energy states. The charge distributions for phenol in the ground and excited state can be obtained from MO calculations. High electron densities are observed on O-atom and ortho-, and para-positions of the benzene ring. This distribution explains the weak acid character of phenol and ortho-, para-directing property of the hydroxyl group. In the first excited singlet state, the charge density on oxygen is reduced and a reasonable charge density is now found on ortho- and meta-positions. Therefore, excited phenol is more acidic and is also ortho-meta-directing towards substitution in benzene ring. The protolytic equilibrium constant pK_a for the reaction is 10.0 and 5.7 in the ground and excited states, respectively, a difference of 4.3 pK units.

$$\text{PhOH} + H_2O \rightarrow \text{PhO}^- + H_3O^+$$

The phenomenon was first observed in 1-naphthylamine-4-sulphonate by Weller who found that the fluorescence of the compound changed colour with pH, although the absorption spectrum was not affected. An explanation of this observation was provided by Förster for a similar observation in 3-hydroxypyrene-5, 8, 10-trisulphonate. Fluorescence characteristic of the ionized species was observed at a pH where the heteroatom is not dissociated in the ground state as indicated by its characteristic absorption. This is a proof of proton transfer in a time period less than the lifetime of the excited state.

The equilibrium between an acid RH and its conjugate base R^- in the ground state is represented as:

$$\text{RH} + H_2O \rightleftharpoons R^- + H_3O^+ \tag{4.18}$$

and the equilibrium constant is expressed as pK_a value ($-\log K_a$). In the excited state, the representation is

$$\text{RH}^* + H_2O \rightleftharpoons R^{-*} + H_3O^+ \tag{4.19}$$

and the constant is designated as pK^*. If $pK^* < pK$, it suggests that the equilibrium is shifted towards the right on excitation:

$$\begin{array}{ccc} \text{RH}^* + H_2O & \underset{pK^*}{\rightleftharpoons} & R^{-*} + H_3O^* \\ h\nu_a \uparrow & & \downarrow h\nu_f \\ \text{RH} + H_2O & \underset{pK}{\rightleftharpoons} & R^- + H_3O^+ \end{array}$$

A molecule which is undissociated in the ground state at a given pH is dissociated on promotion to the higher energy state. If the ion R^{-*} is capable of emitting fluorescence, characteristic fluorescence of the ion is observed but the absorption is characteristic of the undissociated molecule. Therefore, within certain pH range fluorescence due to RH^* and R^{-*} both are observed (Figure 4.11). The pH value at which the two intensities are nearly equal give the approximate value for pK^*. Since absorption wavelengths for the acid and its ion are also different, the ground state pK can be calculated from absorption studies. Absorption and emission characteristics of RH and R^- must be different to utilize this method for the evaluation of pK and pK^*.

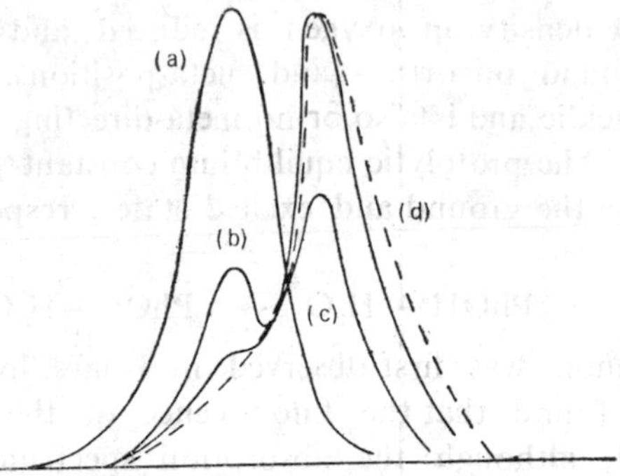

Figure 4.11 Fluorescence spectra of β-napthol as a function pH: (a) pH 13, 0.1 N NaOH; (b) pH 9; (c) pH 7; (d) pH 0, 1 N HCl.

A more general method proposed by Förster is a thermodynamic calculation based on a cycle called after his name (Figure 4.12). If ΔH and ΔH^* are enthalpies of proton dissociation in the ground and the excited states, respectively, then

$$\Delta H = \Delta G + T\Delta S \tag{4.20}$$

$$\Delta H^* = \Delta G^* + T\Delta S^* \tag{4.21}$$

If it is assumed, $\Delta S \approx \Delta S^*$

then, $$\Delta H - \Delta H^* = \Delta G - \Delta G^*$$
$$= - RT(\ln K - \ln K^*) \tag{4.22}$$

where K and K^* are protolytic equilibrium constants in the ground and the excited states respectively. On rearrangement, we have

$$\Delta pK^* = pK - pK^* = \frac{\Delta H - \Delta H^*}{2.3\, RT} \tag{4.23}$$

Depending on the ground state and excited state enthalpies, pK^* can be greater than, equal to or less than pK.
From Figure 4.12 we find,

$$E_1 + \Delta H^* = E_2 + \Delta H \tag{4.24}$$

or, $$E_1 - E_2 = \Delta H - \Delta H^* \tag{4.25}$$

where E_1 and E_2 are O — O absorption energies of the acid RH and its ion R^- respectively. Hence,

$$pK - pK^* = \frac{E_1 - E_2}{2.3\, RT} = \frac{hc\,\Delta\bar{\nu}}{2.3\, RT} \tag{4.26}$$

where $\Delta\bar{\nu}$ is the difference in electronic transition frequencies (in cm^{-1}) of the acid and its conjugate base. The assumption $\Delta S \approx \Delta S^*$ is considered to hold true because pK^* calculated from emission data and absorption data agree within experimental limits.

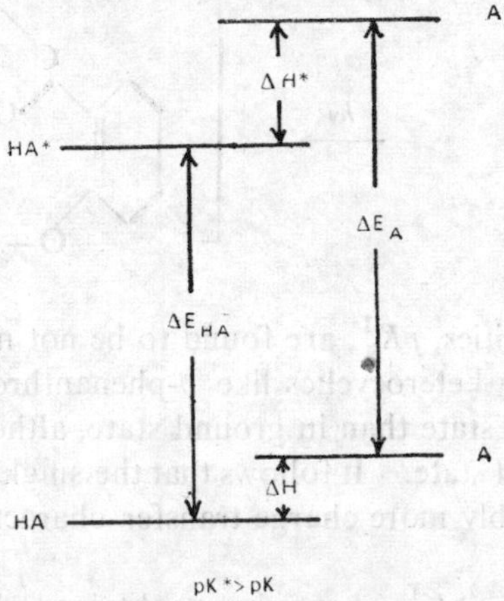

Figure 4.12 Förster's cycles for pK and pK*; HA=undissociated acid; A=conjugated base.

In contrast to alcohols, aromatic carboxylic acids exhibit a decrease in acidity in the first excited singlet state relative to the ground state. For example, benzoic acid can take up a proton

$$C_6H_5C\begin{matrix}\nearrow O\\ \searrow OH\end{matrix} + H_3O^+ \rightleftharpoons C_6H_5C\begin{matrix}\nearrow OH^+\\ \searrow OH\end{matrix} + H_2O$$

only at pH of -7.3 expressed as Hammett's acidity function H_0. But in the excited state, protonation occurs at pH zero suggesting that benzoic acid is more basic or less acidic in the excited state.

In general, phenols, thiols and aromatic amines become much stronger acids on excitation, while carboxylic acids, aldehydes, ketones with lowest (π, π^*) singlet, nitrogen and sulphur heterocyclics become stronger bases. If the protolytic equilibrium is established at a very fast rate, the change of pH values within which the fluorescence colour changes are observed, is small. But if the proton dissociation is a slow process, of the order of lifetime τ_f, the pH range may be considerable.

Intramolecular proton transfer has been observed in molecules like salicyclic acids. This compound exhibits large Stokes' shift. But in the corresponding ether, the shift is small. The unusually large shift in salicyclic acid is rationalized if we assume that upon excitation, carboxyl group becomes more basic and the hydroxylic group becomes more acidic. As a consequence, internal proton transfer occurs when carboxyl and hydroxyl groups are hydrogen bonded internally. In ether such hydrogen bonding is not possible. The effect is not observed in ortho or para isomers of salicyclic acid.

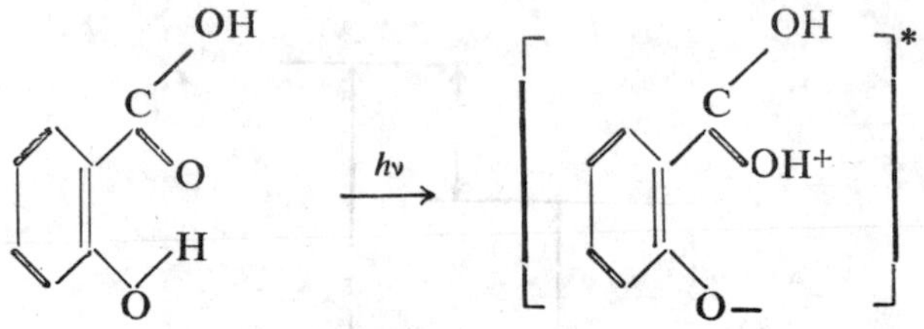

The triplet state acidities, pK_a^T, are found to be not much different from ground state pK_a's. But heterocyclics like *o*-phenanthroline and quinoline are more basic in triplet state than in ground state, although less basic than in the first excited singlet state. It follows that the singlet transition in these molecules has considerably more charge transfer character than the transition in the triplet state.

Triplet state acidities, pK_a^T, have been obtained by Porter and his school from phosphorescence studies, using flash photolysis techniques. The singlet, triplet and ground state acidity constants of some organic molecules are given in Table 4.2.

TABLE 4.2
Singlet, triplet and ground state acidity constants of some organic molecules

	pK_a	pK_a^*	pK_a^T (flash spectroscopy)	pK_a^T (phosphorescence)
Phenol	10.0	4.0	—	8.5
2-Naphthol	9.5	3.1	8.1	7.7
2-Naphthoic acid	4.2	10–12	4.0	4.2
1-Naphthoic acid	3.7	10–12	3.8	4.6
Acridinium ion	5.5	10.6	5.6	—
Quinolinum ion	5.1	—	6.0	5.8
2-Naphthylammonium ion	4.1	—2	3.3	3.1

The pK^* values of phenols in singlet and triplet states are valuable guide to substituent effect in the excited states, specially for the aromatic hydrocarbons. In general, the conjugation between substituents and π-electron clouds is very significantly enhanced by electronic excitation without change in the direction of conjugative substituent effect. The excited state acidities frequently follow the Hammett equation fairly well if 'exalted' substituent constants σ^* are used.

Example

(J. E. Kuder and W. Wychich—Chem. Phys. Litt *24* (1974) 69)

For merocyanine dye 4′—hydroxy—1—methyl stilbazolium betain, following resonance structures are possible:

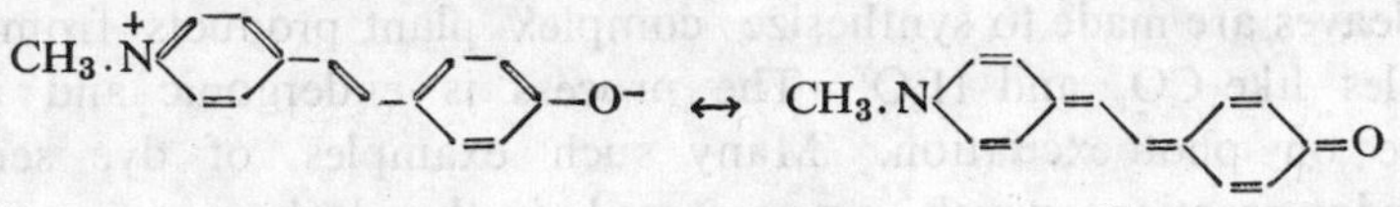

Ia Ib

The wavenumbers of absorption and fluorescence maxima at two differen acidities are:

	IH+ (in acid)	I (pH 9.9)
	$\nu_a \times 10^{-3}$ cm^{-1}	$\nu_b \times 10^{-3}$ cm^{-1}
Absorption max.	26.9	22.6
Fluorescence max.	19.4	17.2
Average	23.1	19.9

From Förster's cycle

$$\Delta pK_a^* = pK_a - pK_a^* = \frac{hc\,(\nu_a - \nu_b)}{2.303\,RT}$$

At T = 298 K, calculated values from absorption and emission data are:

$\Delta pK_a^* = 9.03$ (from absorption data)

$\Delta pK_a^* = 4.62$ (from fluorescence data)

$\Delta pK_a^* = 6.72$ (average value)

This difference in ΔpK^* values as obtained from absorption and emission data arise from differences in solvation energies of the ground and excited states. If the ground state $pK_a = 8.57$, then taking the average value, the excited state $pK_a^* = 1.85$.

4.9 EXCITED STATE REDOX POTENTIAL

Examples of striking differences in redox potentials between ground and first triplet states are observed in the oxidizing powers of methylene blue and thionine. In the ground state, the dyes cannot oxidize Fe^{2+} to Fe^{3+}. But when excited in the presence of Fe^{2+}, the colour of the dye is bleached. The colour returns on removal of the excitation source. In the long lived triplet state, oxidation potential of the dye increases and the ferrous iron is oxidized to ferric state, the dye itself being reduced to the leuco form.

$$\text{Me} + 2\text{Fe}^{++} + 2\text{H}^+ \underset{\text{dark}}{\overset{\text{light}}{\rightleftharpoons}} \underset{\text{leucodye}}{\text{MeH}_2} + 2\text{Fe}^{+++}$$

The electron transfer in the excited state is reversed when the molecule returns to the ground state, the leuco dye being oxidized back by ferric ion. A similar system includes mixed inorganic solutions such as (I_2/I^-) + (Fe^{3+}/Fe^{2+}). In such electron transfer reactions, the acceptor has a much lower electron affinity in the ground state than the donor. Thus, the oxidation-reduction reactions on excitation goes against the gradient of electrochemical potential $\Delta G > 0$. Photochemical redox reactions are the basic photosynthetic mechanism in plants, whereby electrons promoted to higher level by absorption of sunlight by green chlorophyll molecules of the leaves are made to synthesize complex plant products from simple molecules like CO_2 and H_2O. The process is endergonic and is only possible on photoexcitation. Many such examples of dye sensitized photoredox reactions which are reversed in the dark, are available in living systems. They provide a mechanism for photochemical storage of light energy.

The oxidation potentials of the short-lived dye intermediates may be estimated with reference to polarographic half-wave potentials, if it is assumed that the free energy changes are mainly due to the occupation of the highest filled and lowest unfilled molecular orbitals. The electronic configurations of dye in its excited states, singlet and triplet, and in oxidized and reduced forms are given in Figure 4·13.

The computed values of oxidation potentials for some dyes are given in Table 4.3. The potentials E' (volts) correspond to pH 7 reference to which

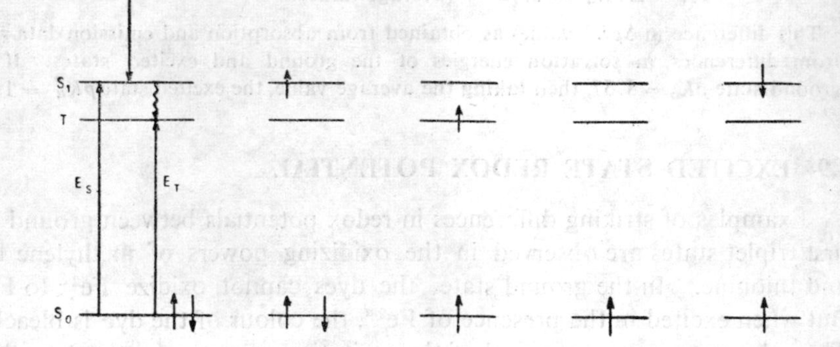

Figure 4.13 Electron configuration of dyes in ground excited singlet and triplet states and in oxidized and reduced forms: (S) ground singlet, (S^-) reduced form, (T) triplet state, (S^+) oxidized form and (S_1) excited singlet state.

standard hydrogen electrode has a value -0.42. $E'(S/S^-)$ are polarographic half wave potentials. E_T and E_S are, respectively, triplet and singlet state energies of the dyes expressed in kJ mol^{-1}.

TABLE 4.3

Oxidation Potentials E' at pH 7 of some typical dyes

Dye	E_T kJ mol^{-1}	E_S kJ mol^{-1}	$E'(S/S^-)$ volt	$E'(T/S)$ volt	$E'(S^+/S)$ volt	$E'(S^+/T)$ volt
Acriflavin	214.1	247.3	+0.61	−1.62	−1.96	+0.27
Acridine Orange	206.2	234.6	+0.67	−1.47	−1.77	+0.37
Fluorescein	197.7	230.4	+0.70	−1.36	−1.70	+0.36
Eosin Y	190.6	220.0	+0.61	−1.35	−1.67	+0.29
Thionine	163.4	201.1	−0.08	−1.79	−2.18	−0.53

Grossweiner, L.I. & Kepka, A.G., *Photochem. Photobiol*, **16** (1972), 305.

Electron-transfer was suggested by Weiss as a mode of quenching of fluorescence and it appears to be a more general mechanism than considered earlier.

4.10 EMISSION OF POLARIZED LUMINESCENCE

From the mechanism of absorption process (Section 3.1). it is evident that a molecule will absorb a photon of proper frequency only if the incident radiation is polarized with its electric vector parallel to the chromophor axis, i.e. the direction of transition moment. For an *isotropic* absorber or oscillator, i.e. an oscillator which vibrates with the same frequency in all directions, the emitted radiation is also plane polarized. But if the oscillators are *anisotropic* and randomly distributed in space, the degree of polarization of emitted radiation is much reduced. In case of solutions of such molecules, the fluorescence has been found to be completely depolarized in solvents of low viscosity but partially polarized in more viscous solvents. A limiting value is obtained in rigid glassy solvents. In crystalline solids with fixed molecular orientations, the degree of polarization depends on the angle of orientation of the molecules in the crystal lattice.

A beam of radiation is said to be depolarized if the electric vectors are oriented in all directions in space (Figure 4.14 a), perpendicular to the

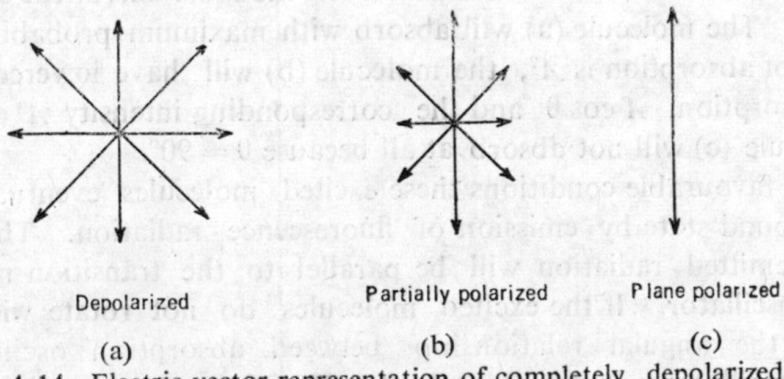

Figure 4.14 Electric vector representation of completely depolarized, partially polarized and completely polarized radiation.

direction of propagation, which is considered perpendicular to the plane of the paper. In a partially polarized beam, the majority of waves will vibrate with electric vector in one plane (Figure 4.14 b) and in a completely polarized beam all the vibrations are unidirectional with respect to a defined coordinate system (Figure 4.14 c).

Let the coordinate system be such as that given in Figure 4.15. The electric vectors of a plane polarized radiation vibrate along *OZ* in the *ZX* plane and *OX* is the direction of propagation of the plane polarized wave. When a solution of anisotropic molecules is exposed to this plane polarized radiation, the electric vector will find the solute molecules in random orientation. Only those molecules absorb with maximum probability which have their transition moment oriented parallel to *OZ* (*photoselection*). Those molecules which are oriented by an angle θ to this direction will have their absorption probability reduced by a factor $\cos\theta$, and the intensity of absorption by $\cos^2\theta$. Finally, the molecules oriented perpendicular to the electric vector will not absorb at all. These statements are direct consequences of directional nature of light absorption

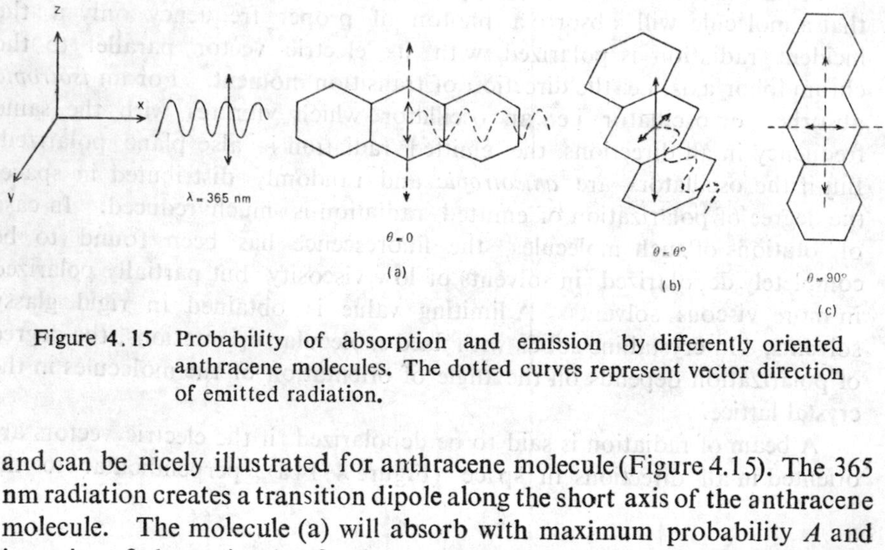

Figure 4.15 Probability of absorption and emission by differently oriented anthracene molecules. The dotted curves represent vector direction of emitted radiation.

and can be nicely illustrated for anthracene molecule (Figure 4.15). The 365 nm radiation creates a transition dipole along the short axis of the anthracene molecule. The molecule (a) will absorb with maximum probability *A* and intensity of absorption is A^2, the molecule (b) will have lowered probability of absorption, $A\cos\theta$ and the corresponding intensity $A^2\cos^2\theta$ and the molecule (c) will not absorb at all because $\theta = 90°$.

Under favourable conditions these excited molecules eventually return to the ground state by emission of fluorescence radiation. The electric vector of emitted radiation will be parallel to the transition moment of emission oscillator. If the excited molecules do not rotate within their lifetimes, the angular relationship between absorption oscillator and emission oscillator will be maintained. Therefore, the electric vector of the molecule (a) in emission will be parallel to the electric vector of incident radiation, that of molecule (b) will maintain an angle and the molecule (c) will not emit as it did not absorb (Figure 4.15). The emission takes place uniformly all around with the molecule as the centre of disturbance. If all the molecules were aligned as in (a), the emitted radiation would have been completely polarized. But the oriented molecules introduce some depolarizing effects. The extent of this effect can be estimated by decomposing the emission electric vector into parallel ($I_{\parallel}$) and perpendicular ($I_{\perp}$) components (Figure 4.16) (App. 1). If the emission is

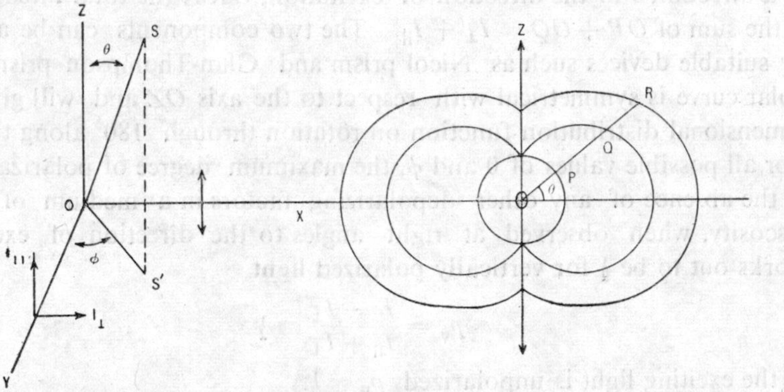

Figure 4.16 Polar curve for fluorescence emission from randomly oriented molecules excited by polarized radiation. Polarization direction *OZ*, propagation direction *OX*.

observed along *OY* direction, a perpendicular component $I_{\perp}$, (perpendicular to the incident electric vector) will be introduced in the emitted radiation causing reduction in the degree of polarization. The greater the angle θ, the larger is the perpendicular component in the emitted radiation causing depolarization. The degree of depolarization is given as the ratio $I_{\parallel}/I_{\perp}$ and the degree of polarization is defined as

$$p = \frac{I_{\parallel} - I_{\perp}}{I_{\parallel} + I_{\perp}} \tag{4.27}$$

where $I_{\parallel}$ and $I_{\perp}$ are the intensities of the components of the beam, parallel and perpendicular to the incident electric vector direction respectively. The polarization of fluorescence emitted by an individual molecule depends on the properties of the molecule and the direction of observation. Its orientation with respect to the direction of polarization of primary light controls the probability of absorption and hence the intensity of fluorescence.

The polarization of the total fluorescence radiation, as it is measured by an observer, results from superposition of all the elementary waves emitted by all the randomly oriented excited molecules. The integrated effect of all the molecules will result in a polar curve given in the Figure 4.16 where θ and ϕ are as defined, and

$$I_{\parallel} = \cos^2\theta \tag{4.28}$$

$$I_{\perp} = \sin^2\theta \cos^2\phi \tag{4.29}$$

In the figure the sample under study is placed at O. The inner curve corresponds to the intensity distribution for the perpendicular component ($I_{\perp}$), the second curve is that for parallel component ($I_{\parallel}$) and the outer curve gives the total intensity in any given direction of observation. Thus, in a direction θ to the direction of excitation, OR is the total intensity and is the sum of $OP + OQ = I_{\perp} + I_{\parallel}$. The two components can be analyzed by suitable devices such as Nicol prism and Glan-Thompson prism. The polar curve is symmetrical with respect to the axis OZ and will give three dimensional distribution function on rotation through 180° along this axis. For all possible values of θ and ϕ, the maximum degree of polarization p_0, in the absence of any other depolarizing factors in a medium of infinite viscosity, when observed at right angles to the direction of excitation, works out to be $\frac{1}{2}$ for vertically polarized light.

$$p_0 = \frac{I_{\parallel} - I_{\perp}}{I_{\parallel} + I_{\perp}} = \tfrac{1}{2} \tag{4.30}$$

If the exciting light is unpolarized, $p_0 = \frac{1}{3}$.

In the system discussed above, the transition moment for absorption coincides with that for emission, that is, the emitting state is the same as the absorbing state. But if the molecule is excited to a higher energy state, but emission is from a lower energy state, for example, $S_2 \leftarrow S_0$ in absorption and $S_1 \rightarrow S_0$ in emission or $S_1 \leftarrow S_0$ in absorption and $T_1 \rightarrow S_0$ in emission, the absorption and emission oscillators may have different transition moment directions. A negative value for the degree of polarization may be observed. The p_0 value is given by

$$p_0 = \frac{3\cos^2\beta - 1}{\cos^2\beta + 3} \tag{4.31}$$

for vertically polarized light where β is the angle between absorption and emission transition moments. For unpolarized exciting light

$$p_0 = \frac{3\cos^2\beta - 1}{7 - \cos^2\beta} \tag{4.32}$$

With vertically polarized exciting light, $p_0 = \frac{1}{2}$ when $\beta = 0$. But when $\beta = \pi/2$, p_0 becomes negative and is equal to $-1/3$. The values are $+1/3$ and $-1/7$ for unpolarized radiation. Thus negative polarization appears when θ is small, i.e. absorption probability is high and the transition moment in emission is perpendicular to that in absorption. These observations provide a suitable method for assigning the polarization directions of transition moments in different absorption bands of a given molecule from *polarization of the fluorescence excitation spectra.*

Suppose a molecule gives rise to two absorption bands *a* and *b* corresponding to transition in two different energy levels S_1 and S_2 (Figure 4.17A). The transition moment in *a* is aligned along the vertically polarised exciting light ($\beta = 0$) and in *b* it is at right angles to it ($\beta = 90°$). The molecule is dissolved in a rigid glassy medium of infinite viscosity. A plot of p_0, the degree of polarization of fluorescence as a function of wavelength or wavenumber of exciting light, gives the *fluorescence polarization spectrum.* In the band *a*, the maximum polarization will be 0.5 and will remain at this level over the absorption band. In the second band *b*, with $\beta = 90°$, the maximum polarization p_0 will be -0.33, since emission always occurs from the lowest excited state. In the region of overlap between the two bands a sharp change in polarization from positive to negative value will be observed. The fluorescence polarization spectrum can be used to identify overlapping bands and can help to resolve hidden transitions.

The fluorescence polarization spectrum of Rhodamin B and the corresponding absorption spectrum are given in the Figure 4.17B. The transition in bands 1 and 5 have the same polarization direction, bands 3 and 4 are polarized almost perpendicular to 1 and the polarization of 2 is at some intermediate angle. These reflect the relative orientation of the transition moments in the respective bands.

4.10.1 Rotational Depolarization of Fluorescence

In all the above arguments, it has been assumed that the molecules are fixed in position, although randomly distributed, in an infinitely viscous medium. If a less viscous medium is used, this condition will no longer hold. The translational component of thermal agitation will have no effect on the degree of polarization of emitted radiation. But the rotational component will change the relative orientation of the transition moment during emission from that at the instant of absorption, and hence the polarization of fluorescence. Obviously, the extent of rotation during the lifetime of the excited molecule only will be effective.

The frequency of rotation in a medium of low viscosity is of the order of 10^{11} s^{-1}, whereas the lifetime of the excited molecule is of the order of 10^{-8} s. Thus, in such a medium, each excited molecule on the average can rotate about 1000 times before it fluoresces. As a result, the fluorescence is completely depolarized and spherically distributed in space. But if rotation is hindered by putting the molecule in a viscous medium like glycerol or sugar solutions, the emitted radiation may not be completely depolarized if the period of rotation is greater than the lifetime τ, of the molecule. The greater the viscosity η of the solvent, greater will be the degree of polarization for a given lifetime τ. For long-lived states (τ large), the degree of polarization will be small. Therefore, the degree of polarization depends on the average period of rotation

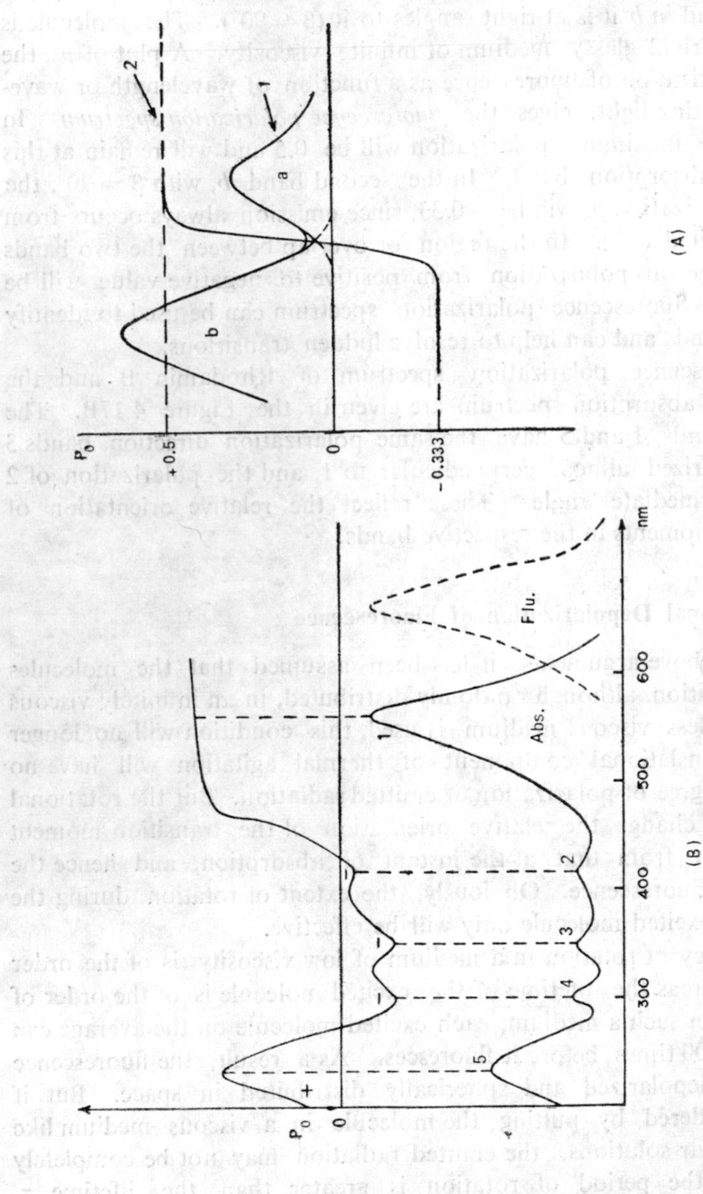

Figure 4.17 *A*. Fluorescence polarization spectrum. A general case of a molecule with two absorption bands *a* and *b* with transition moment vectors at right angles to each other. (1) Absorption spectrum; (2) Polarization of flourescence excitation spectrum. *B*. Fluorescence polarization spectrum and absorption spectrum of Rhodamin B.

and the radiative lifetime of the molecule.

According to the theory developed by Smoluchowski and by Einstein, if a spherical particle of radius r rotates in a liquid of viscosity η, in a short time Δt, by an angle $\Delta\alpha$, then the mean value of angular rotation $\overline{\Delta\alpha^2}$ is given by the Brownian equation for rotational motion:

$$\overline{\Delta\alpha^2} = \frac{kT}{4\pi r^3 \eta} \Delta t$$

$$= \frac{RT}{3\eta V} \Delta t = \frac{\Delta t}{\rho} \tag{4.33}$$

where, $V = \frac{4}{3}\pi r^3 N$, is the molar volume and $\rho = 3\eta V/RT$ is the average period of rotation or the *rotational relaxation time.* By inserting sufficiently short lifetime τ for Δt in this equation, the observed degree of polarization p, which results from rotation of all the individual oscillators around statistically oriented axes, has been obtained by Perrin, for vertically polarized exciting light, as

$$\left(\frac{1}{p} - \frac{1}{3}\right) = \left(\frac{1}{p_0} - \frac{1}{3}\right)\left(1 + \frac{3\tau}{\rho}\right) \tag{4.34}$$

On rearrangement and substitution, we get

$$\frac{1}{p} = \frac{1}{p_0} + \left(\frac{1}{p_0} - \frac{1}{3}\right)\frac{RT}{\eta V}\tau \tag{4.35}$$

giving a linear relationship between $1/p$ and fluidity of the solvent medium. Thus, if the polarization of a solution is measured as a function of temperature under conditions where lifetime τ does not vary with temperature, the plot of $1/p$ vs T/η should be linear. From the intercept, $1/p_0$ and the slope, $RT\,(1/p_0 - 1/3)/V$, the lifetime τ can be calculated if molar volume V is known and molar volume can be calculated if τ is known from other sources. Both these alternatives have been found useful in specific cases. Such plots known as Perrin's plots can be obtained at constant temperature by using solvents of similar chemical nature but of different viscosity, or mixtures of two solvents such as water and glycol to give a series of solutions of variable viscosity. The molar volume V, includes the solvation envelope and may change in different solvents. An uncertainty is that here η is the *microscopic viscosity* of the medium which may be different from *macroscopic viscosity.*

The fluorescence depolarization measurements have found application in biochemical studies to measure the size of large molecules such as protein. The usual method is to tag the molecule with a fluorescent reagent through an isocyanate, isothiocyanate or sulphonyl chloride group. These tagged molecules are known as *fluorescent protein conjugates.* The lifetimes of the conjugates are usually close to those of the simple fluorescent molecules. Assuming a spherical shape, the molar volume can be calculated.

Another parameter characterizing the polarization state of luminescence radiation is the degree of anisotropy R, defined as

$$R=\frac{I_{\parallel}-I_{\perp}}{I_{\parallel}+2I_{\perp}} \tag{4.36}$$

In terms of angle β between absorption and emission oscillator,

$$R=\tfrac{1}{3}(3\cos^2\beta-1) \tag{4.37}$$

4.10.2 Concentration Depolarization

A further depolarizing factor is introduced in concentrated solutions of given viscosity η. The empirical relationship found by Sveshnikov and Feofilov is

$$\frac{1}{p}=\frac{1}{p_0}+A\tau C \tag{4.38}$$

where A is a constant and C, concentration of the solution.

The concentration depolarization is observed at a concentration $<10^{-3}$M where molecules are apart by an average distance of 7 nm or more. The reason for such depolarizing effect is dipole-dipole interaction between the neighbouring molecules whereby electronic excitation energy is transferred from initially excited molecule to its neighbour at fairly large distances. If the acceptor molecule has a different orientation than the donor molecule, the emitted radiation will be depolarized. Such long-range energy transfer by dipole-dipole mechanism can lead to energy wandering from molecule to molecule until it is finally emitted (Figure 4.18) thereby completely losing its initial polarization direction. The concentration depolarization or self-

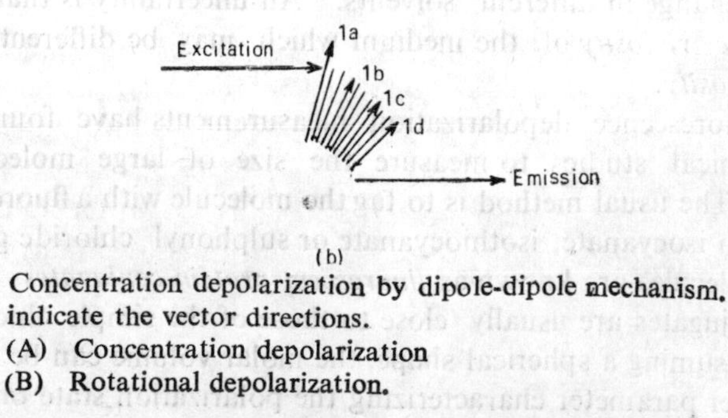

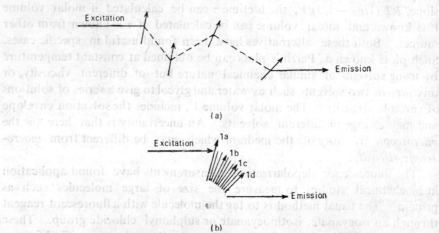

Figure 4.18 Concentration depolarization by dipole-dipole mechanism. Arrows indicate the vector directions.
(A) Concentration depolarization
(B) Rotational depolarization.

depolarization is independent of viscosity of the medium; liquid and solid solutions behave alike. However, this effect is reduced if the lifetime τ, of the excited state is reduced. An expression for reduced degree of polarization due to self-depolarization is given by

$$\left(\frac{1}{p}-\frac{1}{3}\right)=\left(\frac{1}{p_0}-\frac{1}{3}\right)\left[1+\frac{4\pi N R_0^6 \times 10^3}{15(2a)^3}\,C\right] \tag{4.39}$$

where, $2a$ is the molecular diameter, C, concentration in mol litre^{-1} and R_0 is the critical distance between parallel dipoles at which the probability of emission is equal to the probability of energy transfer. Such polarization studies can be used to calculate R_0. The details of energy transfer mechanism will be taken up in Section 6.6.

4.11 GEOMETRY OF SOME ELECTRONICALLY EXCITED MOLECULES

As can be easily visualized from Figure 4.19, formaldehyde molecule has a planar geometry in the ground state. On excitation to the $^1(n, \pi)$ state, an electron localized on O-atom is transferred to an anti-bonding π-MO which has a node between C and O bonds. This destroys the rigidity of the molecule and it can now twist along the $\rangle$C—O bond.

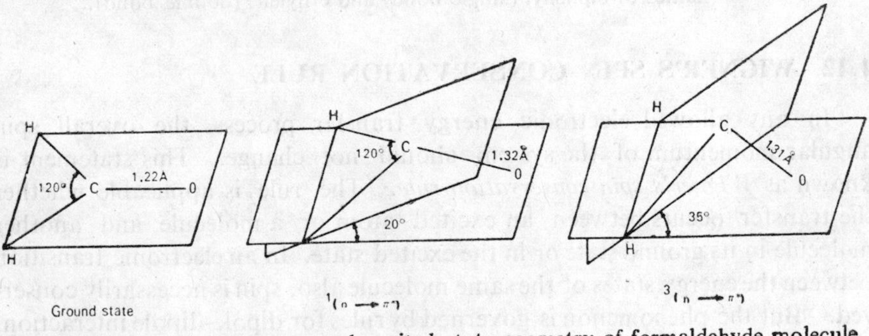

Figure 4.19 Ground state and excited state geometry of formaldehyde molecule.

A similar situation is possible for conjugated systems like ethylene or butadiene. Free rotation in the excited state will lead to cis-trans isomerization, a phenomenon which is utilized by nature for reversible processes like vision, in which changes in the geometry of retinal consequent upon light absorption, can trigger a series of reactions. In the 11-cis configuration, the chromophore snuggly fits into the conformation of the protein opsin. On absorption of radiation it undergoes isomerization to all trans form and is detached from the protein.

Molecules such as CO_2, CS_2, HCN and acetylene are linear in their ground state but bent in their lowest excited states. On the other hand, radicals such as $.NH_2$, .HCO, and possibly NO_2 molecule are bent in their ground state but linear in their lowest excited states.

Those molecules which change their geometry in the excited state show a variation in their potential energy function with the angle of twist. Such variation for a single and a double bonded system is given in Figure 4.20.

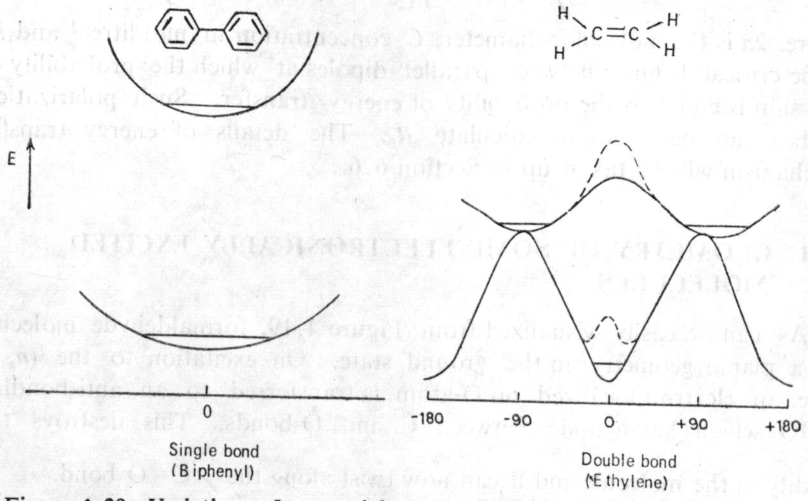

Figure 4.20 Variation of potential energy function for ground and excited states of biphenyl (single bond) and ethylene (double bond).

4.12 WIGNER'S SPIN CONSERVATION RULE

In any allowed electronic energy transfer process, the overall spin angular momentum of the system should not change. This statement is known as *Wigner's spin conservation rule.* The rule is applicable whether the transfer occurs between an excited atom or a molecule and another molecule in its ground state or in the excited state. In an electronic transition between the energy states of the same molecule also, spin is necessarily conserved. But the phenomenon is governed by rules for dipole-dipole interaction.

Wigner's spin conservation rule requires that there must be correlation of spins between the reactants and the products. The possible spin of the transition state for reactants A and B with spins S_A and S_B can be obtained by vector addition rule as $|S_A+S_B|$, $|S_A+S_B-1|$, ..., $|S_A-S_B|$. In order that there is smooth correlation with the products X and Y, the transition state formed by the products must also have total spin magnitude which belongs to one of the above values. This situation allows the reactants and the products to lie on the same potential energy surface. Such reactions, whether physical or chemical, are known as *adiabatic.*

In these reactions a close approach of two reacting partners is necessary. Essentially, Wigner's rule states that the total spins of the α type (↑)

and the β type (↓) in the combined initial and the final states of the system remains the same, no matter how we combine them in the final state. Thus, for example, in singlet-singlet, triplet-triplet and triplet-singlet transfer of energy by exchange mechanism (Section 6.6.7) the following combinations are possible:

Multiplicities of initial state	*Multiplicities of final state*	
Singlet-singlet (↑↓)* + (↑↓)	(↑↓) + (↑↓)*	singlet + singlet
	↑↓↑ + ↓	doublet + doublet
	↑↓ + ↑ + ↓	singlet + doublet + doublet
Triplet–singlet (↑↑)* + (↓↑) (A, B)	↑↓ + (↑↑)* (A, B)	(singlet + triplet*)
	↑↑↓ + ↑	(doublet + doublet)
	↑↑ + ↓ + ↑	(triplet + doublet + doublet)
	↑ + ↑ + ↑↓	(doublet + doublet + singlet)
	(↑↑↓↑)*	triplet*
Triplet-triplet (↑↑)* + (↑↑)*	↑↑ + ↑ + ↑	triplet + doublet + doublet
	↑↑↑ + ↑	quartet + doublet
	(↑↑↑↑)*	quintet*
(↑↑)* (↓↓)*	↑↑↓ + ↓	doublet + doublet
	↑↓ + ↑↓	singlet + singlet

These rules also predict the nature of photoproducts expected in a metal-sensitized reactions. From the restrictions imposed by conservation of spin, we expect different products for singlet-sensitized and triplet-sensitized reactions. The Wigner spin rule is utilized to predict the outcome of photophysical processes such as, allowed electronic states of triplet-triplet annihilation processes, quenching by paramagnetic ions, electronic energy transfer by exchange mechanism and also in a variety of photochemical primary processes leading to reactant-product correlation.

4.13 STUDY OF EXCITED STATES BY FLASH PHOTOLYSIS EXPERIMENTS AND LASER BEAMS

The study of the short-lived excited states is limited by the low concentrations in which they are created on excitation with normal light sources. The use of high intensity sources such as flash lamps with suitable flashing rates (Section 10.3) and laser sources (Section 10.2) have helped in the study of the triplet states of molecules. The triplet-triplet absorption, triplet lifetimes, intersystem crossing rates from triplet to ground singlet state, etc. can be effectively measured because of high concentrations of the triplets generated thereby and because of their long lifetimes. Both these factors are convenient for sequential biphotonic processes.

Using high intensity laser sources, weak emission from upper energy states, higher than the first, has been observed for many systems other than azulene. Azulene which is known to emit from the second singlet state, is found to emit from the first singlet and triplet states following excitation with picosecond (10^{-12} s) laser light pulses, with an efficiency $\simeq 10^{-7}$.

Summary

1. Due to differences in charge distribution in the various energy states of a molecule, the electronically excited species can be completely different in its chemical and physical behaviours as compared to the normal ground state molecule. Molecular geometries and internuclear force constants also change.
2. Variation in intermolecular force constants are reflected in potential energy curves of the various energy states of the molecule. The potential energy as a function of internuclear distances is represented in the form of potential energy (PE) diagrams in two dimensions for diatomic molecules. For polyatomic molecules, polydimensional hypersurfaces are mathematically required, although they cannot be quantitatively obtained in practice.
3. Electronic transitions between two energy states are governed by the Franck-Condon principle. In quantum mechanical terminology, the Franck-Condon overlap integral $\int \chi_f' \chi_i'' \, d\tau_v$ is important. χ_f' and χ_i'' are, respectively, vibration wave functions for v' in the final electronic state, and v'' in the initial electronic state.
4. The shapes of the absorption spectra depend on the shape and disposition of the upper PE curve with respect to the ground state PE curve. In general, vibrational excitation is coupled with electronic excitation and gives rise to structure in the absorption spectra. If the energy of promotion is greater than the dissociation energy of the molecule, a continuum is observed.
5. Emission also follows the Franck-Condon principle. Fluorescence spectrum may appear as mirror image of absorption towards the longer wavelength region for large polyatomic molecules whose equilibrium bond length do not change on excitation. Phosphorescence spectra are shifted at still longer wavelengths and do not obey the mirror image relationship with $S_1 \leftarrow S_0$ transition.
6. Due to differences in charge distribution in the ground and the excited states, the dipole moments and polarizabilities in the two states may be different. Differences in ground state and excited state dipole moments are manifested in differences in solute-solvent interactions, causing blue-shift or red-shift in the absorption and emission spectra.
7. Solute-solvent interactions are of two types: (1) universal interaction, and (2) specific interaction. Universal interaction is due to the collective influence of the solvent as a dielectric medium. It depends on the dielectric constant D and refractive index n of the solvent and the dipole moment μ of the solute molecule. Such interactions are van der Waals type. Specific interactions are short range interactions and involve H-bonding, charge-transfer or exciplex formation. H-bonding ability may change on excitation specially for $n \rightarrow \pi^*$ transitions.
8. A plot of

$$(\bar{\nu}_a - \bar{\nu}_f) \text{ against } \left(\frac{D-1}{2D+1} - \frac{n^2-1}{2n^2+1}\right)$$

gives a positive slope if excited state dipole moment μ_e is greater than ground state dipole moment, μ_g. A negative slope is obtained if $\mu_e < \mu_g$. The relationship is obtained from Onsager's theory of reaction field in solution.

$$\bar{\nu}_a - \bar{\nu}_f = \frac{2}{hc} \frac{(\mu_e - \mu_g)^2}{a^3} \left[\frac{D-1}{2D+2} - \frac{n^2-1}{2n^2+2} \right]$$

$\bar{\nu}_a$ and $\bar{\nu}_f$ are O—O energy in absorption and emission respectively.

9. Singlet excited state acid dissociation constants pK^* can be smaller or greater than the ground state constant pK by as much as 8 units. Phenols, thiols and aromatic amines are stronger acids upon excitation, whereas carboxylic acids, aldehydes and ketones with lowest $^1(\pi, \pi^*)$ states become much more basic. Triplet state constants pK^T are closer to those for the ground state. Förster's cycle may be used to determine $\Delta\, pK\ (=pK^* - pK)$ from fluorescence measurements if proton transfer occurs within the lifetime of the excited molecule.
10. Excited state redox potentials may be different from those of the ground state values. Redox reactions can be initiated on electronic excitation against the electrochemical gradients. Electron transfer in the excited state may be reversed in the ground state. An important example is photosynthesis in plants.
11. When excited with plane polarized light, the degree of polarization p, of emitted radiation from an anisotropic oscillator can be depolarized due to a number of factors: (i) random distribution of molecules such that transition moments of varying orientations are exposed to plane polarized incident radiation; (ii) Brownian motion of rotation which changes the transition moment direction from the instant of absorption to the instant of emission; the effect is viscosity and temperature dependent; and (iii) increase in concentration. If excited to a higher energy state the degree of polarization may be negative.
12. Polarization data can help to assign the transition moment direction in an absorption band.

    The excited state lifetimes, τ, can be calculated from the relationship between the degree of polarization and the viscosity of the medium from the expression,

$$\frac{1}{p} = \frac{1}{p_0} - \left(\frac{1}{p_0} - \frac{1}{3} \right) \frac{RT}{\eta V} \tau$$

    If τ is known, the molar volume V can be calculated.
13. Geometry of the molecules may change on excitation with an important effect on the photochemistry of the molecule.
14. Wigner's spin conservation rules govern energy transfer between two molecules, photodissociation energetics, T—T annihilation energetics and other photophysical and photochemical processes.
15. By using high intensity flash lamps and laser sources, photophysical and photochemical properties of the triplet states can be studied. These sources also help to study emission from upper excited state.

FIVE

Photophysical Processes in Electronically Excited Molecules

5.1 TYPES OF PHOTOPHYSICAL PATHWAYS

A molecule excited to a higher energy state must return to the ground state eventually, unless it gets involved in a photochemical reaction and loses its identity. In monatomic gases at low pressures and temperatures, reverse transition with emission is likely to be the only mode of return. But in condensed systems, e.g. solutions, liquids and solids, in polyatomic molecules and in gases at reasonable pressures, there are more than one pathways available to the excited molecule for dissipation of excitational energy. These different pathways are grouped under photophysical processes in electronically excited molecules. Some are intrinsic properties of the molecule and are unimolecular while some others depend on external perturbations and may involve bimolecular collisions. All these photophysical processes must occur in a time period less than the natural radiative lifetime of the molecule and priorities are established by their relative rate constants.

After the initial act of absorption,

$$A + h\nu \rightarrow A'^* \tag{5.1}$$

where A'^* is either an electronically excited molecule with excess vibrational energy in S_1 state or a molecule excited to higher singlet states S_2, S_3, etc. the various photophysical processes that can occur in a molecule are:

Unimolecular

$A'^* \rightarrow A^* + \text{heat}$	Internal conversion (IC)	$S_n \rightsquigarrow S_1$	(5.2)
$A^* \rightarrow A + \text{heat}$	Internal conversion (IC)	$S_1 \rightsquigarrow S_0$	(5.3)
$A^* \rightarrow A + h\nu_f$	Fluorescence emission	$S_1 \rightarrow S_0$	(5.4)
$A^* \rightarrow {}^3A + \text{heat}$	Intersystem crossing (ISC)	$S_1 \rightsquigarrow T_1$	(5.5)
${}^3A \rightarrow A + h\nu_p$	Phosphorescence emission	$T_1 \rightsquigarrow S_0$	(5.6)
${}^3A \rightarrow A + \text{heat}$	Reverse intersystem crossing (ISC)	$T_1 \rightarrow S_0$	(5.7)

Bimolecular

$A^* + S \rightarrow A + \text{heat}$	Solvent quenching	(5.8)
$A^* + A \rightarrow 2A + \text{heat}$	Self-quenching	(5.9)
$A^* + Q \rightarrow A + Q + \text{heat}$	Impurity quenching	(5.10)
$A^* + B \rightarrow A + B^*$	Electronic energy transfer (ET)	(5.11)

A^*, ${}^3A$ and A are molecules in first excited singlet state, molecules in triplet state and in the ground state respectively. In radiationless processes such as internal conversion and intersystem crossing the excess energy is lost to the environment as thermal energy. Some of the unimolecular processes are represented by a Jablonski diagram in Figure 5.1. *Radiative transitions* are denoted

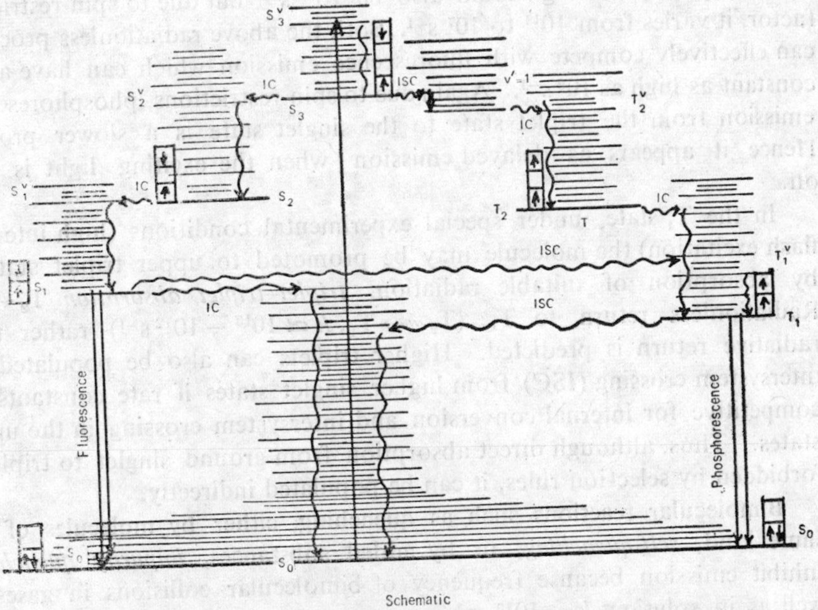

Figure 5.1 Jablonski diagram for photophysical pathways
S_n, S_n^v = nth singlet energy state and vibronically excited state
T_n, T_n^v = nth triplet energy state and vibronically excited state
IC = internal conversion, ISC = intersystem crossing
F = fluorescence emission, P = phosphoresence emission.

by straight arrows $S_1 \rightarrow S_0$ whereas *nonradiative* processes by wavy arrows $S_1 \rightsquigarrow S_0$.

The initial act of absorption may promote a molecule to higher energy states S_2, S_3, etc. or to higher vibrational levels of the S_1 state in a time period 10^{-15} s ($k = 10^{15}$ s^{-1}) obeying Franck-Condon principle. The electronic energy of S_2, S_3 states or excess vibrational energy of S_1 state is quickly lost to the surroundings by a mechanism known as *internal conversion* (IC). The rate constant of internal conversion is $\simeq 10^{13} - 10^{12}$ s^{-1}, the same as vibrational frequencies. Once in the zero vibrational level of the first excited singlet state, the molecule may return to the ground state, in the absence of a photochemical reaction, by radiative *fluorescence emission* $S_1 \rightarrow S_0$, nonradiative *internal conversion*, $S_1 \rightsquigarrow S_0$ or partly radiative and partly nonradiative pathways as represented in the Jablonski diagram. Internal conversions from $S_1 \rightsquigarrow S_0$ have smaller rate constants 10^8 s^{-1} or less, as compared to the same processes in higher energy states because of large energy gap between the two. *Intersystem crossing* (ISC), involves nonradiative transition from singlet to triplet state, $S_1 \rightsquigarrow T_1$, generating ^3A which can then decay by radiative *phosphorescence emission*, $T_1 \rightarrow S_0$ or by nonradiative *reverse intersystem crossing*, $T_1 \rightsquigarrow S_0$ processes. Rate constant for intersystem crossing should also be 10^{12} s^{-1} but due to spin restriction factor, it varies from 10^{11} to 10^7 s^{-1}. Both the above radiationless processes can effectively compete with fluorescence emission which can have a rate constant as high as 10^9 s^{-1}. Again due to spin restrictions, phosphorescence emission from the triplet state to the singlet state is a slower process. Hence it appears as delayed emission when the exciting light is shut off.

In the T_1 state, under special experimental conditions (high intensity flash excitation) the molecule may be promoted to upper triplet state T_2 by absorption of suitable radiation, *triplet–triplet absorption* $T_2 \leftarrow T_1$. Radiationless return to T_1, ($T_2 \rightsquigarrow T_1$; $k \simeq 10^{13} - 10^{12}$ s^{-1}) rather than radiative return is predicted. Higher triplets can also be populated by intersystem crossing (ISC) from higher singlet states if rate constants are competitive for internal conversion and intersystem crossing in the upper states. Thus, although direct absorption from ground singlet to triplet is forbidden by selection rules, it can be populated indirectly.

Bimolecular reactions such as quenching, either by molecules of the same kind, *self-quenching*, or by added substances, *impurity quenching*, inhibit emission because frequency of bimolecular collisions in gases as well as in solution, $k \simeq 10^{10}$ s^{-1} can compete with fluorescence emission. Solvent quenching may involve other physical parameters as well such as solute-solvent interactions. Since the solvent acts as the medium in which the solute molecules are bathed, solvent quenching may be classified under unimolecular processes and a clear distinction between it and internal conversion $S_1 \rightarrow S_0$ is difficult.

A very important bimolecular deactivation process is the *electronic energy transfer* (ET). In this process, a molecule initially excited by absorption of radiation, transfers its excitation energy by nonradiative mechanism to another molecule which is transparent to this particular wavelength. The second molecule, thus excited, can undergo various photophysical and photochemical processes according to its own characteristics.

Under certain conditions, a few other processes may be initiated, such as

I.	$A^* + A \rightarrow AA^*$	Excimer formation	(5.12)
	$AA^* \rightarrow 2A + h\nu_E$	Excimer emission	(5.13)
II.	$A^* + Q \rightarrow AQ^*$	Exciplex formation	(5.14)
	$AQ^* \rightarrow A + Q + h\nu_{EC}$	Exciplex emission	(5.15)
III.	$T^* + \text{heat} \rightarrow A^*$	Thermal excitation of triplet to singlet	(5.16)
	$A^* \rightarrow A + h\nu_{ED}$	E-type delayed emission of fluorescence	(5.17)
IV.	$T + T \rightarrow A^* + A$	Triplet-triplet annihilation to excited singlet	(5.18)
	$A^* \rightarrow A + h\nu_{PD}$	P-type delayed emission of fluorescence	(5.19)

The processes III and IV termed as *E-type* and *P-type delayed emissions* have emission spectra identical with that of the normal fluorescence but with longer radiative lifetime. The long life is due to the involvement of the triplet state as an intermediate. Hence the short-lived direct fluorescence emission from the S_1 state is referred to as *prompt fluorescence.* E-type delayed fluorescence was called α-phosphorescence by Lewis in his early works.

These photophysical processes often decide the photochemical behaviour of a molecule and reduce the quantum yield of a photochemical reaction to much less than unity. A molecule in the singlet state is a different chemical species from that in the triplet state and may initiate different chemistry. Therefore, for a complete understanding of a photochemical reaction, a clear knowledge of various photophysical processes, that is, how the absorbed quantum is partitioned into different pathways is essential. This account keeping of the absorbed quanta, so to say, may help modify a given chemical reaction if it is so desired. We shall discuss each of these processes one by one.

5.2 RADIATIONLESS TRANSITIONS—INTERNAL CONVERSION AND INTERSYSTEM CROSSING

A polyatomic molecule in condensed system when excited to a higher vibrational level of the first excited state, loses its excess vibrational energy to the surroundings in a time period $\simeq 10^{-13}-10^{12}$ s, the time for a molecular vibration. This radiationless cascade of energy is known as *internal conversion.* Even if the excitation is to an energy state higher than S_1, the molecule tumbles down quickly to the zero vibrational level of the first excited state S_1, losing all its excess electronic and vibrational energy within 10^{-12} s. Due to large energy gaps, transition from S_1 to S_0 is not always probable by radiationless mechanism. Under these circumstances the molecule has two alternatives: (i) to return to the ground state by fluorescence emission, or (ii) to cross over to the lowest triplet state non-radiatively. This nonradiative transfer from singlet excited to triplet state is known as *intersystem crossing.* In these radiationless processes, the environment acts as a heat sink for dissipation of extra energy as thermal energy. In a polyatomic molecule with 3N–6 modes of vibrations such loss in energy is observed even in the vapour phase at very low pressures where collision frequencies are likely to be less than the rates for radiationless conversion. It follows that nonradiative conversion is an *intrinsic* property of polyatomic molecules.

Thus, there are two major types of radiationless or nonradiative transitions: (i) internal conversion, and (ii) intersystem crossing. The *internal conversion* is so called because the nonradiative loss of energy occurs between electronic energy manifold of the same spin type: singlet-singlet or triplet-triplet, $S_j \rightsquigarrow S_k$ or $T_j \rightsquigarrow T_k$. The *intersystem crossing* involves nonradiative energy loss between energy states of two different spin manifolds $S_j \rightsquigarrow T_k$ or $T_k \rightsquigarrow S_p$.

From kinetic considerations each can be further subdivided according to observed values of rate constants: k_{IC}, the rate constant for internal conversion and k_{ISC}, the rate constant for intersystem crossing.

(ia) $S_j \rightsquigarrow S_1$; $k_{IC} \simeq 10^{13}-10^{12}\ s^{-1}$

(ib) $S_1 \rightsquigarrow S_0$; $k_{IC} \simeq 10^{8}\ s^{-1}$ or less

(iia) $S_1 \rightsquigarrow T_1$; $k_{ISC} \simeq 10^{11}-10^{6}\ s^{-1}$

(iib) $T_1 \rightsquigarrow S_0$; $k_{ISC}^{T} \simeq 10^{4}-10^{-2}\ s^{-1}$

5.2.1 Theory of Radiationless Transitions

Radiationless transition between two electronic states may be represented as occurring at the point of intersection of potential energy surfaces. The phenomenon is similar to the one encountered in predissociation spectra of diatomic molecules. In an N-atomic molecule with 3N–6 modes of vibration there will be 3N–6 polydimensional hypersurfaces describing the potential energy functions for each mode. There will be many points of crossing, or points of near-crossing amongst them. A crossing point is the point of equal energy for both the curves. The transfer occurs irreversibly at this *isoenergetic point* to the high vibrational level of the lower energy state and the excess vibrational energy rapidly cascades down the vibrational manifold. Thus, the radiationless conversion of energy involves *two steps*: (i) the vertical transfer of energy at the isoenergetic point from the zero-point level of higher electronic energy state to the high vibrational level of the lower electronic state, and (ii) the rapid loss of excess vibrational energy after transfer. The first step is the rate determining step and is of main interest. The second is merely *vibrational relaxation.*

Various theories have been proposed for *horizontal transfer* at the isoenergetic point. Gouterman considered a condensed system and tried to explain it in the same way as the radiative mechanism. In the radiative transfer, the two energy states are coupled by the photon or the radiation field. In the nonradiative transfer, the coupling is brought about by the phonon field of the crystalline matrix. But this theory is inconsistent with the observation that internal conversion occurs also in individual polyatomic molecules such as benzene. In such cases the medium does not actively participate except as a heat sink. This was taken into consideration in theories proposed by Robinson and Frosch, and Siebrand and has been further improved by Bixon and Jortner for isolated molecules, but the subject is still imperfectly understood.

In the theory of radiative transition, *the dipole moment operator* $\hat{\mu}$ couples the two electronic energy states and the Franck-Condon overlap integral determines the *vertical transfer probability* $(\int \Psi_f \hat{\mu} \Psi_i \, d\tau)^2$ between the vibronic wave functions of the two states. In the theory of non-radiative transition, the two states are coupled by an operator called the *nuclear kinetic energy operator* $\hat{J}_N$ and the Franck-Condon overlap integral determines the probability of *horizontal transfer* between the potential functions of the two electronic states. The operator $\hat{J}_N$ is effective on Born-Oppenheimer states only in which nuclear and electronic motions can be separated. Hence, if Ψ_i and Ψ_f are wave functions of two combining states, the initial and the final, then under the perturbation $\hat{J}_N$ the probability of energy transfer between these two states is:

$$\text{Probability} \simeq \left[\int \Psi_f^* \hat{J}_N \Psi_i \, d\tau \right]^2$$

$$\simeq \left[\int \Psi_f \hat{J}_N \Psi_i \, d\tau_e \int \chi_f^{v'} \chi_i^{v''} \, d\tau_v \int S_f S_i \, d\tau_s \right]^2 \qquad (5.20)$$

where $d\tau_e$, $d\tau_v$ and $d\tau_s$ are the configuration spaces for electronic, vibrational and spin motions respectively. The perturbation acts on the electronic wave function only, which is the first term in the above expression. The last term is the spin integral. It is unity when the spin functions of the two states are the same. Otherwise it is zero but may have nonzero value under spin-orbit coupling interactions. The middle term defines the overlap criterion for vibrational wave functions of the two combining states and is the familiar *Franck-Condon* integral. A large overlap integral indicates a high transfer efficiency. A qualitative understanding of the concept can be obtained from the potential energy surfaces for a ground state S_0 and two upper electronic states S_1 and S_2 of a simple diatomic molecule (Figure 5.2).

Energy difference between S_1 and S_0 is generally much larger than that between higher energy states S_2, S_1, S_n, etc. Therefore, zero-vibrational level of S_1 ($v' = 0$) state can overlap only with a high vibrational level of the ground energy state S_0. In this diagram, $v'' = 12$, and the Franck-Condon integral is

$$\int \chi_1^{v'=0} \chi_0^{v''=12} \, d\tau_v$$

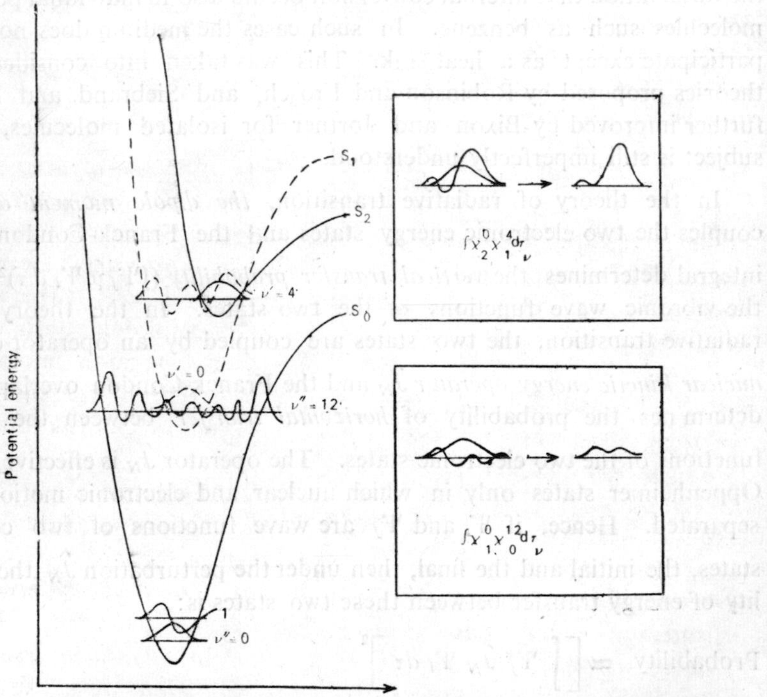

Figure 5.2 Potential energy surface for S_0, S_1 and S_2 and Franck-Condon overlap integral. Inset: (A) Overlap integral between $S_0^{v=12}$ and $S_1^{v=0}$ and (B) $S_1^{v=4}$ and $S_2^{v=0}$ states for horizontal transfer of energy.

The two curves are nearly parallel and they overlap near the equilibrium nuclear geometry where kinetic energy is large. On the other hand, the dispositions of PE surfaces of S_2 and S_1 states are such that the two curves intersect at a point $v=0,\ v'=4$. At the point of intersection the lower curve is at the extreme position of vibrational oscillation where energy is all potential and the probability function is large. The overlap integral is

$$\int \chi_2^{v=0}\, \chi_1^{v'=4}\, d\tau_v$$

Pictorially, the overlap integrals for the two cases can be represented as shown in the inset of Figure 5.2. Since only the overlap regions need be considered, by simple superposition principle, we find that $S_1 \rightsquigarrow S_0$ has very poor overlap integral (Figure 5.2a) because the higher vibrational energy states of S_0 have low probability distribution function in the centre. Such a situation is likely to be obtained when the (O – O) energy gap between the two combining states is large. For $S_2 \rightsquigarrow S_1$, the two potential functions intersect (Figure 5.2b). The overlap is good mainly because the wave functions have large values at the extrema as expected from a classical description of harmonic oscillation. It follows that the larger the energy gap between S_1 and S_0 states, the smaller will be the overlap integral and the smaller will be the transfer efficiency. Equation (5.20) thus predicts low probability of internal conversion between S_1 and S_0 states.

Since the higher energy states are closer in energy, there is always a possibility of potential energy surfaces crossing at some point. Transfer occurs at the crossing points which are isoenergetic for the two combining states. The transfer is further facilitated by momentary freezing of the nuclear coordinates at the vibrational turning points. This is the rate determining step and must occur before the molecule starts oscillating, i.e. within $10^{-13} - 10^{-12}$ s. The large Franck-Condon integral is not always the sole criterion for efficient cross-over from one energy state to the other. Symmetry restrictions and spin multiplicity rules impose their own inefficiency factors.

The transfer is in general irreversible and is immediately followed by very fast vibrational relaxation phenomenon. The irreversibility of transfer is not due to any difference in the probabilities of forward $S_2 \rightsquigarrow S_1$ and the reverse $S_2 \leftsquigarrow S_1$ nonradiative transfer but due to the difference in the densities of energy states in the initial and the final states. The *density of energy state* ρ_E is defined as *the number of vibrational levels per unit energy interval (in cm*$^{-1}$*) at the energy of the initial state*. From Figure 5.3, it is observed that since the initial state is at or near the zero-point energy of S_2, the energy levels are very sparsely spaced. But the state to which the energy is transferred is associated with high vibrational quantum numbers and hence have very closely spaced vibrational energy states as expected for an anharmonic oscillator. Furthermore, in solution, specially in a rigid matrix, the medium may provide a background of its own energy states which are nearly degenerate and in resonance with the

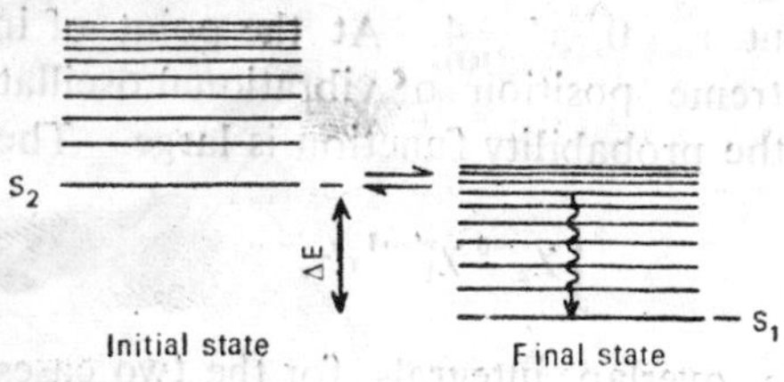

Figure 5.3 Rate of horizontal transfer from initial to final states as a function of density of energy states.

initial state. Therefore, the final state is a *quasicontinuum*. The energy finds a large number of exit paths of varied vibrational modes but very few reentry paths. From the theories of unimolecular reactions we know that once the energy is transferred to different vibrational modes, it is impossible to collect it back to the initial high energy vibrational state of a complex molecule.

The quasidegeneracy of the isoenergetic energy state E is described by

$$E = \Delta E \pm \tfrac{1}{2}\,\rho_E \tag{5.21}$$

where ΔE is the O—O energy difference between S_2 and S_1. The energy $E = \sum_n v_n h\nu_n$, where v_n is the nth vibrational quantum number and ν_n the corresponding vibrational frequency in the final (lower electronic) state. Increase in ΔE implies that a higher vibronic level of the lower electronic state is isoenergetic with zero-point energy of the upper state. Since the vibrational modes increase, the permutation possibility of this energy in various vibrational modes also increases. This results in large transfer efficiency in the forward direction and large probability of vibrational relaxation. But increase in ΔE can be helpful only to a limited extent. Further increase in the energy gap between the pure electronic states tends to remove the system from the condition of resonance. As a consequence, the transfer probability decreases.

The role played by the medium in radiationless transitions is not yet very clear. It certainly acts as a heat sink. It may provide many closely spaced energy levels which are in near resonance with vibrational levels for heat flow into the sink. Solute-solvent interactions may further perturb the potential energy leading to crossing of surfaces, if such a crossing point was not present initially, thereby promoting nonradiative transitions. But in the absence of specific solute-solvent interactions, the medium probably has very little effect on the rates of radiationless transitions in polyatomic molecules. In condensed systems like solid crystalline states, the crystal fields may perturb the vibronic energy states enhancing the probability of radiationless transitions.

To sum up, radiationless transitions are governed by the following three factors:

(1) Density of states, ρ_E, the vibronic energy levels;
(2) Energy gap ΔE between the interacting electronic states; and
(3) Vibronic overlap or the Franck-Condon factor.

These three factors are included in the expression derived by Robinson and Frosch from time-dependent perturbation theory, for the nonradiative energy transfer or radiationless transition probability k_{NR} per unit time,

$$k_{NR} = \frac{4\pi^2}{h} \rho_E V^2 \tag{5.22}$$

$$= \left[\frac{4\pi^2}{h}\rho_E\right] \quad [\beta^2] \quad \left[\int \chi_f^{v'} \chi_i^{v''} d\tau_v\right]^2 \tag{5.23}$$

$$= \begin{matrix}\text{density of} \\ \text{state} \\ \text{factor}\end{matrix} \times \begin{matrix}\text{electronic} \\ \text{interaction} \\ \text{factor}\end{matrix} \times \begin{matrix}\text{Franck-Condon} \\ \text{overlap} \\ \text{factor}\end{matrix}$$

Here β^2 is the interaction energy between the two states. It includes interaction with all the vibronic levels in the final state close to the initially populated state and, therefore, depends on the O–O energy gap ΔE.

These discussions provide an explanation for the fact that fluorescence emission is normally observed from the zero vibrational level of the first excited state of a molecule (*Kasha's rule*). The photochemical behaviour of polyatomic molecules is almost always decided by the chemical properties of their first excited state. Azulenes and substituted azulenes are some important exceptions to this rule observed so far. The fluorescence from azulene originates from S_2 state and is the mirror image of $S_2 \leftarrow S_0$ transition in absorption. It appears that in this molecule, $S_1 \leftarrow S_0$ absorption energy is lost in a time less than the fluorescence lifetime, whereas certain restrictions are imposed for $S_2 \rightsquigarrow S_0$ nonradiative transitions. In azulene, the energy gap ΔE, between S_2 and S_1 is large compared with that between S_1 and S_0. The small value of ΔE facilitates radiationless conversion from S_1 but that from S_2 cannot compete with fluorescence emission. Recently, more sensitive measurement techniques such as picosecond flash fluorimetry have led to the observation of $S_1 \rightarrow S_0$ fluorescence also. The emission is extremely weak. Higher energy states of some other molecules have been observed to emit very weak fluorescence. The effect is controlled by the relative rate constants of the photophysical processes.

The rate determining step in *intersystem crossing* is the transfer from the thermally relaxed singlet state to the vibronically excited triplet state $S_j \rightsquigarrow T_k$ $(j \geq k)$. This is followed by vibrational relaxation. The spin-orbital interaction modifies the transition rates. A prohibition factor of $10^{-2} - 10^{-6}$ is introduced and the values of k_{ISC} lie between 10^{11} and 10^7 s^{-1}. The reverse transfer from the relaxed triplet to vibronically excited singlet is also possible.

Of special interest are the transfers $S_1 \rightsquigarrow T_1$ and $T_1 \rightsquigarrow S_0$. The rate constants for these transfers are important parameters in photochemistry. The lowest energy excited state is the T_1 state and is fairly long lived in the time scale of excitation process. Therefore, it is the favoured seat of photochemical reactions. Since direct excitation to a triplet state is a forbidden process by electric dipole mechanism, intersystem crossing is the pathway by which the reactive triplet state of an organic molecule can be populated under normal conditions. If $S_1 - T_1$ energy gap is small, as it is in some cases, the molecule can gain thermal energy from the surrounding and repopulate the S_1 level, subsequently emitting *E-type delayed fluorescence* (Section 5.9).

There is a large difference in the rate constants k_{ISC} ($S_1 \rightsquigarrow T_1$) and k_{ISC}^T ($T_1 \rightsquigarrow S_0$). The ratio of the two rates may be as high as 10^9 in certain molecules. This is due to the differences in the *zero-point energy* ΔE_{ST} between the combining singlet and triplet states which is normally much less than that between T_1 and S_0, i.e. ΔE_{TS}. In some molecules another higher triplet state T_j may intervene between S_1 and T_1 states, further reducing the energy gap ΔE_{ST}. The overall transfer S_1 to T_1 may in such cases proceed as $S_1 \overset{\text{ISC}}{\rightsquigarrow} T_j \overset{\text{IC}}{\rightsquigarrow} T_1$, where $\Delta E_{S_1 T_j}$ is the controlling factor. The sensitivity of cross-over rates on the energy gap is

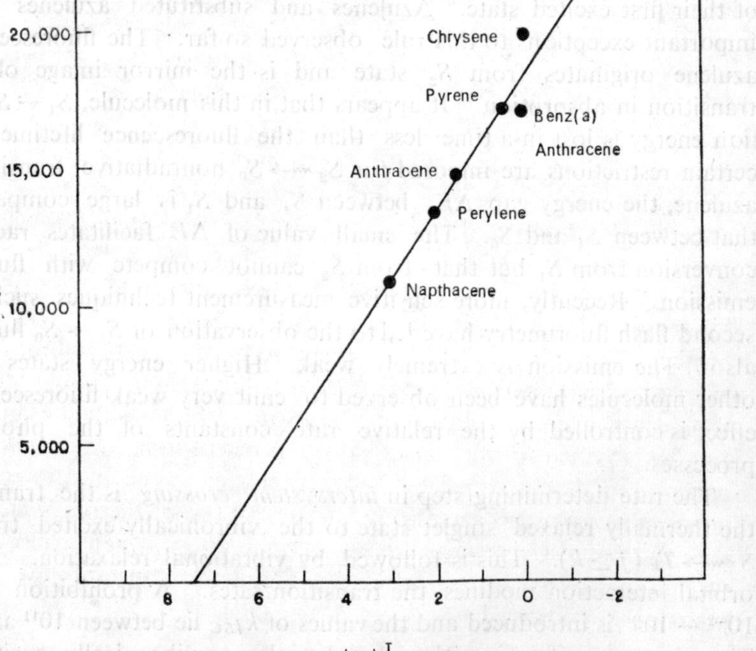

Figure 5.4 Plot of $\Delta E_{T_1 - S_0}$ *vs* log k_{ICS}^T for a number of aromatic hydrocarbons.

nicely demonstrated in Figure 5.4. The formal explanation of such a dependence on ΔE_{ST} is again found to lie in the Franck-Condon overlap integral.

In aromatic hydrocarbons, the radiationless transitions from the triplet to the ground state are dominated by CH stretching vibrational modes. In these hydrocarbons perdeuteration reduces k_{ISC}^T and consequently enhances k_p, the rate constant for phosphorescence emission. The lifetime τ_p, is considerably reduced. The reduction of k_{ISC}^T ($T_1 \rightsquigarrow S_0$) which competes with radiative constant k_p ($T_1 \rightarrow S_0$) is as expected because on perdeuteration, zero-point energy and vibrational energy differences are reduced. Consequently, the cross-over point now lies at higher quantum vibronic state implying reduced FC factor.

5.2.2 Selection Rule for Radiationless Transitions

The selection rules for radiationless transitions are just the opposite of those for radiative transitions. The nuclear kinetic operator $\hat{J}_N$ is symmetric. The symmetric aromatic molecules normally have symmetrical ground state and antisymmetrical excited state. Therefore, allowed transitions are:

$$g \rightsquigarrow g, \; u \rightsquigarrow u \;\text{ and }\; S_1 \rightsquigarrow T_1 \tag{5.24}$$

and forbidden transitions are:

$$u \rightsquigarrow g, \; S_1 \rightarrow S_0 \;\text{ and }\; T_1 \rightsquigarrow S_0. \tag{5.25}$$

In heteroaromatic systems, transitions between $^1(n, \pi^*) \rightsquigarrow {}^3(\pi, \pi^*)$ states and $^1(\pi, \pi^*) \rightsquigarrow {}^3(n, \pi^*)$ states occur more readily if energy relationships are good. $^1(n, \pi^*) \rightsquigarrow {}^3(n, \pi^*)$ and $^1(\pi, \pi^*) \rightsquigarrow {}^3(\pi, \pi^*)$ transitions are less probable and, therefore, are slower processes as shown by El Sayed.

5.3 FLUORESCENCE EMISSION

The mechanism of fluorescence emission has been discussed in Section 4.5. Fluorescence emission is normally observed from the first excited singlet state of the molecule. Even if the electronic transition promotes the molecule to an energy state higher than S_1, the excess energy in condensed systems is dissipated away to the surroundings as thermal energy and the molecule comes to stay in the lowest excited state. In this state it has a lifetime governed by the transition probabilities of absorption. In the absence of any deactivating perturbations, the molecule has a natural radiative lifetime, τ_N, inversely related to the integrated absorption intensity as given by equation (3.97). All those factors which effectively alter the radiative lifetime are the same as those which cause variation in oscillator strength in absorption. The rate constant for fluorescence emission k_f is defined as

$$k_f = \frac{1}{\tau_N} = \frac{1}{\tau_f^0} \tag{5.26}$$

In the presence of other competitive deactivating processes, the average lifetime is much reduced and actual lifetime τ_f is

$$\tau_f = \frac{1}{k_f + \sum k_i} \tag{5.27}$$

where Σk_i is the rate constant for the ith competitive process, assumed to be unimolecular.

Due to the working of the Franck-Condon principle and thermal relaxation of vibrational modes, the fluorescence spectrum is always observed on the red side of the absorption spectrum, in approximate mirror image relationship for polyatomic molecules. This red shift of the fluorescence implies that the emitted quanta are of lower energy than the absorbed quanta, i.e. $h\nu_f < h\nu_a$, *Stokes' shift.* Under certain conditions at high temperatures, when higher vibrational levels of the ground state are thermally populated, anti-Stokes effect ($h\nu_f > h\nu_a$) may be observed. Generally, there is an overlap region between absorption and emission spectra (Figure 5.5). The intensity distribution in fluorescence spectra is

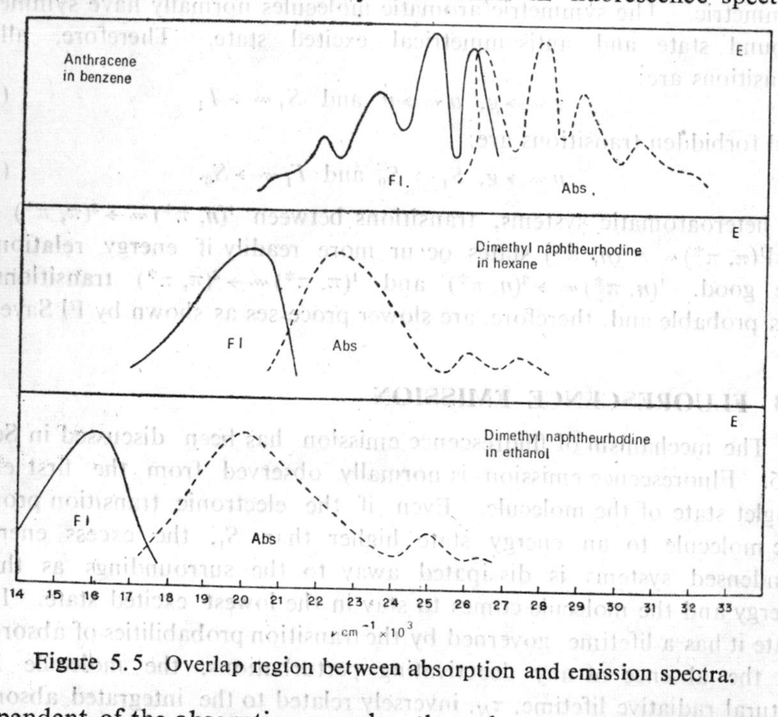

Figure 5.5 Overlap region between absorption and emission spectra.

independent of the absorption wavelength as long as the emission occurs from the vibrationally relaxed upper state, emphasizing the Boltzmann distribution of molecules in the emitting state. Mirror image relationship (Levschin's rule) in fluorescence spectra is not observed if the excited state has very different geometry from that of the ground state as is usually the case for small unconjugated molecules.

The *quantum efficiency of fluorescence* ϕ_f is defined as

$$\phi_f = \frac{\text{number of quanta emitted/s/cm}^2}{\text{number of quanta absorbed/s/cm}^2} = \frac{n_f\,(h\nu_f)}{n_a\,(h\nu_a)}$$

$$= \frac{\text{rate of emission}}{\text{rate of absorption}} = \frac{k_f\,[S_1]}{I_a} \tag{5.28}$$

$$= \frac{\text{intensity of emission}}{\text{intensity of absorption}} = \frac{F \text{ einstein } s^{-1}\text{ cm}^{-2}}{I_a \text{ einstein } s^{-1}\text{ cm}^{-2}} \tag{5.29}$$

$[S_1]$ is the concentration of the lowest excited singlet molecules. It is independent of the exciting wavelength except when chemical changes occur but is lowered by competing deactivating processes as described in Section 5.1. In condensed systems, radiationless pathways are most common deactivating mechanisms. Since these deactivating processes also reduce the radiative lifetime, the measured lifetime τ_f is related to ϕ_f through τ_f^0, the intrinsic lifetime as

$$\tau_f = \tau_f^0\,\phi_f \tag{5.30}$$

Because of the independence of quantum efficiency on wavelength it is possible to obtain a *fluorescence excitation spectrum* which is defined and derived as follows:

Intensity of fluorescence F varies as fractional light absorption I_a, that is

$$F = \phi_f\,I_a \tag{5.31}$$

$$= \phi_f\,I_0\,(1 - e^{-2.303\,\epsilon\,Cl}) \tag{5.32}$$

where ϕ_f is fluorescence quantum efficiency; I_0 is incident light intensity; ϵ, molar extinction coefficient; C, concentration in moles per litre and l, optical path length.

For small fractional absorption, $e^{-2.303\,\epsilon\,Cl}$ can be expanded and

$$F = \phi_f\,I_0\,(1 - 1 + 2.303\,\epsilon\,Cl + \text{higher powers})$$

$$= \phi_f\,I_0\,2.303\,\epsilon\,Cl \tag{5.33}$$

Since ϕ_f is independent of wavelength, if F is measured as a function of wavelength of exciting light of constant intensity I_0, the resulting variation in fluorescence intensity called fluorescence excitation spectrum, will reflect the variation in ϵ, molar extinction coefficient, as a function of wavelength or wavenumber. The fluorescence excitation spectrum should, therefore, reproduce the absorption spectrum of the molecule. A fluorescence excitation spectrum is a more sensitive method of obtaining the absorption spectrum and is of great analytical value in systems containing a number of fluorescent species or a large background absorption. From equation (5.33) it is also seen that for low fractional absorption, fluorescence intensity is directly proportional to concentration, thus providing a sensitive method for *quantitative fluorimetry*.

5.4 FLOURESCENCE AND STRUCTURE

Although all molecules are capable of absorption, fluorescence is not observed in large number of compounds. Fluorescence is generally observed in those organic molecules which have rigid framework and not many loosely coupled substituents through which vibronic energy can flow out. Following structures are called fluorophores in analogy with chromophores and they are part of the molecules with conjugated system of double bonds. These are:

$$C = C, \quad N = O, \quad -N = N, \quad -C = O, \quad -C = N, \quad -C = S,$$

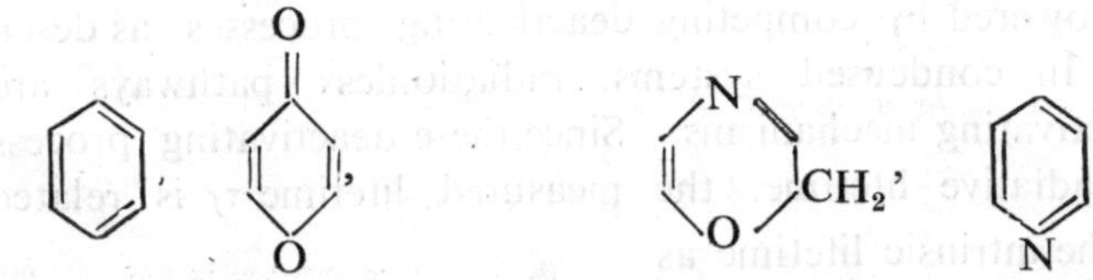

Some substituents tend to enhance fluorescence. They are known as *fluorochromes* in the same sense as auxochromes. In general, these are electron donors such as — OH, NH_2, etc. which enhance the transition probability or intensity of colour, e.g. acridine and acridine orange.

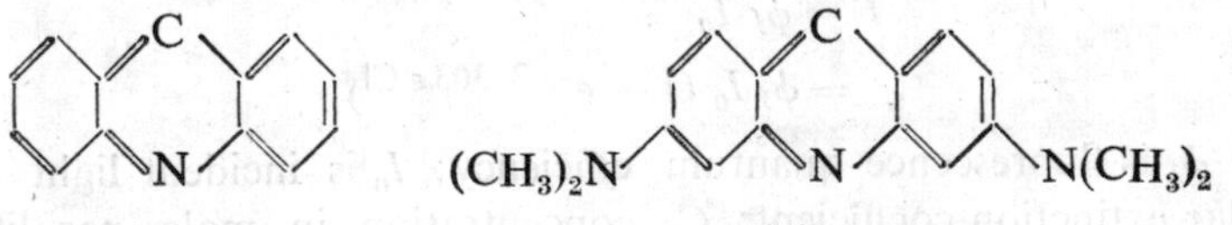

acridine
(non fluorescent)

acridine orange
(fluorescent)

On the other hand, electron-withdrawing substituents tend to diminish or inhibit fluorescence completely. In dilute solutions in water, aniline is about 40 times more fluorescent than benzene whereas benzoic acid is nonfluorescent.

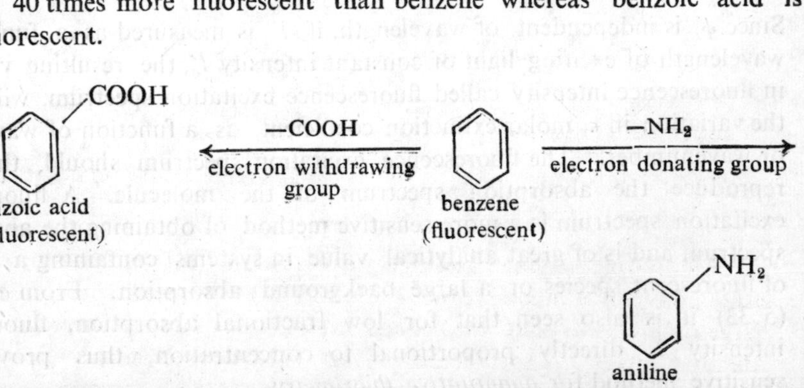

benzoic acid
(nonfluorescent)

$\xleftarrow{\text{electron withdrawing group}}$ —COOH

benzene
(fluorescent)

—NH_2 $\xrightarrow{\text{electron donating group}}$

aniline
(highly fluorescent)

The classic example of influence of rigidity of a molecule on its capacity to fluoresce is found in the pair phenolphthalein and fluorescein.

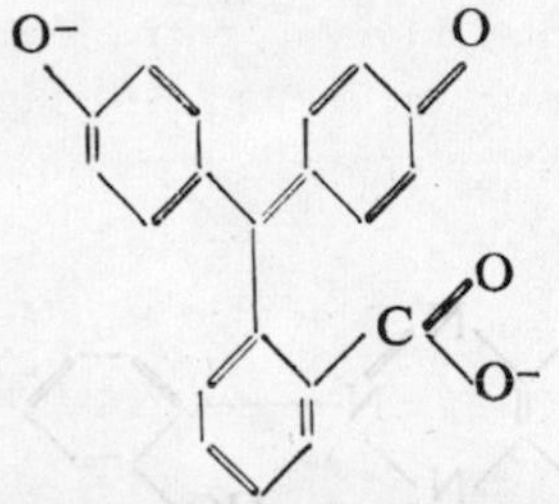

phenolphthalein (nonfluorescent)

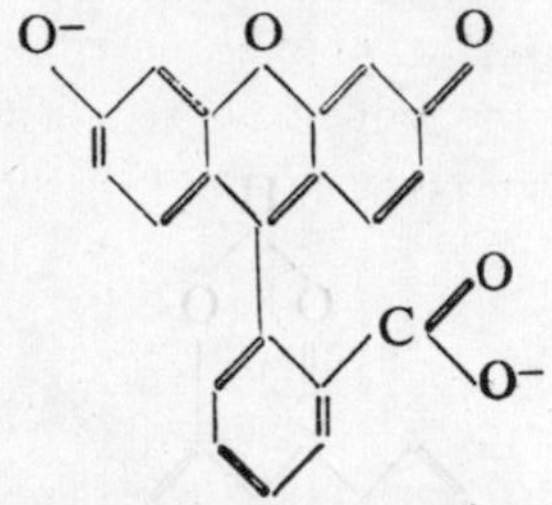

fluorescein (fluorescent)

The rigidity introduced by O-bridge in fluorescein makes the molecule highly fluorescent. Similarly, azobenzene is nonfluorescent, whereas diazaphenanthrene is capable of fluorescence.

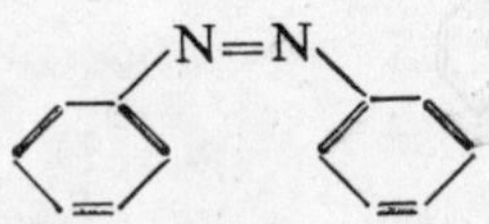

azobenzene (nonfluorescent)

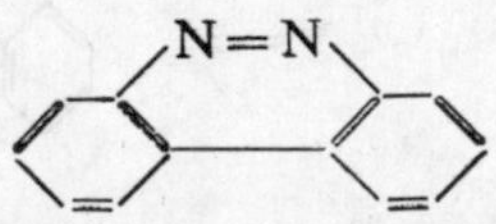

azaphenanthrene (fluorescent)

Naphthalene with same number of double bonds is nearly five times more fluorescent than vitamin A.

naphthalene (fluorescent)

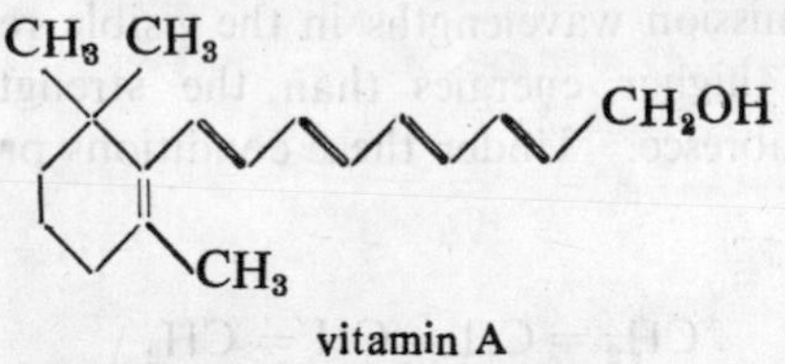

vitamin A
(fluorescence 1/5 that of naphthalene)

Anthracene crystals are highly fluorescent ($\phi_f = 1.0$) but in dissolved state emission is much reduced ($\phi_f = 0.25$). A recent explanation of this large difference is that the second triplet state T_2 of anthracene lies above the first singlet in anthracene crystal but below it in the dissolved state thereby enhancing the nonradiative dissipative processes. Molecular adsorption on a substrate also enhances the fluorescence. Hydrogen bonding in some cases can activate fluorescence, in others it helps to degrade the energy. Some ortho hydroxy-benzophenones are used as energy degraders in plastic materials and printed clothes to protect them from damaging effects of sunlight.

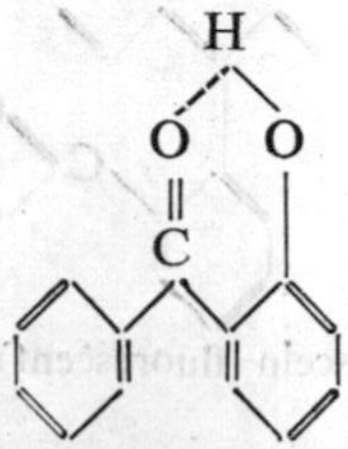

o-hydroxy benzophenones
($R=H$, Me, C_8H_{17}, $C_{10}H_{21}$ etc.)

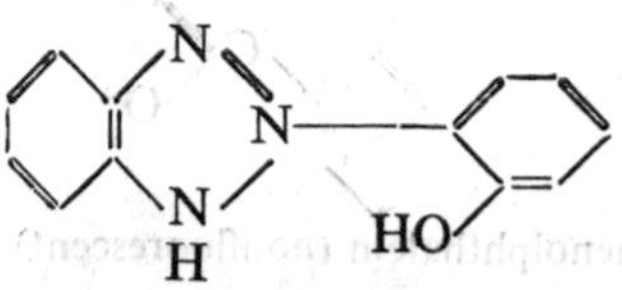

benzotriazole

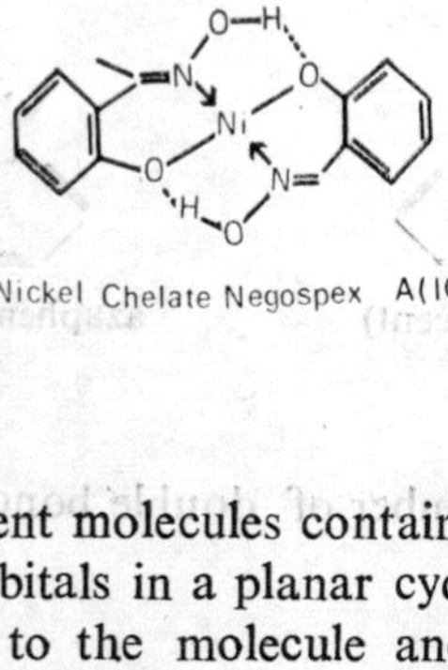

Nickel Chelate Negospex A(ICI)

Most organic fluorescent molecules contain conjugated system of double bonds with extended π-orbitals in a planar cyclic structure. Such structure imparts certain rigidity to the molecule and shifts the absorption and emission wavelengths in the visible region. The compounds which absorb at higher energies than the strengths of their weakest bonds do not fluoresce. Under these conditions predissociation is favoured.

	$CH_2=CH-CH=CH_2$ butadiene	$\phi CH=CH-CH=CH\phi$ 1, 4 diphenylbutadiene
λ_{max} (alcohol) :	210 nm ($\simeq$ 590 kJ mol^{-1})	350 nm ($\simeq$ 336 kJ mol^{-1})
least stable bond cleavage energy (in alcohol)	525 kJ mol^{-1} (nonfluorescent)	525 kJ mol^{-1} (fluorescent)

Large energy separation between excited singlet S_1 and triplet T_1 states should favour fluorescence since intersystem crossing is less probable in such systems. For this reason aromatic hydrocarbons are normally fluorescent. Fluorescence efficiency increases with increase in the number of condensed ring, e.g. benzene $<$ naphthalene $<$ anthracene, etc. Linear ring systems are good fluorescers as compared to angular compounds like phenanthrene. Planarity of the ring is also an important criterion. Substituents which produce steric hindrance to planarity thus reducing π-electron mobility, lower the fluorescence efficiency of the molecule.

From the principle of microscopic reversibility it may be inferred that what is probable in absorption should also be probable in emission. A high emission probability for (π, π^*) state is predicted. The rate constant for emission is observed to be large for dipole allowed $\pi^* \rightarrow \pi$ transitions. As a consequence, other deactivating processes cannot compete with the radiative process, and a high fluorescence efficiency for such a system is usually observed.

In heteroaromatic molecules (π, π^*) and (n, π^*) energy states have different emission probabilities. Since fluorescence is generally observed from the lowest energy excited state of the same multiplicity as the ground state, the nature of this state will dictate the chemistry of the molecule. If for a heterocyclic molecule, $^1(n, \pi^*)$ state is the lowest excited state, because of the space forbidden character of this transition, a longer life and hence a smaller rate constant for emission is expected. Any other process with higher rate constant can compete with radiative transition and completely inhibit fluorescence radiation. That is the reason why pyridine, quinoline and acridine which are isoelectronic N-analogs of benzene, naphthalene and anthracene, respectively, are nonfluorescent in hydrocarbon solvents. All these compounds have $^1(n, \pi^*)$ as the lowest singlet excited state. This explanation is substantiated by the fact that, in those compounds in which $^1(\pi, \pi^*)$ and $^1(n, \pi^*)$ are not very far apart, polar solvents can bring about the change in the order of the energy levels, making $^1(\pi, \pi^*)$ state as the lowest excited level with subsequent appearance of fluorescence. For example, the relative quantum yields of fluorescence of quinoline in benzene, ethanol and water are 1 : 30 : 1000, evidently due to such solvent effects on the energy states. In Figure 5.6, the effect of polar solvent on energy levels of chlorophyll molecule is given. Chlorophyll is fluorescent in *polar* solvent.

Substituents have considerable influence on emission characteristics of aromatic compounds. Heavy atom substituents tend to reduce the fluorescence quantum yield ϕ_f in favour of phosphorescence emission ϕ_p. In halogen series the effect increases in the order F $<$ Cl $<$ Br $<$ I. In Table 5.1 are recorded experimental data for halogen substituted naphthalenes.

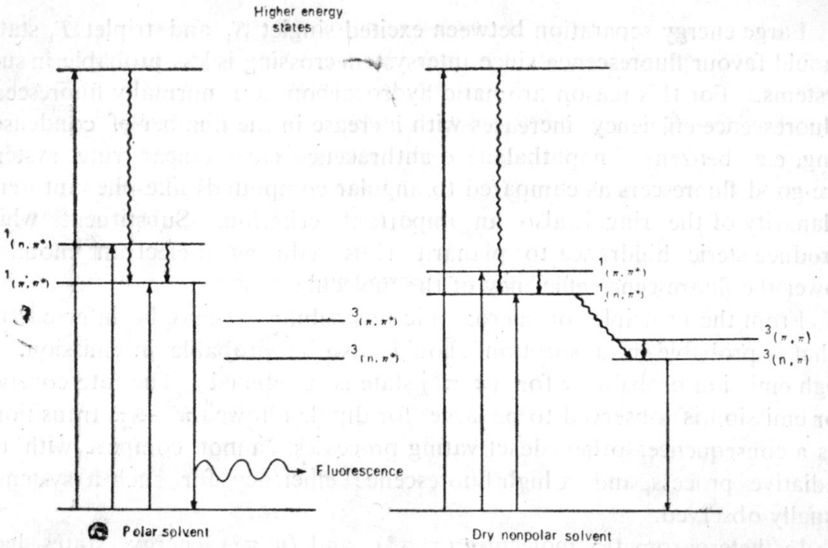

Figure 5.6 Energy levels of chlorophyll *a* in polar and nonpolar solvents.

TABLE 5.1

Effect of halosubstitution upon emission characteristics of naphthalene

Compound	ϕ_p/ϕ_f	ν_f, cm^{-1}	$\bar{\nu}_p$, cm^{-1}	τ_p
Naphthalene	0.093	31750	21250	2.5
1-Fluoronaphthalene	0.068	31600	21150	1.4
1-Chloronaphthalene	5.2	31360	20700	0.23
1-Bromonaphthalene	6.4	31280	20650	0.014
1-Iodonaphthalene	> 1000	not observed	20500	0.0023

In dianions of xanthene dyes although halosubstituents decrease the quantum efficiency of fluorescence, phosphorescence efficiency is not increased proportionately. The phosphorescence lifetime decreases with ϕ_f. It is suggested that in these dyes, besides enhancement of $S_1 \rightsquigarrow T_1$ intersystem crossing rates, $S_1 \rightsquigarrow S_0$ nonradiative transition is promoted by heavy atom substitution.

5.5 TRIPLET STATES AND PHOSPHORESCENCE EMISSION

The heterocyclics pyridine, quinoline and acridine, although nonfluorescent at room temperature, emit at longer wavelength with increased decay constant at liquid nitrogen temperature, i.e. 77 K. It was first suggested by Jablonski that the emitting energy state for this longwave emission component is a metastable state lying just below the singlet S_1 state. This metastable state was later identified with the triplet state by Lewis and his associates. Identity of the phosphorescent state with the triplet state was conclusively established by them for acid fluorescein dye in boric acid glass. A large number of molecules were transferred to the triplet state via intersystem crossing from the initially excited singlet, using high intensity excitation source. A triplet state with unpaired spin should have residual magnetic field. The paramagnetic susceptibility of the dye under this condition was found to correspond to a pair of electrons with parallel spins.

The appearance of phosphorescence in the above mentioned heterocyclic aromatic compounds at low temperature implies that the triplet state becomes populated by intersystem crossing from the thermally equilibrated lowest excited singlet state. Nonappearance of any emission at room temperature suggests that phosphorescence is inhibited by competitive nonradiative processes such as impurity quenching or internal conversion to the singlet ground state. Because of spin forbidden character, phosphorescence has a low probability of emission, hence a long radiative lifetime. In rigid glassy deoxygenated solutions, all deactivating collisions are prevented and the molecule is able to emit. At low temperature and in rigid solvents, emission is a rule rather than exception.

After crossing over to the triplet energy state at the isoenergetic point by nonradiative mechanism, the molecule quickly relaxes to the zero vibration level of the triplet state. From this level it can return to the ground state either by *phosphorescence* emission (radiative pathway) or by *intersystem crossing* to the higher vibrational level of S_0 with fast dissipation of excess vibrational energy (nonradiative pathway). Phosphorescence is always at a longer wavelength than fluorescence. The transition $T_1 \leftarrow S_0$ is forbidden by spin selection rule and is not observed under ordinary conditions. The transition probability can be enhanced by spin-orbit coupling interactions (Section 3.7.3) promoted by a number of techniques described below.

(i) *External heavy atom perturbation.* The effect was first discovered by Kasha who observed that a mixture of 1-chloronaphthalene and ethyl iodide, each colourless alone, is yellow and that the colour is due to an increase in the absorption intensity of the $T_1 \leftarrow S_0$ transition. This increase is due to an increase in spin-orbit coupling when the π-electrons of naphthalene penetrate near the large nuclear field of the iodine atom. The intensity of the transition increases with the atomic number of the halogen substituent in the solvent. The external heavy atom perturbations can bring about regular increase in the phosphorescence to fluorescence ratio for naphthalene on going to solvents with more and heavier atoms (Figure 3.8).

(ii) A second way in which singlet-triplet transitions can be enhanced is through the presence of *paramagnetic molecules* such as oxygen and nitric oxide. Evans has shown that $T_1 \leftarrow S_0$ transition of a large number of organic molecules can be observed if a concentrated solution of the substrate is saturated with O_2 at higher pressures. The enhancement observed in the presence of oxygen disappears when oxygen is removed. Exciplex formation is probably involved.

For aromatic hydrocarbons, the quantum efficiencies of fluorescence and phosphorescence in low temperature glasses (Appendix II) such as EPA (ether : isopentane : ethyl alcohol in the ratio 2 : 2 : 5) add up to unity. This suggests that direct nonradiative decay from $S_1 \rightsquigarrow S_0$ is of very low probability. All the nonradiative paths to the ground state are coupled via the triplet state. The sequence of transfers is:

$$S_1^o \rightsquigarrow T_1^v \rightsquigarrow T_1^o \rightsquigarrow S_0^v \rightsquigarrow S_0^o$$

where T_1^v and S_0^v are vibrationally excited T_1 and S_0 state, respectively. The coupling transition in the horizontal direction at the isoenergetic point is the rate determining step; subsequent thermal equilibration is much faster (Section 5.2.1). In such cases:

$$\phi_f + \phi_p + \phi_{ISC}^T (T_1 \rightsquigarrow S_0) \simeq 1 \qquad (5.34)$$

If a quantum efficiency less than unity is observed, even in the absence of a photochemical reaction, a direct internal conversion from $S_1 \rightsquigarrow S_0$ via higher vibrational level of the ground state is proposed.

If the lower energy state intersects at a point (Figure 5.1) above the zero vibrational level of the transferring state, a temperature dependent factor $e^{-W/kT}$ may be involved in the rate constant for intersystem crossing. The energy term W corresponds to the activation energy needed to raise the molecule from the zero point to the point of intersection.

The *quantum yield of phosphorescence*, ϕ_p is defined as

$$\phi_p = \frac{\text{number per second of quanta emitted as phosphorescence in } T_1 \rightarrow S_0 \text{ transition}}{\text{number per second of quanta absorbed in } S_1 \leftarrow S_0 \text{ transition}}$$

$$= \frac{\text{rate of phosphorescence}}{\text{rate of absorption}} = \frac{k_p\,[T_1]}{I_a} \qquad (5.35)$$

$$= \frac{\text{intensity of phosphorescence emission}}{\text{intensity of absorption}} = \frac{\text{P einstein s}^{-1}}{I_a \text{ einstein s}^{-1}} \qquad (5.36)$$

$[T_1]$ is the concentration of the triplet molecules.

Because of the forbidden nature of $T_1 \leftarrow S_0$ transition, T_1 is long lived and subjected to rapid collisional deactivation and thermal relaxation. As a result phosphorescence is not observed at room temperature except for a few cases. It is observed at low temperatures only in rigid glassy solutions.

The intrinsic lifetime of triplet T_1 state τ_p^0, is the reciprocal of the rate constant k_p, for phosphorescence emission and the actual lifetime τ_p, is the reciprocal of the sum of all the steps which deactivate the triplet.

$$\tau_p^0 = \frac{1}{k_p}\,;\ \tau_p = \frac{1}{k_p + k_{ISC}^T} \tag{5.37}$$

Hence,

$$\frac{\tau_p}{\tau_p^0} = \frac{k_p}{k_p + k_{ISC}^T} = \frac{k_p/I_a}{(k_p + k_{ISC}^T)/I_a} = \frac{\phi_p}{\phi_T} \tag{5.38}$$

where k_{ISC}^T is the rate constant for intersystem crossing from triplet to ground singlet. ϕ_p and ϕ_T are quantum efficiencies of phosphorescence emission and triplet formation respectively. ϕ_T is identical with intersystem crossing efficiency ϕ_{ISC}, assuming that all those molecules which do not fluoresce are transferred to the triplet state. It follows that

$$\tau_p^0 = \tau_p \frac{\phi_T}{\phi_p} = \tau_p \cdot \frac{1 - \phi_f}{\phi_p} \tag{5.39}$$

The phosphorescence lifetimes may vary from 10^{-6} second to more than a second.

5.6 EMISSION PROPERTY AND THE ELECTRONIC CONFIGURATION

The ground states of most molecules are singlet and contain a pair of electrons with antiparallel spins. On electronic excitation, the ground and the excited electronic configurations are delinked and singlet and triplet states are possible.

$$(\phi)^2 \xrightarrow{h\nu} (\phi)^1 (\phi^*)^1$$

and

$$\psi_1 \xrightarrow{h\nu} {}^1\psi^* \text{ or } {}^3\psi^*$$

where ϕs are the electronic orbital wave functions and ψs are the state functions. The singlet and triplet states have different energies because of the electron-electron repulsion in the triplet state. The extent of singlet-triplet split will be governed by the interaction energy as computed from the exchange integral

$$K = \iint \phi\phi^* \frac{e^2}{r_{12}} \phi^* \phi \, d\tau_1 \, d\tau_2$$

$$= \iint \frac{(\phi\phi^* e)(e\,\phi^* \phi)}{r_{12}} \, d\tau_1 \, d\tau_2 \tag{5.40}$$

This is equivalent to coulombic repulsional interaction between two equal charge densities ($\phi\phi^{*}e$). The S – T energy split ΔE_{ST} is equal to 2K. The value of K depends on the overlap between ϕ and ϕ^*; the larger the overlap, the greater is the value of K. For example, n and π^* orbitals do not overlap much as they lie in orthogonal planes, hence the energy separation ΔE_{ST} between $^1(n, \pi^*)$ and $^3(n, \pi^*)$ is small. On the other hand, in $\pi \rightarrow \pi^*$ transitions, π and π^* orbitals occupy more or less the same configuration space. Large interactions result in relatively large ΔE_{ST}. In general, the singlet and triplet states will fall in the following order (Figure 5.7).

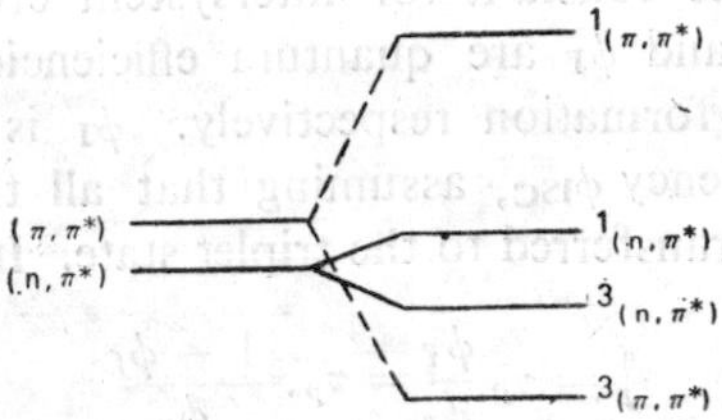

Figure 5.7 Splitting of singlet and triplet energy states of (π, π^*) and (n, π^*) character.

Some possible dispositions of (n, π^*) and (π, π^*) states are given in Figure 5.8 for a few representative molecule types.

A small value of ΔE_{ST} facilitates intersystem crossing. We expect singlet state to be fast depleted along this pathway if the lowest excited state is of (n, π^*) type. This pathway is further promoted due to the fact that $\tau_{n\pi^*} > \tau_{\pi\pi^*}$ by a factor of ten, due to the forbidden character of $n \rightarrow \pi^*$ transition. Fluorescence with decreased rate constant for emission cannot compete efficiently with intersystem crossing. This explains the absence of room temperature emission in heterocyclics like benzophenone, acetophenone, quinoline, acridines, etc. They phosphoresce at low temperatures only.

In aromatic hydrocarbons, short-lived $^1(\pi, \pi^*)$ is the lowest excited S state and energy gap between $^1(\pi, \pi^*)$ and $^3(\pi, \pi^*)$ states is large. Both these factors are conducive to fluorescence emission and in general aromatic hydrocarbons are good fluorescer. Sometimes, the prediction may not come true if a higher triplet state T_2 is available near the S_1 state such as in anthracene. In such cases, fluorescence and phosphorescence both are observed at low temperatures in suitable solvent medium specially when S_1 and T_2 are states of different symmetry type. Some data correlating ΔE_{ST} and ϕ_{ISC} are given in the Table 5.2.

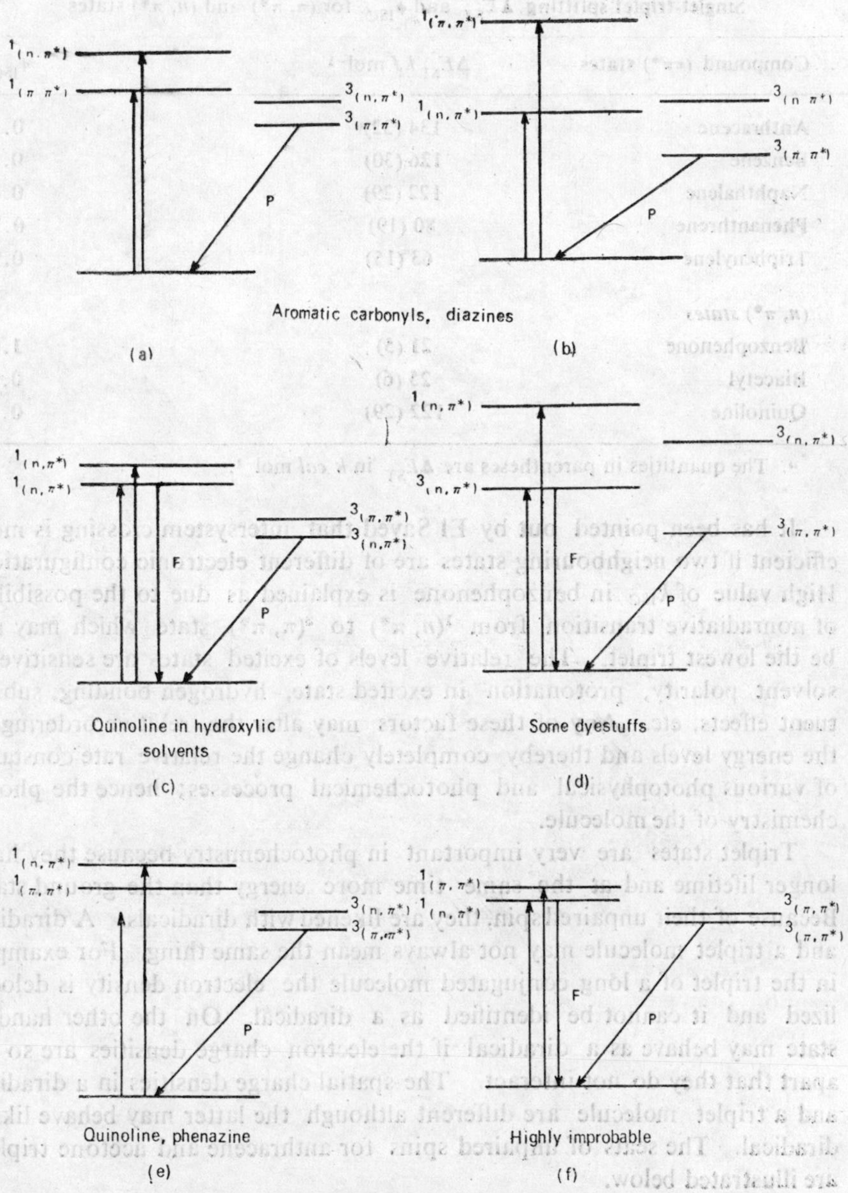

Figure 5.8 Disposition of (π, π*) and (n, π*) singlet and triplet energy states for a few representative types of molecules and their emission characteristics. [F. Wilkinson and A.R. Horrocks in E.J. Bowen (ed), *Luminiscence in Chemistry*, D. Van Norstand, 1968]

TABLE 5.2

Singlet-triplet splitting ΔE_{ST} and ϕ_{ISC} for (π, π^*) and (n, π^*) states

Compound $(\pi\pi^*)$ states	ΔE_{ST} kJ mol^{-1}	ϕ_{ISC}
Anthracene	134 (32)[a]	0.70
Benzene	126 (30)	0.24
Naphthalene	122 (29)	0.40
Phenanthrene	80 (19)	0.76
Triphenylene	63 (15)	0.96
(n, π^*) states		
Benzophenone	21 (5)	1.0
Biacetyl	25 (6)	0.99
Quinoline	122 (29)	0.32

[a]. The quantities in parentheses are ΔE_{ST} in *k cal* mol^{-1}.

It has been pointed out by El Sayed that intersystem crossing is more efficient if two neighbouring states are of different electronic configuration. High value of k_{ISC} in benzophenone is explained as due to the possibility of nonradiative transition from $^1(n, \pi^*)$ to $^3(\pi, \pi^*)$ state which may not be the lowest triplet. The relative levels of excited states are sensitive to solvent polarity, protonation in excited state, hydrogen bonding, substituent effects, etc. Any of these factors may alter the relative ordering of the energy levels and thereby completely change the relative rate constants of various photophysical and photochemical processes; hence the photochemistry of the molecule.

Triplet states are very important in photochemistry because they have longer lifetime and at the same time more energy than the ground state. Because of their unpaired spin, they are likened with diradicals. A diradical and a triplet molecule may not always mean the same thing. For example, in the triplet of a long conjugated molecule the electron density is delocalized and it cannot be identified as a diradical. On the other hand, a state may behave as a diradical if the electron charge densities are so far apart that they do not interact. The spatial charge densities in a diradical and a triplet molecule are different although the latter may behave like a diradical. The seats of unpaired spins for anthracene and acetone triplets are illustrated below.

5.7 PHOTOPHYSICAL KINETICS OF UNIMOLECULAR PROCESSES

Rate constants of unimolecular processes can be obtained from spectral data and are useful parameters in photochemical kinetics. Even the nature of photoproducts may be different if these parameters change due to some perturbations. In the absence of bimolecular quenching and photochemical reactions, the following reaction steps are important in deactivating the excited molecule back to the ground state.

	Step	*Rate*
$S_0 + h\nu \xrightarrow{k_0} S_1$	Excitation	I_a
$S_1 \xrightarrow{k_{IC}} S_0 + \text{heat}$	Internal conversion	$k_{IC}[S_1]$
$S_1 \xrightarrow{k_{ISC}} T_1 + \text{heat}$	Intersystem crossing	$k_{ISC}[S_1]$
$S_1 \xrightarrow{k_f} S_0 + h\nu_f$	Fluorescence	$k_f[S_1]$
$T_1 \xrightarrow{k_p} S_0 + h\nu_p$	Phosphorescence	$k_p[T_1]$
$T_1 \xrightarrow{k^T_{ISC}} S_0 + \text{heat}$	Reverse intersystem crossing	$k^T_{ISC}[T_1]$

k's are respective rate constants and $[S_1]$ and $[T_1]$ are concentrations of excited singlet and triplet molecules, respectively.

Under the condition of photostationary equilibrium, rate of formation of S_1 = rate of deactivation of S_1

$$I_a = \{k_{IC} + k_{ISC} + k_f\}[S_1] \tag{5.41}$$

and

$$[S_1] = \frac{I_a}{k_{IC} + k_{ISC} + k_f} \tag{5.42}$$

Hence the quantum yield of fluorescence ϕ_f° in the absence of any external quenching is, from definition, (5.28)

$$\phi_f^\circ = \frac{k_f[S_1]}{I_a} = \frac{k_f}{k_{IC} + k_{ISC} + k_f} \tag{5.43}$$

Similarly for (T_1)

$$k_{ISC}[S_1] = (k^T_{ISC} + k_p)[T_1] \tag{5.44}$$

and on substituting for S_1,

$$[T_1] = I_a\left(\frac{k_{ISC}}{k_{IC} + k_{ISC} + k_f}\right)\left(\frac{1}{k^T_{ISC} + k_p}\right) \tag{5.45}$$

Hence, the quantum yield for phosphorescence ϕ_p from (5.35) and (5.45) is given as

$$\phi_p = \frac{k_p[T_1]}{I_a} \tag{5.35}$$

$$=\left(\frac{k_{ISC}}{k_{IC}+k_{ISC}+k_f}\right)\left(\frac{k_p}{k_{ISC}^T+k_p}\right) \quad (5.46)$$

$$=\phi_T \frac{k_p}{k_{ISC}^T+k_p}$$

If $k_{ISC} \gg (k_{IC}+k_f)$ as is generally the case for many organic compounds, specially heteroaromatic hydrocarbons, the intrinsic phosphorescence efficiency is given by

$$\phi_p^0 = \frac{k_p}{k_{ISC}^T+k_p} = k_p \tau_p \quad (5.47)$$

The ratio of the two emission efficiencies is

$$\frac{\phi_p^0}{\phi_f^0} = \frac{k_p[T_1]}{k_f\,[S_1]} = \frac{k_{ISC}}{k_f}\left(\frac{k_p}{k_{ISC}^T+k_p}\right) \quad (5.48)$$

There are established techniques for the determination of ϕ_f^0 and ϕ_p^0 (Section 10.2). In this expression, k_f and k_p are reciprocals of the radiative lifetimes of fluorescene and phosphorescence states, respectively. k_f can be obtained experimentally from the integrated area under the absorption curve and k_p is obtained from the measured decay rates for phosphorescence at 77K in EPA. In Table 5.3 the observed quantities, their symbols, relation to rate constants and sources of studies are summarized.

TABLE 5.3

Observable photophysical parameters, their relationship to rate constants of various photophysical processes and sources of their information.

Observable	Symbol	Relation to rate constant	Source
Fluorescence quantum yield	ϕ_f	$\frac{k_f}{k_{IC}+k_{ISC}+k_f}$	Fluorescence spectrum
Phosphorescence quantum yield	ϕ_p	$\frac{k_p}{(k_{ISC}^T+k_p)}\frac{k_{ISC}}{(k_{IC}+k_{ISC}+k_f)}$	Phosphorescence spectrum
Singlet lifetime	τ_f	$\frac{1}{k_{IC}+k_{ISC}+k_f}$	Fluorescence decay
Triplet lifetime	τ_p	$\frac{1}{k_{ISC}^T+k_p}$	Phosphorescence decay
Singlet radiative lifetime	τ_f^0	$\frac{1}{k_f}$	(i) Absorption spectrum (ii) τ_f/ϕ_f

Rate constants for nonradiative steps are difficult to obtain experimentally. The quantum yield for intersystem crossing is expressed as

$$\phi_{ISC} = \frac{k_{ISC}}{k_{IC} + k_{ISC} + k_f} \tag{5.49}$$

A limiting value of k_{ISC} can, however, be obtained by making certain approximations. For example:

(i) If ϕ_p is large and $\phi_p + \phi_f \simeq 1$, then $k_{ISC}^{T} \ll k_p$ and in equation (5.48)

$$\frac{\phi_p}{\phi_f} = \frac{k_{ISC}}{k_f} = k_{ISC} \cdot \tau_f^0 \tag{5.49 a}$$

This can give an estimate of k_{ISC} from the knowledge of ϕ_p, ϕ_f and τ_f^0. In actual practice, the condition $\phi_p + \phi_f \simeq 1$ may not hold.

(ii) If internal conversion $S_1 \rightsquigarrow S_0$ is neglected but k_{ISC}^{T} is reasonable, then

$$\phi_f + \phi_p + \phi_{ISC}^{T} \simeq 1$$

and the quantum yield for nonradiative decay of triplet

$$\phi_{ISC}^{T} = 1 - (\phi_p + \phi_f) \tag{5.50}$$

and

$$\frac{\phi_{ISC}^{T}}{\phi_p} = \frac{k_{ISC}^{T}}{k_p} \tag{5.51}$$

Therefore,

$$k_{ISC}^{T} \simeq k_p \frac{1 - (\phi_p + \phi_f)}{\phi_p} \tag{5.52}$$

$$= \left(\begin{array}{c} \text{rate const. for} \\ \text{phosphorescence decay} \end{array} \right) \left(\frac{\text{fraction that decayed nonradiatively}}{\text{fraction that decayed radiatively}} \right)$$

The value of k_{ISC}^{T} can now be calculated from the knowledge of ϕ_f, ϕ_p and k_p.

Furthermore, since

$$\frac{\phi_p}{\phi_f} = \frac{k_{ISC}}{k_f} \frac{k_p}{k_{ISC}^{T} + k_p} \simeq \tau_f^0 \, k_{ISC} \quad \frac{\phi_p}{1 - \phi_f}, k_{ISC} = \frac{1}{\tau_f^0} \frac{(1 - \phi_f)}{\phi_f} \tag{5.53}$$

Example:

Calculation of radiationless transition probabilities in benzene:

ϕ_f for benzene at 77°C = 0.2

ϕ_p ,, ,, ,, = 0.2

$$k_f = \frac{1}{\tau_N} = \frac{1}{\tau_f^0}$$

$= 2 \times 10^6 \text{ s}^{-1}$ (calculated from integrated area under the absorption spectrum)

$k_p = 3.5 \times 10^{-2} \text{ s}^{-1}$ (calculated from phosphorescence decay and eq. (5.39))

$$k_{ISC} = \frac{1 - \phi_f}{\phi_f} k_f = \frac{1 - 0.2}{0.2} \times 2 \times 10^6 \text{ s}^{-1}$$

$$= 8 \times 10^6 \text{ s}^{-1}$$

$$k_{ISC}^{T} = 3.5 \times 10^{-2} \text{ s}^{-1} \times \frac{1 - (0.2 + 0.2)}{0.2}$$

$$= 1.0 \times 10^{-2} \text{ s}^{-1}$$

The rate constants for unimolecular photophysical processes in few representative organic molecules are given in the Table 5.4. These give an idea of the order of magnitude expected for various processes.

TABLE 5.4
Rate constants for unimolecular photophysical processes in some organic molecules

Compound	ϕ_f at 77 K	ϕ_p at 77 K	k_f 10^{-6} s^{-1}	k_p[a] s^{-1}	k_{ISC}^T[b] s^{-1}	k_{ISC}[c] s^{-1}
Naphthalene	0.55	0.05	1	0.044	0.35	1×10^6
1-Chloronaphthalene	0.06	0.54	3	1.7	1.4	5×10^7
1-Bromonaphthalene	0.002	0.55	3	28	22	2×10^9
1-Iodonaphthalene	10^{-4}	0.70	2	350	150	3×10^{10}
Benzophenone	10^{-4}	0.90	1	160	18	1×10^{10}
Acetophenone	10^{-4}	0.63	0.4	76	50	5×10^9
Biacetyl	0.001	0.25	0.1	125	375	2×10^7
Quinoline	0.10	0.20	10	0.15	0.5	8×10^7

[a] calculated from expression $k_p = \dfrac{1}{\tau_p^0} = \dfrac{1}{\tau_p}\left(\dfrac{\phi_p}{1-\phi_f}\right)$; equations (5.39) and (5.37)
[b] calculated from equation (5.52).
[c] calculated from equation (5.53).

5.8 STATE DIAGRAMS

All the information collected from spectroscopic data such as energies, lifetimes and populations of the S_1 and T_1 states can be incorprorated in the construction of a Jablonski type *state diagram* for a molecule. Such a diagram can be of immense help in predicting the photochemical behaviour of a molecule.

The energy difference between the ground and the first excited levels S_1 can be obtained from the O-O band in absorption and/or fluorescence. The O-O frequency in fluorescence is a better measure. For a T_1 state, the O-O band or the blue edge of the phosphorescence spectrum is considered. Other parameters are:

τ_f^0 : intrinsic radiative lifetime; from integrated absorption intensity using the formula given in Section 3.9.

ϕ_f^0 : intrinsic fluorescence quantum yield; from integrated fluorescence spectrum of a dilute solution expressed as number of quanta emitted per unit wavenumber *vs* wavenumber. It is compared with a standard of known ϕ_f, under equal light absorption. A simple way is to measure integrated fluorescence of the sample and the standard under identical experimental conditions such as geometry of experimental set up and optical densities of two solutions (Section 10.2).

τ_f: actual radiative lifetime $= \phi_f \tau_f^0$ (equation 5.30).

ϕ_p^0: quantum efficiency of phosphorescence emission measured at 77K in EPA solvent using a phosphorimeter.

τ_p^0: intrinsic lifetime of phosphorescence emission; on the assumption that all the deactivation occurs radiatively either from singlet or from triplet states, natural phosphorescence lifetime

$$= \tau_p \frac{1-\phi_f}{\phi_p} \quad \text{(equation 5.39)}$$

τ_p: lifetime measured from phosphorescence decay curve.

k_f: rate constant for fluorescence emission $= 1/\tau_f^0$.

k_p: rate constant for phosphorescence emission

$$= \frac{1}{\tau_p^0} = \frac{1}{\tau_p} \frac{\phi_p}{1-\phi_f} \quad \text{(equation 5.37)}$$

k_{ISC}: rate constant for intersystem crossing $S_1 \rightsquigarrow T_1$

$$= \frac{1}{\tau_f^0} \frac{1-\phi_f}{\phi_f} \quad \text{(equation 5.52)}$$

k_{ISC}^T: rate constant for intersystem crossing $T_1 \rightsquigarrow S_0$

$$k_p \frac{1-(\phi_p+\phi_f)}{\phi_p} \quad \text{(equation 5.53)}$$

The state diagram for biacetyl is presented in Figure 5.9 where $E_{S_1} = 259.6$ kJ mol^{-1} and $E_{T_1} = 234.4$ kJ mol^{-1}.

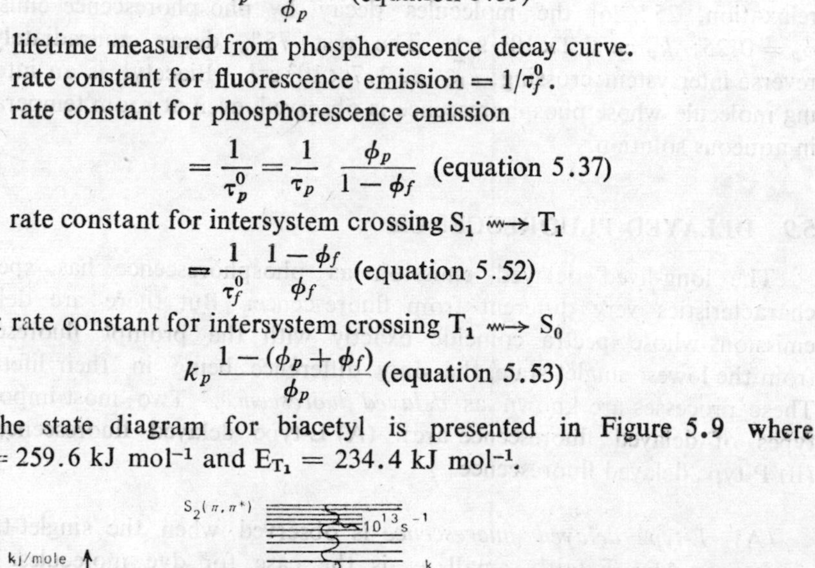

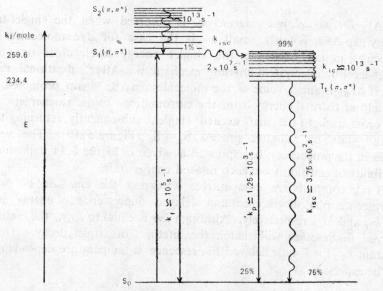

Figure 5.9 State diagram for biacetyl.

Whether the singlet is a (π, π^*) state or an (n, π^*) state can be established from solvent effect. A blue shift in polar solvent suggests an (n, π^*) state, a red shift a (π, π^*) state. A rough guess can be made from the radiative lifetime for fluorescence which is of the order of 10^{-5} s for biacetyl. Hence, it appears that the lowest excited singlet is an $S_1 (n, \pi^*)$ state. From dilute solution value of ϕ_f, we find that only 1% of the initially excited molecules are capable of emission with a rate constant $1 \times 10^5\ s^{-1} = 1/\tau_f^0$. Therefore 99% of the molecules must be transferred to the triplet state, assuming $S_1 \rightsquigarrow S_0$ do not occur. The rate constant for intersystem crossing, k_{ISC} is $2 \times 10^7\ s^{-1}$. After fast ($k \simeq 10^{12}\ s^{-1}$) vibrational relaxation, 25% of the molecules decay by phosphorescence emission, $\phi_p = 0.25$; $k_p \simeq 1.2 \times 10^2\ s^{-1}$. The rest 75% decay nonradiately by reverse intersystem crossing; $k_{ISC}^T \simeq 3.7 \times 10^2\ s^{-1}$. Biacetyl is an interesting molecule whose phosphorescence is observed even at room temperature in aqueous solution.

5.9 DELAYED FLUORESCENCE

The long-lived delayed emission as phosphorescence has spectral characteristics very different from fluorescence. But there are delayed emissions whose spectra coincide exactly with the prompt fluorescence from the lowest singlet state, the only difference being in their lifetimes. These processes are known as *delayed fluorescence*. Two most important types of delayed fluorescence are: (A) E-type delayed fluorescence and (B) P-type delayed fluorescence.

(A) *E-type delayed fluorescence* is observed when the singlet-triplet energy gap ΔE_{ST} is fairly small, as is the case for dye molecules. The molecules, initially excited to the singlet energy level, cross over to the triplet level by intersystem crossing mechanism. After vibrational relaxation, if ΔE_{ST} is small, some of the molecules may be again promoted, with the help of thermal energy from the surrounding, to the isoenergetic point and cross back to the first excited singlet, subsequently returning to the ground state by radiative process $S_1 \rightarrow S_0$ (Figure 5.10). The various stages in the process can be expressed as given in Figure 5.11 indicating the possibility of emission from each relaxed energy state.

If rate constants are competitive, we expect the emission of prompt fluorescence, phosphorescence and delayed fluorescence of energy quanta $h\nu_f$, $h\nu_p$ and $h\nu_{ED}$ respectively. Although $h\nu_f$ is equal to $h\nu_{ED}$, the lifetime of delayed fluorescence will match the lifetime of triplet decay. The rate constant k_{ED} for E-type delayed fluorescence is temperature dependent and can be expressed as

$$k_{ED} = A \exp(-\Delta E_{ST}/RT) \tag{5.54}$$

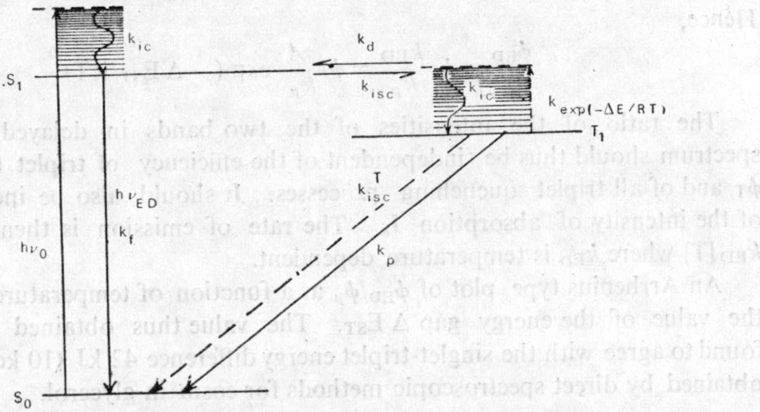

Figure 5.10 Jablonski diagram for E-type delayed fluorescence.

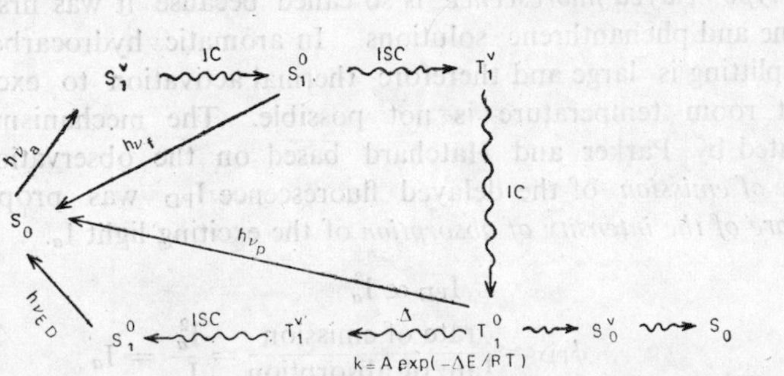

Figure 5.11 E-type delayed fluorescence pathways indicating possibility of emission from each relaxed state.

where A is the frequency factor and ΔE_{ST}, the activation energy equal to O – O difference between the singlet and the triplet level. This type of emission was first observed in deoxygenated solutions of eosin in glycerol and ethanol at room temperature and hence it is designated as E-type or *eosin type*. In the early work of Lewis and Kasha, direct phosphorescence was called β-phosphorescence and E-type delayed fluorescence as α-phosphorescence.

The ratio of quantum yields ϕ_p to ϕ_{ED} of phosphorescence and E-type delayed emissions respectively can be deduced as follows.

If ϕ_T is the quantum yield of triplet formation, then

$$\phi_p = \phi_T \frac{k_p}{k_p + k^T_{ISC} + k_{ED}} \tag{5.55}$$

On thermal excitation to the S_1 state, fluorescence will occur with usual fluorescence efficiency ϕ_f. Therefore, the efficiency of delayed fluorescence ϕ_{ED} is given as

$$\phi_{ED} = \phi_T\,\phi_f \frac{k_{ED}}{k_p + k^T_{ISC} + k_{ED}} \tag{5.56}$$

Hence,

$$\frac{\phi_{ED}}{\phi_p} = \phi_f \frac{k_{ED}}{k_p} = \phi_f \frac{A}{k_p} \exp(-\Delta E_{ST}/RT) \qquad (5.57)$$

The ratio of the intensities of the two bands in delayed emission spectrum should thus be independent of the efficiency of triplet formation ϕ_T and of all triplet quenching processes. It should also be independent of the intensity of absorption I_a. The rate of emission is then given as k_{ED} [T] where k_{ED} is temperature dependent.

An Arrhenius type plot of ϕ_{ED}/ϕ_p as a function of temperature will give the value of the energy gap ΔE_{ST}. The value thus obtained has been found to agree with the singlet-triplet energy difference 42 kJ (10 kcal) mol^{-1} obtained by direct spectroscopic methods for eosin in glycerol.

(B) *P-type delayed fluorescence* is so called because it was first observed in pyrene and phenanthrene solutions. In aromatic hydrocarbons singlet-triplet splitting is large and therefore thermal activation to excited singlet state at room temperature is not possible. The mechanism was first formulated by Parker and Hatchard based on the observation that the *intensity of emission* of the delayed fluorescence I_{PD} was proportional to the *square of the intensity of absorption* of the exciting light I_a.

$$I_{PD} \propto I_a^2$$

$$\phi_{PD} = \frac{\text{rate of emission}}{\text{rate of absorption}} = \frac{I_a^2}{I_a} = I_a \qquad (5.58)$$

Since ϕ_f is independent of I_a, the ratio of the intensity of P-type delayed emission to that of prompt fluorescence should show linear dependence on the intensity of absorption. The *square law* dependence indicates the necessity of two photons for the act of delayed emission and is hence known as *biphotonic* process. It has been observed in fluid solutions of many compounds and also in the vapour state.

The mechanism proposed for this biphotonic delayed emission is that encounter between two triplet molecules gives rise to an intermediate species X, which subsequently dissociates into an excited and a ground state singlet molecule. The excited singlet molecule finally relaxes by emission of radiation. The rate constant of emission is not that of fluorescence but is governed by the rate of formation of triplet molecules. The various steps leading to emission process are:

	Rate
$S_0 \rightarrow S_1 \rightarrow T$	$I_a\phi_T$
$T \rightarrow S_0$	$\Sigma k_T[T]$
$T + T \rightarrow X$	$k_X[T]^2$
$X \rightarrow$ Product	$k_r[X]$

$$X \rightarrow S_1 + S_0 \qquad k_s[X]$$

$$S_1 \rightarrow S_0 + h\nu_f \qquad k_f[S_1]$$

in which Σk_T represents the sum of all first order processes by which the triplet state is depleted. At low light intensities triplet-triplet interaction is less probable than the first order decay. Hence, under steady state conditions.

$$I_a \phi_T = \Sigma k_T [T] = \frac{1}{\tau_T} [T] \qquad (5.59)$$

where τ_T is the lifetime of the triplet. If the dissociation of the intermediate X is also fast, then the rate controlling step is the formation of X by *triplet-triplet annihilation process*. Hence, the intensity I_{PD} of P-type delayed emission is given as:

$$I_{PD} = \tfrac{1}{2} k_X [T]^2 \phi_f \frac{k_s}{k_s + k_r} \qquad (5.60)$$

since $\phi_{PD} = I_{PD}/I_a$, on transformation and substituting for $[T]^2$ from equation (5.59), we have,

$$\phi_{PD}/\phi_f = \tfrac{1}{2} I_a (\phi_T \tau_T)^2 k_X \frac{k_s}{k_s + k_r} \qquad (5.61)$$

In the pyrene system, the intermediate X has been identified with an excited dimer or excimer, since characteristic excimer emission is also observed. The process of *triplet-triplet annihilation* then consists of transfer of energy from one triplet to another to form excited dimeric species $(S)_2^*$, which dissociates into an excited and a ground state singlets:

$$\begin{array}{ccccc} T + T & \rightarrow & (S)_2^* & \rightarrow & S_1 + S_0 \\ & & \downarrow & & \downarrow \\ & & S_0 + S_0 + h\nu_E & & S_0 + h\nu_{PD} \\ & & \text{excimer emission} & & \text{delayed emission} \end{array}$$

The generation of an excited and a ground state singlets from two triplets is a spin allowed exchange mechanism:

$$T(\uparrow\uparrow) + T(\downarrow\downarrow) \rightarrow S_0(\uparrow\downarrow) + S_1(\downarrow\uparrow)$$

Such bimolecular quenching reactions occur at diffusion controlled rates. A probability P for reaction per encounter may be included. The rate constant terms in (5.61) become

$$k_X \frac{k_s}{k_s + k_r} = P k_{diff} \qquad (5.62)$$

and

$$\phi_{PD}/\phi_f = \tfrac{1}{2} P k_{diff} I_a (\phi_T \tau_T)^2 \qquad (5.63)$$

The intensity of P-type delayed fluorescence should decay exponentially with a lifetime equal to *one-half* of that of phosphorescence. Since in fluid solutions, phosphorescence is much weaker than delayed fluorescence, the decay constant of the latter may be used for the determination of the triplet state lifetimes;

$$\tau_{PD} = \frac{\tau_T}{2} = \frac{\tau_p}{2} \tag{5.64}$$

The intensity of P-type delayed fluorescence should decrease with viscosity and should be observed, if at all, with very low efficiency in rigid media.

(C) *Recombination luminescence.* Still another kind of delayed emission observed in dye solutions in low temperature glasses is by cation-electron recombination mechanism. Under high intensity irradiation and in rigid media, a dye molecule can eject an electron which is trapped in suitable sites. When this electron recombines with the dye cation, an S_1 state is generated and a photon is emitted as fluorescence.

A similar phenomenon of cation-anion recombination may generate an excited state which eventually can relax by emission of radiation. The process can be explained as follows: a radical anion carries the extra electron in the next higher energy state, whereas a radical anion has an electron deficiency in the ground state. When the two are brought together, singlet, excited and ground state, molecules are generated. Experimentally, the radical anions can be produced at the cathode by electrolysis of the organic compound. The anion A^- so generated will diffuse into the solution. Now if the polarity of the electrodes is reversed A^+ will be generated. This will also diffuse into solution. Under favourable conditions, the two may combine to generate A* and A and cause *chemiluminescence*, i.e. luminescence initiated by chemical reaction and not by irradiation. This is also known as electrochemiluminescence (ECL).

5.10 THE EFFECT OF TEMPERATURE ON EMISSION PROCESSES

In general, the natural radiative lifetimes of fluorescence and phosphorescence should be independent of temperature. But the emission intensities may vary due to other temperature dependent and competitive rate constants.

Interesting examples are found in substituted anthracenes. The lifetimes and quantum yields of fluorescence of substituted anthracenes show different dependencies on temperature. The position of the substituent is more important than its nature. For 9- and 9,10-substituted anthracenes, fluorescence quantum yields increase steeply with decrease of temperature, while side-substituted derivatives have low yield and small temperature dependence. The variation is of the form

$$\ln\left(\frac{1}{\phi_f} - A\right) = \frac{B}{T} \tag{5.65}$$

where A and B are constants, $A = 1 + k_1\tau$, k_1 being the rate constant for crossing over from zero-vibrational level of S_1 to T and τ, the radiative lifetime of S_1 state. Evidently cross-over point lies above $S_1^{v=0}$. At the same time the quantum yield of production of triplet decreases with decrease of temperature.

Anthracene has a triplet state T_2 which is about 650 cm^{-1} below the S_1 level. The T_2 and S_1 states are affected to different extents by substitution, whereas T_1 is nearly unaffected.

The temperature sensitivity arises due to disposition of T_2 state with respect to S_1 state. If T_2 is considerably above S_1, transfer to T_1 is less probable because of unfavourable Franck-Condon factor. As a consequence, fluorescence is the easiest way for deactivation and fluorescence yield is nearly unity. No dependence on temperature is expected. On the other hand, if T_2 is sufficiently below S_1, so that the density of state is high at the crossing point, fluorescence quantum yield should be less than unity as triplet transfer is fecilitated. Again no temperature dependence is observed. But if T_2 is nearly at the same energy as S_1, a barrier to intersystem crossing is expected and fluorescence yield will show temperature dependence.

A temperature dependent intersystem crossing rate will be obtained whenever the triplet state crosses the PE surface of S_1 state at a position slightly above the zero-point energy of S_1 state. This energy barrier to transition is of the form $e^{-\Delta E/kT}$ where ΔE is the height of the energy barrier. In general, ΔE is found to be in the range of 17–24 kJ mol^{-1} (4–7 kcal), which is equivalent to 1–2 quanta of vibrational energy of simple vibrational modes. The variation in fluorescence yield may also arise due to change in the viscosity of the medium with temperature. The dependence is of the same form as already mentioned, $\log (1/\phi_f - A)$ varies as $1/T$. But the energy of activation obtained in this case is of the order of temperature coefficient for the viscous flow. The role of viscosity in such cases is apparently to decrease energy degrading vibrational-rotational motions.

The temperature sensitivity of phosphorescence mainly arises from fast impurity quenching processes. At low temperatures and rigid glassy medium, emission is a rule rather than exception.

Summary

1. After an initial act of photon absorption, an atom or a molecule, if not consumed by photochemical reaction, can return to the ground state by a number of photophysical pathways, some of which are radiative and others nonradiative.
2. In radiationless or nonradiative transitions, electronic energy is dissipated as small vibrational quanta in which the environment acts as the heat sink. If the transition is within the states of the same multiplicity such as singlet-singlet, $S_j \rightsquigarrow S_k$ or triplet-triplet, $T_j \rightsquigarrow T_k$, it is termed internal conversion (IC), if it is

between the states of different multiplicities $S_j \rightsquigarrow T_k$, or $T_k \rightsquigarrow S_p$, it is termed intersystem crossing (ISC). The rate constants for internal conversion are of the order of vibrational frequencies, $k_{IC} \simeq 10^{13} - 10^{12}\ s^{-1}$. That for intersystem crossing k_{ISC} is reduced by a factor of $10^2 - 10^4$, due to violation of spin conservation rule. The transitions from the lowest excited state to the ground state k_{IC} ($S_1 \rightsquigarrow S_0$) and k_{ISC}^T ($T_1 \rightsquigarrow S_0$) are less efficient due to larger energy gap ΔE between the two combining states.

3. The theory of radiationless transition considers the transition to occur in two steps: (i) horizontal transition from one energy state to the other at the isoenergetic point, for the two combining states and (ii) vibrational relaxation of the lower energy state. The step (i) is the rate determining step. The rate constant k_{NR} is given by the theory of Robinson and Frosch as,

$$k_{NR} = \frac{4\pi^2 \rho_E}{hN} \beta^2 \int \chi_f^{v'} \chi_i^{v''} \, d\tau_v$$

where ρ_E is the density of vibronic energy levels of the lower energy state E_1, which are in near resonance with the zero-point energy of the transferring state E_2, β^2 is the interaction energy which couples the Born-Oppenheimer satets and is proportional to Δ E; and $\int \chi_f^{v'} \chi_i^{v''} \, d\tau_v$ is the Franck-Condon overlap integral.

4. The selection rules for radiationless transitions are just the opposite of those for radiative transitions. Allowed transitions are:

$$g \rightsquigarrow g, \quad u \rightsquigarrow u \quad \text{and} \quad S_1 \rightsquigarrow T_1$$

Transitions $\quad {}^1(n, \pi^*) \rightsquigarrow {}^3(\pi, \pi^*), \quad {}^1(\pi, \pi^*) \rightsquigarrow {}^3(n, \pi^*)$

are more probable than

$${}^1(\pi, \pi^*) \rightsquigarrow {}^3(\pi, \pi^*) \quad \text{or} \quad {}^1(n, \pi^*) \rightsquigarrow {}^3(n, \pi^*).$$

5. The radiative transitions can occur (a) from the lowest excited singlet S_1 to the ground state S_0 and (b) from the lowest triplet T_1 to ground singlet S_0. The radiative transition between $S_1 \rightarrow S_0$ leads to fluorescence emission and that from $T_1 \rightarrow S_0$ to phosphorescence emission.

Fluorescence always occurs from the lowest singlet state even if the initial excitation is to higher energy state (Kasha's rule). Azulene and some of its derivatives are exceptions to this rule. Because of vibrational relaxation of initially excited vibronic state, the fluorescence spectrum may appear as a mirror image of the absorption spectrum for large polyatomic molecules. The shape of the emission spectrum is independent of the exciting wavelength.

6. The fluorescence yield ϕ_f is defined as

$$\phi_f = \frac{\text{intensity of emission } (S_1 \rightarrow S_0)}{\text{intensity of absorption } (S_1 \leftarrow S_0)} = \frac{F \text{ einstein s}^{-1}}{I_a \text{ einstein s}^{-1}}$$

The rate constant k_f for fluorescence emission is related to natural radiative lifetime τ_N as

$$k_f = \frac{1}{\tau_N} = \frac{1}{\tau_f^0}$$

In the presence of other competitive deactivating processes, actual lifetime is given as

$$\tau_f = \frac{1}{k_f + \Sigma k_i} \quad \text{and} \quad \phi_f = \frac{\tau_f}{\tau_f^0} = \frac{k_f}{k_f + \Sigma k_i}$$

Furthermore, $\quad F = \phi_f I_a = \phi_f I_0 (1 - e^{-2 \cdot 303\, \epsilon\, Cl})$

In the case of small fractional absorption, $F = \phi_f^\circ . I_0 . 2.303\, \epsilon c l$, two observations can be made:

(i) If fluorescence intensity is measured as a function of wavelength or wavenumber of excitation radiation of constant incident intensity I_0 and under fixed experimental conditions, the variation of fluorescence intensity will simulate variation of ϵ, the molar extinction coefficient of the emitter, as a function of λ or $\bar{\nu}$. This is known as *fluorescence excitation spectrum* and, in general, should be identical with absorption spectrum.

(ii) For low absorption intensities, fluorescence should be a linear function of concentration and hence a sensitive analytical tool.

7. Fluorescence appears mostly in molecules with conjugated system of double bonds and rigid structure. Heavy atom substituents like halogens, enhance $S_1 \rightsquigarrow T_1$, intersystem crossing and hence reduce $S_1 \rightarrow S_0$, radiative transition probabilities. Organic molecules with $^1(\pi, \pi)^*$ state as the lowest excited state are fluorescent, but those with $^1(n, \pi^*)$ state as the lowest excited state are, in general, nonfluorescent due to longer radiative lifetimes and enhanced intersystem crossing rates. These compounds emit phosphorescence radiation at low temperatures in rigid glassy medium.

8. The radiative transition $T_1 \rightarrow S_0$, from the lowest triplet state T_1 gives rise to phosphorescence emission. The quantum yield of phosphorescence emission ϕ_p is expressed as

$$\phi_p = \frac{\text{Intensity of phosphorescence emission } (T_1 \rightarrow S_0)}{\text{Intensity of absorption } (S_1 \leftarrow S_0)}$$

The phosphorescence lifetime τ_p^0 in the absence of nonradiative transitions is related to observed lifetime τ_p, as

$$\tau_p^0 = \tau_p \frac{1-\phi_f}{\phi_p}$$

assuming that all those molecules which do not fluoresce are transferred to the triplet state. Furthermore, τ_p^0 is inversely related to k_p, the rate constant for phosphorescence emission in the absence of all nonradiative deactivation pathways for triplet state. The actual lifetime τ_p is

$$\tau_p = \frac{1}{k_p + k_{ISC}^T}$$

where k_{ISC}^T is the rate constant for intersystem crossing, $T_1 \rightsquigarrow S_0$, and

$$\frac{\tau_p}{\tau_p^0} = \frac{k_p}{k_p + k_{ISC}^T} = \frac{\phi_p}{\phi_T}$$

where ϕ_T is quantum yield of triplet formation. Hence,

$$k_p = \frac{1}{\tau_p} \frac{\phi_p}{(1-\phi_f)} \tau_p^0$$

is generally much greater than τ_f^0 because of the forbidden character of $T_1 \rightarrow S_0$ transition.

9. The singlet-triplet energy gap ΔE_{ST} is governed by the electron-electron repulsion term. The interaction energy is computed from the exchange integral K which depends on the overlap between the single electron orbitals, ϕ and ϕ^*. For $\pi \rightarrow \pi^*$ transition, ΔE_{ST} is large because π and π^* orbitals occupy more or less the same configuration space. For $n \rightarrow \pi^*$ transition, ΔE_{ST} is small because of the small overlap between n and π^* orbitals. Therefore, general ordering of

energy states in an organic molecule with heteroatom is

$$^1(\pi, \pi^*) > {}^1(n, \pi^*) > {}^3(n, \pi^*) > {}^3(\pi, \pi^*).$$

10. Rate constants for photophysical unimolecular radiative processes can be obtained from spectral data and k_{ISC} and k_{ISC}^T computed therefrom. The rate constants for radiationless processes are important parameters in photochemistry because the lowest singlet and triplet states are seats of photochemical reactions.

11. Besides long lived phosphorescence emission, which has spectral characteristics different from fluorescence, there are few other delayed emission processes which have same spectral characteristics as prompt fluorescence.
    (i) E-type delayed fluorescence is observed in dyes which have small ΔE_{ST}. The molecules transferred to the triplet state nonradiatively are promoted with the help of thermal energy, to the singlet state from where they decay radiatively.
    (ii) P-type delayed fluorescence is caused by triplet-triplet annihilation process which generates an excited singlet and a ground state molecule. The excited singlet then decays radiatively. This is a biphotonic process and emission intensity is proportional to the square of the incident intensity.
    (iii) Electrochemiluminescence is observed when radical-electron recombination reactions occur in rigid matrix or radicals A^+ and A^- generated in electrolysis of organic compounds combine to form A^* and A.

12. Temperature has no effect on intrinsic transition probabilities of fluorescence or phosphorescence but emission characteristics may be affected by other temperature dependent nonradiative competitive processes.

SIX

Photophysical Kinetics of Bimolecular Processes

6.1 KINETIC COLLISIONS AND OPTICAL COLLISIONS

The primary condition for a bimolecular process is close approach of two molecules, which in normal terminology is called a *collision*. The '*closeness*' of approach is a variable term and depends on the nature of interaction between the two colliding partners. For ground state molecules, the collisions are defined by gas kinetic theory based on a hard sphere model. The number of collision per second between unlike molecules A and B and between molecules of the same kind are, respectively,

$$\left[\frac{8\pi RT}{\mu}\right]^{1/2} r_{AB}^2 [A][B] \quad \text{and} \quad 2\left[\frac{\pi RT}{M_A}\right]^{1/2} r_A^2 [A]^2 \qquad (6.1)$$

where μ, the reduced mass $= M_A M_B/(M_A + M_B)$, M_A and M_B being the molecular weights of A and B, respectively; $r_{AB} = (r_A + r_B) =$ the sum of the radii of molecules A and B in centimetres and defines the collision cross-section $\sigma_k \ (= \pi r_{AB}^2)$ in cm^2, when two spheres are in contact. Collision numbers or frequencies Z_{AB} or Z_{AA} are defined for unit concentrations of [A] and [B] and for molecules of common diameters and weights are in the range $10^9 - 10^{10}$ 1 mol^{-1} s^{-1} at ordinary temperature.

For a molecule in the excited state, direct contact between two interacting partners may not be necessary and effective cross-section for *optical* collisions may be much larger than those for kinetic collisions (Figure 6.1). The *effective crosssection* $\sigma = (\pi R_{AB}^2)$ for *optical* collisions is proportional to *the square of distance* R_{AB} *over which the excited molecule can*

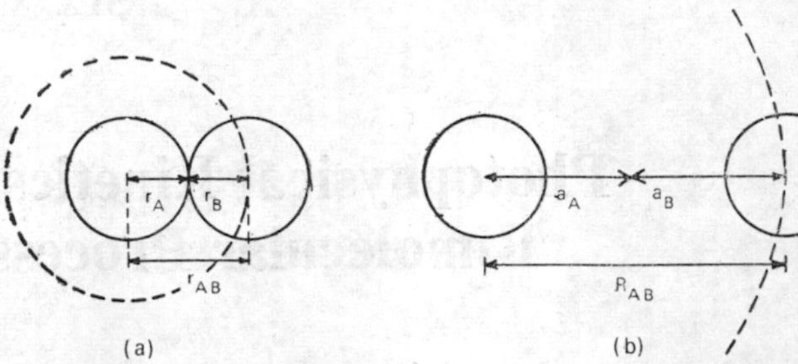

Figure 6.1 Schematic representation of (a) kinetic collision and (b) optical collision.

interact with another molecule to bring about a physical or chemical change. It is different for different types of reactions and for different types of molecules. This approach is normally used for atomic gases and vapours which will be discussed first because it has provided a wealth of information about the mode of excited state interactions.

6.2 BIMOLECULAR COLLISIONS IN GASES AND VAPOURS AND THE MECHANISM OF FLUORESCENCE QUENCHING

The quenching of resonance radiation of mercury from excited 6^3P_1 state by large number of added gases has been studied. This state is populated through $6^3P_1 \leftarrow 6^1S_0$ transitions by mercury atoms and emits strong radiation at 253.7 nm. It is a spin forbidden transition and hence has a long lifetime ($\tau_0 = 110$ ns).

The mechanism of quenching is not the same for all the added gases. The basic mechanisms may be identified as:

(i) *Electronic energy transfer to the added gas molecules resulting in excitation of the latter ($E \rightarrow E$ transfer).* Such transfer may result in the characteristic emission from the added gas with concomitant quenching of mercury resonance lines. Examples are: mixtures of Hg + Tl and Hg + Na. These energy transfer reactions have a large cross-section for quenching and will be discussed in detail at a later stage.

(ii) *Energy transfer in the vibrational modes of the quencher ($E \rightarrow V$ transfer).* In the classical work by Zeemansky, a correlation was established between the quenching cross-section and the close matching of energy difference, Hg (6^3P_1) — Hg (6^3P_0), and the vibrational energy spacing Δv of $v = 0 \rightarrow v = 1$ transition in the quencher. Simple molecules like NO, CO, N_2 were found to be very efficient quenchers (Table 6.1).

TABLE 6.1

Quenching crosssection of various gases for mercury resonance radiation

Gas	H_2O	D_2O	NH_3	NO	CO_2	N_2	H_2
σ in $cm^2.10^{16}$	1.43	0.46	4.2	35.3	3.54	0.25	8.6

But more recent work using flash excitation has shown that Hg (6^3P_0) may not be the final state in all cases. It is true for N_2, but CO deactivates the excited mercury to ground state Hg (6^1S_0), itself being transferred to high vibrational level ($v = 9$) through the intermediary of mercury-carbonyl complex formation.

$$Hg(6^3P_1) + CO\ (X^1\Sigma) \rightarrow HgCO\ (^3\Sigma)\text{ or }(^3\Pi) \qquad \textit{complex formation}$$

$$HgCO\ (^3\Sigma) \rightarrow Hg\ CO(^1\Sigma) \qquad \textit{intersystem crossing}$$

$$HgCO\ (^1\Sigma) \rightarrow Hg\ (6^1S_0) + CO\ (X^1\Sigma)\ v' \leqslant 9 \qquad \textit{dissociation}$$

A similar mechanism is now suggested for quenching by NO. Thus a strong interaction between the excited atom and the quencher appears essential for quenching to occur.

(iii) *Quenching by inert gases like argon and xenon.* These gases are known to be good quenchers of excited states. It is expected that in such cases electronic excitation energy can only be dissipated as translational energy ($E \rightarrow T$ transfer) because energy levels are much above Hg (6^3P_1). Conversion of a large amount of electronic energy into translational energy is not a very efficient process as formulated in Franck-Condon principle. Xenon is observed to quench by forming a complex which dissociates on emission of radiation.

(iv) *Quenching due to abstraction of* H *atom.* Such a process results in the formation of free radicals and consequent chemical reactions. Mercury photosensitized reactions are an important class of gas phase reactions. The simplest s quenching by H_2, which probably proceeds as follows:

$$Hg(6^3P_1) + H_2\ (X^1\Sigma_g^+) \rightarrow Hg\,H(^2\Sigma) + H(1^2S_{1/2}) \rightarrow Hg(6^1S_0) + 2H(1^2S_{1/2})$$

Both saturated and unsaturated organic compounds have been found to be good quenchers. In many cases, H atom abstraction may be a very probable mechanism. In others, some form of association involving donation ɹo π-electrons into vacant atomic orbital of mercury seems likely. There is also the possibility of energy transfer to the triplet level of the quencher.

Some of these mechanisms may be explained by the potential energy surfaces for interaction between an excited molecule A and a quencher B which may have following dispositions, Figure 6.2 (a, b, c, d):

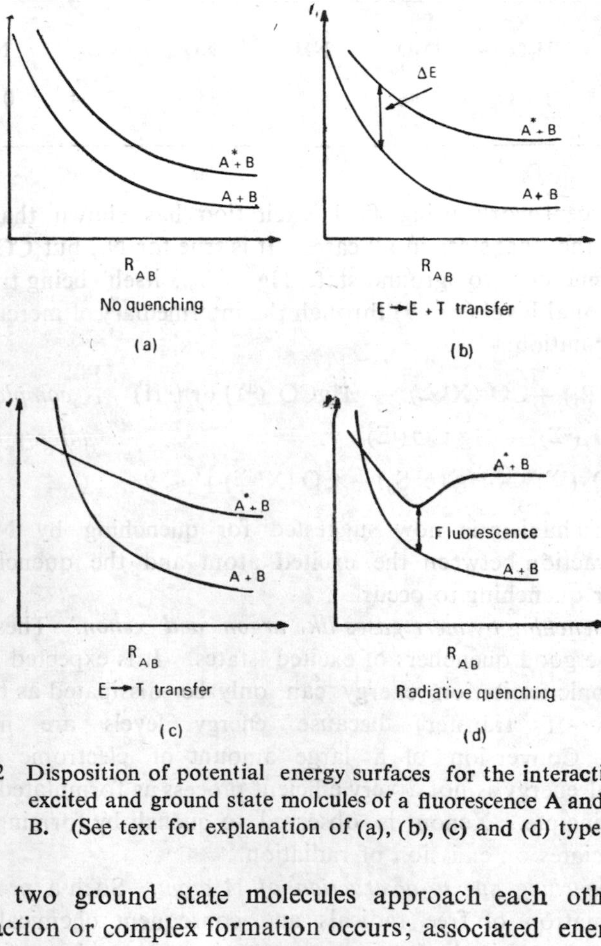

Figure 6.2 Disposition of potential energy surfaces for the interaction between excited and ground state molcules of a fluorescence A and a quencher B. (See text for explanation of (a), (b), (c) and (d) types.)

(a) The two ground state molecules approach each other and no chemical reaction or complex formation occurs; associated energy changes are repulsional all the time. The interaction between one excited and the other normal molecule are also repulsional and PE surfaces run parallel to each other. No quenching is visualized. (b) The two curves tend to approach each other, whereby the electronic or vibrational transfer mechanism ($E \rightarrow E$ or $E \rightarrow V$) is possible. After fruitful collision any excess energy is converted into translational energy of the partners. (c) The two repulsional curves cross at a point. The high kinetic energy of collision between A* and B pushes the colliding partners at this point of intersection, the two eventually separate as ground state molecules with large kinetic energy of translation. (d) A shallow minimum appears in the upper PE surface suggestive of complex formation between A* and B. A radiative mechanism leads to dissociation of the complex. Characteristic emission from the complex may be identifiable. Such excited state complexes, now termed *exciplexes*, are discussed later.

Similar quenching studies for more complex systems have established that quenching is efficient when electronic energy is transferred as electronic energy, vibrational energy as vibrational energy and rotational energy as rotational energy, with the minimum conversion to kinetic energy of translation. This is, again, a statement of *the Franck-Condon principle*. A polyatomic molecule with many more vibrational modes is a much better quencher than a simple molecule. The reason is that any one of these modes may be in energy resonance with some vibrational energy level of the excited molecule facilitating quenching by energy transfer. Many collisions are necessary to deactivate a molecule with high vibrational energy because only one or two vibrational quanta can be transferred at a time ($\Delta v = \pm 1$ for a harmonic oscillator, > 1 for an unharmonic oscillator). Eventually, the system attains thermal equilibrium when all the excess energy is dissipated to the surrounding. Potential energy representations of molecular collisions go some way to interpreting quenching but the complexity of the field effects makes them necessarily non-quantitative.

Collisions in gas phase need not always cause fluorescence quenching. If a possibility of predissociation exists due to crossing of a repulsional PE surface with that of the first excited state at a slightly higher vibrational level, collisional deactivation may prevent transfer of the excited molecule to the repulsional state. Once brought down to the zero vibrational level of the first singlet state, fiuorescence emission is facilitated. β-naphthylamine is an instance in point, which in the gas phase is observed to be fluorescent in presence of deactivating gases.

6.3 COLLISIONS IN SOLUTION

Gas collision theory cannot be applied to liquid solutions because of the Franck-Rabinowitch 'cage' effect. Two solute molecules close together and hemmed in by solvent molecules make many repeated collisions with each other before drifting away. This is called an *encounter*, usually involving 20–100 collisions. Solute molecules, initially far apart, can only come together by a relatively slow diffusion process. If the number of collisions two molecules need to make before interacting is small compared with the number they make in a solvent cage, the bimolecular reaction rate is limited by the rate of formation of new encounters through diffusion. The liquid viscosity η becomes a controlling factor. By a balance of factors, in common liquids of low viscosity, the interaction rate here happens to lie in the same range as the gas mixtures, $k_2 \simeq 10^9 - 10^{10}$ $l\ mol^{-1}\ s^{-1}$. The theory of such diffusion controlled reactions derived by Debye from Smoluchowski equation for rapid coagulation of colloids is based on Fick's law of diffusion. According to this expression, the rate constant k_2 for bimolecular quenching reaction is given by

$$k_2 = p \frac{4\pi R_{AB} N (D_A + D_B)}{1000} \text{ l mol}^{-1} \text{ s}^{-1} \tag{6.2}$$

This involves the diffusion coefficients D_A and D_B for the two colliding partners and the encounter radii $R_{AB} = a_A + a_B$, sum of the interaction radii (Figure 6.1), p is a probability factor per encounter. The diffusion coefficients are given by the Stokes-Einstein equation

$$D = \frac{kT}{6\pi \eta r}$$

where k is the Boltzmann constant and r the kinetic radius of the diffusing molecule. Let r_A and r_B be the radii of the two diffusing partners, then

$$D_A + D_B = \frac{kT}{6\pi \eta}\left(\frac{1}{r_A} + \frac{1}{r_B}\right) \tag{6.3}$$

If it is assumed that (i) $a_A = a_B = a$ so that $R_{AB} = 2a$, and (ii) $r_A = r_B = r$ then on substitution in (6.2), we have

$$k_2 = p \frac{8 \text{ RT}}{3{,}000\ \eta} \frac{a}{r} \text{ l mol}^{-1} \text{ s}^{-1}$$

On further assumption that the interaction radius and the kinetic radius are equal, i.e. $a = r$, and the probability factor $p = 1$, we have the final form of the equation for efficient reaction,

$$k_2 = \frac{8\text{RT}}{3{,}000\ \eta} \text{ l mol}^{-1} \text{ s}^{-1} \tag{6.4}$$

Thus, bimolecular rate constant depends only on the viscosity and the temperature of the solvent. The calculated rate constants for diffusion-controlled bimolecular reactions in solution set the upper limit for such reactions.

However, if the diffusing molecules are much smaller than the solvent molecules, the coefficient of sliding friction is zero, i.e. the solute molecules can move freely in contact with the solvent molecule. In such cases k_2 is given by

$$k_2 = \frac{8 \text{ RT}}{2000\,\eta} \text{ l mol}^{-1} \text{s}^{-1} \tag{6.5}$$

The oxygen quenching of aromatic hydrocarbons in viscous solutions is better approximated by this expression. For ionic solutions of charges Z_A and Z_B in a medium of dielectric constant ϵ, a coulombic interaction term f is included in the denominator, where

$$f = \frac{\delta}{e^{\delta} - 1} \text{ and } \delta = \frac{Z_A Z_B e^2}{\epsilon k T r_{AB}}$$

6.4. KINETICS OF COLLISIONAL QUENCHING : STERN-VOLMER EQUATION

A quenching process is defined as one which competes with the spontaneous emission process and thereby shortens the lifetime of the emitting molecule. Basically, these quenching reactions are *energy transfer or electron transfer processes.* In some cases transient complex formation in the excited state may be involved (*exciplex formation*). These *excited state complexes* may or may not emit their own characteristic emission which is likely to be different from that of the original molecule. In solution, the possibility of weak complex formation in the ground state is also present. In the limit this may just be juxtaposition of fluorescer and quencher molecules within the solvent cage at the moment of excitation. This may cause immediate loss of electronic energy and is called *static quenching*. Since there is no competition with emission processes, the lifetime is not affected. Ground state complex formation reduces fluorescence intensity by competing with the uncomplexed molecules for the absorption of the incident radiation. This is called the *inner filter effect*, similar to the presence of other added absorbing molecules.

Under steady illumination, the rate of formation of an excited molecule A* is equal to its rate of deactivation and the concentration [A*] remains constant:

$$\frac{d[A^*]}{dt}=0 \tag{6.6}$$

The concentration of A* in the absence of any bimolecular quenching step is given as before (Section 5.7).

$$[A^*]^\circ=\frac{I_a}{k_f+k_{IC}+k_{ISC}}=\frac{I_a}{k_f+\Sigma k_i} \tag{6.7}$$

where I_a is the rate of absorption or the rate of formation of the activated molecule, k_f is the rate constant for fluorescence and Σk_i, the sum of rate constants for all the unimolecular deactivating steps such as internal conversion k_{IC} and intersystem crossing k_{ISC} which originate from this energy state.

If another molecule Q is added to the solution which quenches the fluorescence by a bimolecular quenching step, than

$$A^*+Q \rightarrow A+Q \qquad\qquad \text{Rate}=k_q[A^*][Q]$$

then the concentration of the fluorescer [A*] in presence of the quencher is given as

$$[A^*]=\frac{I_a}{k_f+\Sigma k_i+k_q(Q)}$$

If [A*]° and [A*] are fluorescer concentrations in absence and in presence of the quencher, the respective quantum yields are

$$\phi_f{}^\circ = \frac{k_f[\mathrm{A}^*]^\circ}{\mathrm{I}_a} = \frac{k_f}{k_f + \Sigma k_i} \tag{6.8}$$

$$\phi_f = \frac{k_f[\mathrm{A}^*]}{\mathrm{I}_a} = \frac{k_f}{k_f + \Sigma k_i + k_q[\mathrm{Q}]} \tag{6.9}$$

The ratio of the two yields is

$$\frac{\phi_f^0}{\phi_f} = \frac{k_f + \Sigma k_i + k_q[\mathrm{Q}]}{k_f + \Sigma k_i} = 1 + \frac{k_q[\mathrm{Q}]}{k_f + \Sigma k_i}$$

$$= 1 + k_q\tau[\mathrm{Q}] = 1 + \mathrm{K_{SV}}[\mathrm{Q}] \tag{6.10}$$

This expression is known as the *Stern-Volmer equation* and $\mathrm{K_{SV}}$ as *Stern-Volmer constant*. $\mathrm{K_{SV}}$ is the ratio of bimolecular quenching constant to unimolecular decay constant and has the dimension of litre/mole. It implies a competition between the two decay pathways and has the character of an equilibrium constant. The Stern-Volmer expression is linear in quencher concentration and $\mathrm{K_{SV}}$ is obtained as the slope of the plot of $\phi_f{}^\circ/\phi_f$ vs [Q], if the assumed mechanism of quenching is operative. Here, τ is the actual lifetime of the fluorescer molecule in absence of bimolecular quenching and is expressed as

$$\tau = \frac{1}{k_f + \Sigma k_i}$$

If τ is measured independently then from the knowledge of $\mathrm{K_{SV}}$, the rate constant k_q for the bimolecular quenching step can be determined. For an efficient quencher, $\mathrm{K_{SV}} \simeq 10^2 - 10^3$ l mol^{-1} and if $\tau \simeq 10^{-8}$ s, then

$$k_q = \frac{\mathrm{K_{SV}}}{\tau} \simeq \frac{10^2}{10^{-8}} \text{ l mol}^{-1}\text{ s}^{-1} \tag{6.11}$$

$$\simeq 10^{-10} \text{ l mol}^{-1}\text{ s}^{-1}$$

which is of the same order as the encounter frequency. In such cases quenching is diffusion controlled and is obtained directly from the expression (6.4). Therefore, an upper limit for k_q may thus be obtained. (Table 6.2).

TABLE 6.2

Values of $k_q \times 10^{-10}$ for anthracene fluorescence: A for O_2 quenching, B for concentration quenching, C calculated from equation (6.4).

Solvent	Viscosity in cP at 25°C	A	B	C
Benzene	0.65	4.0	0.93	1. 0
Chloroform	0.54	3.4	0.97	1. 2
Kerosene	2.0	2.9	1.17	0.31

The quenching constant can also be calculated from the condition of 50% quenching. If $[Q]_{1/2}$ is the concentration of the quencher when the solution is half-quenched, then

$$\frac{\phi_0}{\phi} = 2 = 1 + K_{SV}[Q]_{1/2}$$

or
$$K_{SV} = k_q \tau = \frac{1}{[Q]_{1/2}} = \frac{1}{C_h} \quad (6.12)$$

K_{SV} is the reciprocal of half-quenching concentration or *half-value concentration* C_h. If k_q is assumed to be diffusion controlled, the lifetime τ of unquenched molecule can be obtained from a knowledge of K_{SV}. For a strongly absorbing solution with fluorescence lifetime $\simeq 10^{-8}$ s and assuming diffusion controlled quenching, the half-quenching concentration, $[Q]_{50\%}$ is of the order of 0.01 M and for 10% quenching, $[Q]_{10\%} \simeq 0.001$ M. But if $\tau = 10^{-2}$ s, $[Q]_{50\%} = 10^{-8}$ M. This explains the sensitivity of the long-lived phosphorescent state to traces of impurities.

For weak quenchers, all the encounters may not be fruitful and collisional quenching efficiency p is less than unity, so that

$$K_{SV} = p\,\tau\,k_q$$

In systems where quenching is much smaller than that predicted by diffusion-controlled encounter frequencies, the reason for inefficiency may be that either a heat of activation or an entropy of activation is necessary. The dependence of K_{SV} on solvent viscosity then disappears. For example, bromobenzene is a weak quencher for fluorescence of aromatic hydrocarbons, the quenching constant being nearly the same in hexane as in viscous paraffins.

For ionic solutions, ionic strength is an important parameter, affecting the quenching constant. The rate constant for bimolecular quenching collisions should be corrected to the limiting value k^0_q according to Brönsted theory

$$\log k_q = \log k^0_q + 0.5\,\Delta z^2 \sqrt{\mu} \quad (6.13)$$

where μ is the ionic strength, $\Delta z^2 = z^2_{AQ} - (z^2_A + z^2_Q)$ and z_{AQ}, z_A, z_Q are number and nature of charges on the intermediate complex AQ, the fluorescer A and the quencher Q, respectively. With rise in ionic strength the quenching action may increase, decrease or remain unchanged depending on the sign of Δz^2.

On the other hand, if the rate constant for the quenching step exceeds that expected for a diffusion-controlled process, a modification of the parameters in the Debye equation is indicated. Either the diffusion coefficient D as given by the Stokes-Einstein equation is not applicable because the bulk viscosity is different from the microviscosity experienced by the quencher (e.g. quenching of aromatic hydrocarbons by O_2 in paraffin solvents) or the encounter radius R_{AB} is much greater than the gas-kinetic collision radius. In the latter case a long-range quenching interaction such as electronic energy transfer is suggested. Therefore, simple quenching experiments can provide useful information regarding the quenching mechanism. Hence, they are powerful tools for mechanistic studies of photochemical reactions.

6.4.1 Deviations from Stern-Volmer Equation

(i) Deviations from Stern-Volmer relationship may arise if static quenching is present. The dynamic and static quenching may be distinguished by the following mode of action :

$$\begin{array}{lll} \text{dynamic} & A^* + Q \rightleftharpoons (A.Q)^* \rightarrow & \left[\begin{array}{l} A + Q + h\nu' \\ A + Q + \Delta \end{array}\right. \\ & \quad h\nu \uparrow \qquad\quad \uparrow h\nu & \\ \text{static} & A + Q \rightleftharpoons (AQ) & \end{array}$$

If K_0 is the molar equilibrium constant for the formation of the complex in the ground state

$$K_0 = \frac{[AQ]}{[A][Q]} \tag{6.14}$$

then the fraction α, of incident light absorbed by the complex AQ is given by

$$\alpha = \frac{\epsilon' K_0 [Q]}{\epsilon + \epsilon' K_0 [Q]} \simeq \frac{\epsilon'}{\epsilon} K_0 [Q] = K[Q] \tag{6.15}$$

where ϵ and ϵ' are the molar extinction coefficients of A and AQ, respectively at the wavelength of excitation. If the complex is non-fluorescent, only the fraction $(1 - \alpha)$ is effectively used for fluorescence emission and for small α, and $\epsilon = \epsilon'$,

$$\frac{F_0 - F}{F} = (K_{SV} + K_0)[Q] + K_{SV} K [Q]^2$$

so that

$$\frac{F_0 - F}{F}\Big/[Q] = (K_{SV} + K_0) + K_{SV} K [Q] \tag{6.16}$$

where F_0 and F are fluorescence intensities in absence and in presence of quencher, respectively. In this expression, the quantity on the left is the apparent quenching constant and varies linearly with [Q] with a slope given by the product of collisional (K_{SV}) and static (K_0) quenching constants, which is identical with the equilibrium constant. The intercept corresponds to the sum of the two. The expression becomes more elaborate if the ground state complex also emits after excitation or fluorescent complexes are formed in the excited state (exciplex formation and emission).

(ii) In the absence of static quenching a small positive deviation from the

Stern-Volmer equation may be introduced by the transient component of dynamic quenching. If the quencher molecules are present near the fluorescent molecule at the moment of excitation, the initial quenching before a steady state is achieved, leads to the non-steady state term in the quenching expression. A sphere of transient quenching of volume v, may be defined as

$$v = 4\pi(pR)^2\sqrt{D\tau} \tag{6.17}$$

where R is the encounter distance, p the probability of quenching and τ the fluorescence lifetime and $D = D_A + D_B$. On this model,

$$\frac{F}{F_0} = e^{-K'[Q]} \tag{6.18}$$

where $K' = vN = 4\pi N(pR)^2\sqrt{D\tau} \times 10^{-3}$

The total quenching consisting of transient and steady-state contributions is given by

$$\frac{F}{F_0} = \frac{e^{-K'[Q]}}{1 + K_{SV}[Q]} \tag{6.19}$$

For weak quenching, the expression reduces to the Stern-Volmer type relation,

$$\frac{F}{F_0} = \frac{1}{1 + K''[Q]} \quad \text{where } K'' = K' + K_{SV}$$

and

$$\frac{F_0 - F}{F} = K''[Q]$$

$$= (K_{SV} + K')[Q]$$

$$= \frac{4\pi N R_{AB} D\tau}{1000}\left(1 + \frac{R_{AB}}{\sqrt{D\tau}}\right)[Q] \tag{6.20}$$

assuming $p = 1$. The transient quenching accounts for the deviation from the Stern-Volmer relation in a manner similar to the static quenching but K' is proportional to $\sqrt{D}$ and lifetime changes with quenching although the simple relationship, $F_0/F = \tau_0/\tau$ does not hold. For static quenching the constant is independent of D and the lifetime of the fluorescer molecule is unaffected.

6.5 CONCENTRATION DEPENDENCE OF QUENCHING AND EXCIMER FORMATION

The fluorescence intensity F per unit volume is proportional to the fractional light absorption I_a per unit volume per unit time.

$$F = \phi_f I_a$$

$$= \phi_f I_0 (1 - e^{-2.3 \epsilon Cl}) \tag{6.21}$$

where ϕ_f is the proportionality constant termed as the *quantum yield of fluorescence*. In an actual experiment, a factor G, which expresses the effect of geometry of the experimental setup on the measurements of fluorescence and absorption intensities must be considered. For dilute solutions, the term inside the bracket can be expanded. If identical experimental conditions are maintained fluorescence intensities and concentrations are linearly related according to the expression:

$$F = \phi_f\, 2.3\, I_0 \epsilon C l \qquad (6.22)$$

where the symbols have their usual meanings.

On the other hand, at high concentrations where all the incident light is completely absorbed, the exponential term in the expression (6.21) becomes negligible and $F = \phi_f I_0$. The fluorescence intensity should remain constant for any further increase in concentration. A plot of F vs C should be as given in Figure 6.3a. But for many compounds, the intensity

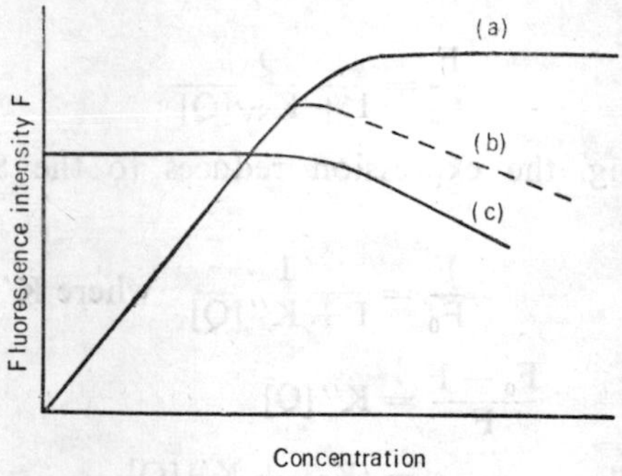

Figure 6.3 Plot of F vs C with (a) geometrical error, (b) without geometrical error, (c) ϕ_f vs C.

starts decreasing after reaching a critical concentration even if geometrical effects are avoided in the experimental setup (Figure 6.3b). The quantum yield ϕ_f, which is expected to be independent of concentration, starts decreasing at the same concentration (Figure 6.3c). The decrease is due to the quenching of fluorescence by molecules of the same kind and is known as *concentration quenching*. It follows the Stern-Volmer equation, [Q] representing the concentration of the fluorescer itself.

In some cases, simultaneously with the quenching of the normal fluorescence a new structureless emission band appears at about 6000 cm^{-1} to the red side of the monomer fluorescence spectrum (Figure 6.4). This phenomenon was first observed in pyrene solution by Förster and was explained as due to transitory complex formation between the ground and the excited state molecules since the absorption spectrum was not modified by increase in concentration. Furthermore, cryoscopic experiments gave negative results for the presence of ground state dimers. These short-lived excited state dimers are called *excimers* to differentiate them from

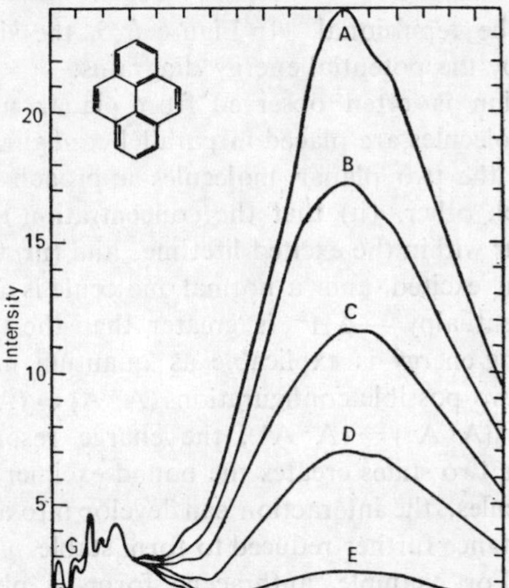

Figure 6.4 Variation in intensity of monomer and excimer emission spectra with concentration.

electronically excited dimeric ground state species. The excimers dissociate when they return to the ground state by emission, giving rise to a structureless envelope for the emission spectrum. This shows that the ground state

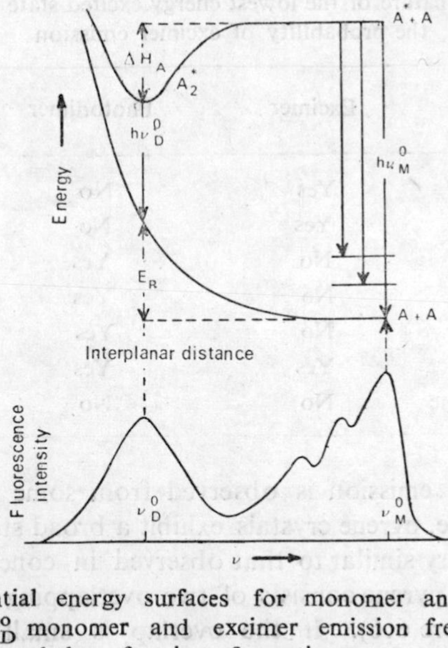

Figure 6.5 Potential energy surfaces for monomer and excimer emission. ν_M^o, ν_D^o monomer and excimer emission frequency, A_2^* excimer, ΔH_A enthalpy of excimer formation.

PE surface must be repulsional. In Figure 6.5, the situation is qualitatively illustrated by the potential energy diagrams.

Excimer emission is often observed from planar molecules when the two component molecules are placed in parallel configuration. The conditions are that (i) the two planar molecules approach within a distance $\simeq 0.35$ nm of each other, (ii) that the concentration is high enough for interaction to occur within the excited lifetime, and (iii) that the interaction energy between an excited and a normal molecule is attractive such that the excited state enthalpy $-\Delta H^*$ is greater than the thermal energy RT. The excimer binding energy is explicable as quantum mechanical interaction between the two possible configurations $(A^* A) \rightleftharpoons (AA^*)$, known as an exciton state and $(A^+ A^-) \rightleftharpoons (A^- A^+)$, the charge resonance state. The mixing between the two states creates the bound excimer state.

In some molecules, the interaction can develop into a stronger force and the interplanar distance further reduced to form stable *photodimers* through covalent bonds. For example, anthracene forms a photodimer and no excimer emission is observed, whereas some of its derivatives with bulky substituents which hinder close approach give excimer fluorescence. In 9-methylanthracene both photodimer formation and excimer emission is observed. 9, 10–diphenylanthracene neither forms a photodimer nor emits excimer fluorescence due to steric hindrance. These observations are tabulated in the Table 6.3, which shows that the nature of the excited state is also important.

TABLE 6.3

The nature of the lowest energy excited state and the probability of excimer emission

	Excimer	Photodimer	Lowest energy excited state
Pyrene	Yes	No	^1L_b
Naphthalene	Yes	No	^1L_b
Anthracene	No	Yes	^1L_a
Naphthacene	No	Yes	^1L_a
2-methylanthracene	No	Yes	^1L_a
9-methylanthracene	Yes	Yes	^1L_a
9, 10-diphenylanthracene	No	No	^1L_a

A similar type of emission is observed from some crystalline hydrocarbons. For example, pyrene crystals exhibit a broad structureless band in the visible region, very similar to that observed in concentrated solutions. The crystal lattice of pyrene consists of two overlapping molecules (Type B_1 crystal lattice) (Figure 6.6). If the overlap is small as in anthracene

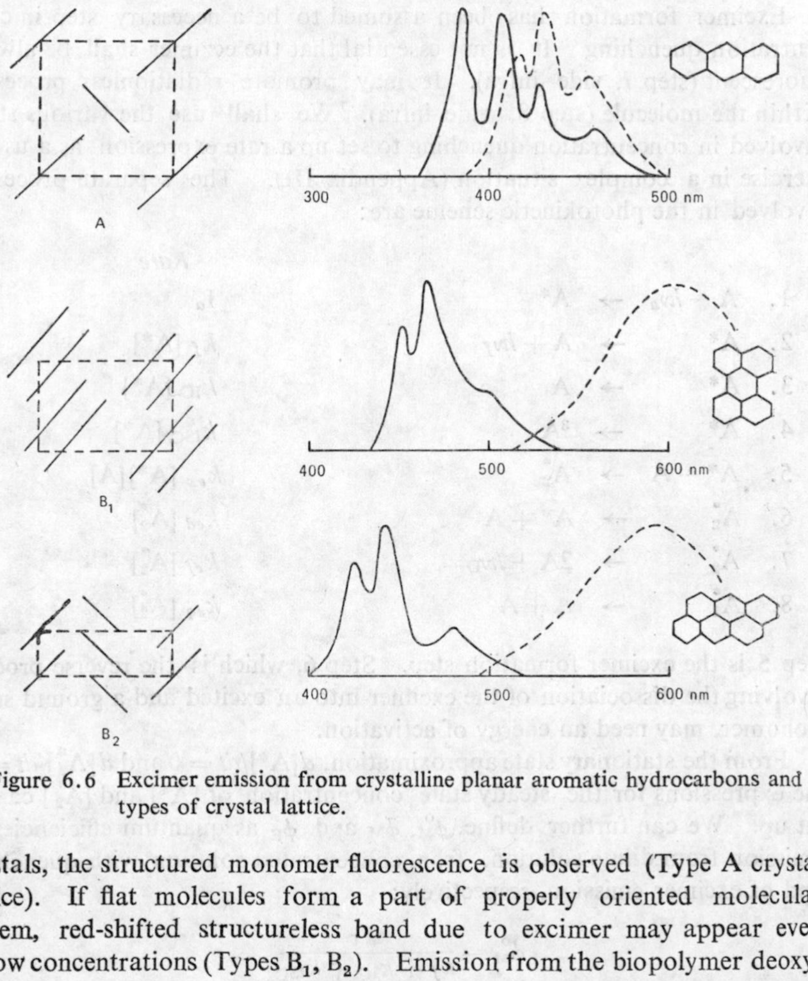

Figure 6.6 Excimer emission from crystalline planar aromatic hydrocarbons and types of crystal lattice.

crystals, the structured monomer fluorescence is observed (Type A crystal lattice). If flat molecules form a part of properly oriented molecular system, red-shifted structureless band due to excimer may appear even at low concentrations (Types B_1, B_2). Emission from the biopolymer deoxyribonucleic acid (DNA) is due to excimer formation between the constituent bases. Excimer emission is known to occur in inert gases like He and Xe also. The utility of xenon arc as a source of continuous high intensity radiation is due to this emission.

Besides the prompt excimer emission, *delayed excimer emission* has been observed. The mechanism is through *triplet-triplet annihilation process* (Section 5.9B):

$$^3A + {}^3A \rightarrow \underset{\displaystyle h\nu_{EX}}{\overset{}{A_2^*}\downarrow} \rightarrow \underset{\displaystyle h\nu_{DF}}{A^*\downarrow} + A$$

The excimer is formed as an intermediate which may dissociate into an excited singlet and a ground state singlet. The delayed excimer emission is a *biphotonic* process in contrast to the prompt excimer emission which is *monophotonic*.

Excimer formation has been assumed to be a necessary step in concentration quenching. It is not essential that the eczimer shall be always fluorescent (step 7, vide infra). It may promote radiationless processes within the molecule (step 8, vide infra). We shall use the various steps involved in concentration quenching to set up a rate expression as a useful exercise in a complex situation (Appendix III). The separate processes involved in the photokinetic scheme are:

				Rate
1.	$A + h\nu_a$	$\rightarrow$	A^*	I_a
2.	A^*	$\rightarrow$	$A + h\nu_f$	$k_f\,[A^*]$
3.	A^*	$\rightarrow$	A	$k_{IC}\,[A^*]$
4.	A^*	$\rightarrow$	^3A	$k_{ISC}\,[A^*]$
5.	$A^* + A$	$\rightarrow$	A_2^*	$k_{ex}\,[A^*][A]$
6.	A_2^*	$\rightarrow$	$A^* + A$	$k_{ed}\,[A_2^*]$
7.	A_2^*	$\rightarrow$	$2A + h\nu_D$	$k_{ef}\,[A_2^*]$
8.	A_2^*	$\rightarrow$	$A + A$	$k_{eq}\,[A_2^*]$

Step 5 is the excimer formation step. Step 6, which is the reverse process involving the dissociation of the excimer into an excited and a ground state monomer, may need an energy of activation.

From the stationary state approximation, $d[A^*]/dt = 0$ and $d[A_2^*]/dt = 0$, the expressions for the steady state concentration of $[A^*]$ and $[A_2^*]$ can be set up. We can further define, ϕ_M^o, ϕ_M and ϕ_E as quantum efficiencies of emission from dilute solution, from concentrated solution with quenching and of excimer emission, respectively:

$$\phi_M^o = \frac{k_f}{k_f + k_{IC} + k_{ISC}}$$

$$\phi_M = \frac{k_f a}{ab - k_{ex} k_{ed}\,[A]}$$

$$\phi_E = \frac{k_{ef} k_{ex}\,[A]}{ab - k_{ex} k_{ed}\,[A]}$$

where, $$a = k_{ed} + k_{ef} + k_{eq} = \frac{1}{\tau_E^o}$$

$$b = k_f + k_{IC} + k_{ISC} + k_{ex}\,[A] = \frac{1}{\tau_M}$$

and $$k_f + k_{IC} + k_{ISC} = \frac{1}{\tau_M^o}$$

We get

$$\frac{\phi_E}{\phi_M} = \frac{k_{ef} k_{ex}\,[A]}{k_f\,(k_{ed} + k_{ef} + k_{eq})} = K\,[A] \qquad (6.23)$$

and

$$\frac{\phi_M^\circ}{\phi_M} = 1 + \left(\frac{k_{ex}}{k_f + k_{IC} + k_{ISC}}\right)\left(1 - \frac{k_{ed}}{k_{ed} + k_{ef} + k_{eq}}\right)[A]$$

$$= 1 + k_{ex}.\tau_M^\circ (1 - k_{ed}\ \tau_E^\circ)[A] \qquad (6.24)$$

The final expression is in the form of the Stern-Volmer equation, where τ_M° is the monomer lifetime in dilute solution and τ_E° is assumed to be excimer lifetime in infinitely dilute solution.

For 50% quenching

$$C_h = [A]_{50\%} = \frac{1}{k_{ex}\,\tau_M^\circ (1 - k_{ed}\ \tau_E^\circ)} \qquad (6.25)$$

$$\frac{1}{C_h} = k_{ex}\ \tau_M^\circ\ (1 - k_{ed}\ \tau_E^\circ)$$

where C_h is the 'half-value concentration'. The Stern-Volmer expression can be rewritten as

$$\frac{\phi_M^\circ}{\phi_M} = 1 + \frac{C}{C_h} \qquad (6.26)$$

C is the concentration of the monomer and we have,

$$\phi_M = \phi_M^\circ \left(\frac{1}{1 + \dfrac{C}{C_h}}\right) \qquad (6.27)$$

Also

$$\phi_E = \phi_E^\infty \frac{1}{1 + \dfrac{C_h}{C}} \qquad (6.28)$$

ϕ_E^∞ is the maximum fluorescence efficiency of the excimer at high concentrations. $1/C_h$ has been found to be viscosity dependent confirming that k_{ex} indeed is diffusion-limited. Because of the diffusion-controlled kinetics, excimers are not observed in dilute solid solutions. Mixed excimers of the type (AB)* have been detected in liquid solutions.

In contrast to flat aromatic hydrocarbons, many dyes form ground state dimers and sometimes larger complexes. In such cases, modification of the absorption spectra takes place as a function of concentration and Beer's law is not obeyed. But fluorescence spectra are unchanged although lowered in intensity. The exciton interaction which provides part of the binding energy for excimer formation is also responsible for absorption changes in the ground state stable dimers of dyes. The concentration quenching constant can be calculated from the Stern-Volmer equation for not very concentrated solutions of dyes in which fractional absorption by the monomeric dye molecules is high. Some cyanine dyes exhibit both monomer and dimer emission, e.g. 1, 1-diethyl-2, 2-pyridocyanine iodide. In general, aggregation of dyes inhibits fluorescence emission but enhances low temperature (77K) phosphorescence. This increase is due to increase in intersystem crossing probability because of reduction in $S_1 - T_1$ energy gap brought about by exciton splitting of singlet S_1 state (Section 6.6.12). Some distinctive properties of excimers and excited state dimers are tabulated in Table 6.4. The term excimer was coined by B. Stevens to differentiate it from excited state of dimeric species.

TABLE 6.4
Distinctive properties of excimer and excited state dimers

Excimer	Excited state dimers
1. Dissociated in the ground state.	Stable in the ground state at room temperature.
2. Dimerization follows act of light absorption. $A + A \xrightarrow{h\nu} A^* + A \rightarrow (AA^*)$	Dimerization precedes act of light absorption. $A + A \rightarrow (AA) \xrightarrow{h\nu} (AA)^*$
3. Radiative transition weak.	Radiative transition strong.
4. No absorption observed.	Intensity of dimer absorption band increases with concentration.
5. Emission characteristics.	Normally nonfluorescent. If fluorescent,
(i) Broad structureless band shifted to about 6000 cm^{-1} to red of the monomer fluorescence band.	(i) Give sharp emission on the long wave side of corresponding dimer absorption.
(ii) Emission concentration dependent.	(ii) Usually independent of concentration.
6. Observed for planar aromatic hydrocarbons.	6. Observed for ionic forms of dyes in solvents of high dielectric constant.

6.6 QUENCHING BY FOREIGN SUBSTANCES

Physical quenching by foreign added substances generally occur by two broad based mechanisms:

(a) Charge transfer mechanism.

(b) Energy transfer mechanism.

6.6.1 Charge Transfer Mechanism: Exciplex Formation and Decay

The formation of excimers between ground and excited states of similar molecules has already been discussed. When two dissimilar molecules collide, attractive tendencies are usually greater, depending on polar and polarizability properties. Interaction involves some degree of charge transfer, and the frequently formed complexes between excited fluorescent molecules and added foreign molecules are called *exciplexes*. Absorption spectra remain unchanged, in contrast to ground state interactions, because of the short life of the complexes, unless high concentrations can be built up by the use of very intense light.

Molecules with heavy atoms act as quenchers by an exciplex mechanism. Quantum considerations show that they facilitate change of electron spin, so that exciplex dissociation occurs with energy degradation via a triplet level :

$$^1A^* + Q \rightarrow (^1AQ)^* \rightarrow {}^3A + Q \rightarrow {}^1A + Q$$

Examples of such quencher molecules are xenon, bromobenzene, bromides and iodides and some rare earth compounds.

In absence of the heavy atom effect and where the extent of charge transfer is large enough, exciplex lifetimes may be long enough for them to degrade by light emission. More commonly, however, dissociation may occur via an ion-pair complex, as was first suggested by Weiss.

$$^1A^* + Q \rightarrow (^1AQ)^* \rightarrow (A^{\pm} \ldots Q^{\mp}) \rightarrow {}^1A + Q$$

Solvation is an important factor affecting such reactions. In some systems recognizably free solvated ions may be detected, as when deoxygenated solution of perylene in a polar solvent containing sufficient concentration of dimethylamine to give 80% quenching is subjected to flash spectroscopic studies. Absorption due to both perylene mononegative ion and perylene triplets is observed. The quenching efficiency increases with decrease in the ionization potential of the amines (Table 6.5).

The solvent-shared ion-pair character of the exciplex increases with the increase in solvent dielectric constant and ionization potential of the donor and electron affinity of the acceptor.

The free energy change ΔG involved in the actual electron transfer process: encounter complex $\rightarrow$ ion-pair, can be calculated according to the expression 6.29.

TABLE 6.5

Quenching of acridine fluorescence in aqueous solution (0.03 M NaOH) at 25°C.

Quencher	Ionization potential I_D (ev)	D^a 10^2 cm^2 s^{-1}	K_{sv} (l mol^{-1})	pR (Å)
NH_3	10.16	3.10	0.38	0.012
CH_3NH_2	8.97	2.35	2.45	0.10
$i-C_3H_7.NH_2$	8.72	1.80	2.4	0.13
$(CH_3)_2.NH$	8.24	2.00	21.5	1.07
$(CH_3)_3.N$	7.84	1.80	25.2	1.40

[a]Sum of the diffusion coefficients of fluorescence and quencher.

$$\Delta G = 23.06\left[E(D/D^{+}) - E(A^{-}/A) - \frac{e^2}{\varepsilon a}\right] - \Delta E_{00} \quad (6.29)$$

where E (D/D^+) and E (A^-/A) are, respectively polarographic oxidation potential of the donor and reduction potential of the acceptor, $e^2/\varepsilon\, a$ is the free energy change gained by bringing two radical ions to encounter distance a in the solvent of dielectric constant ε, ΔE_{00} is the electronic excitation energy. If fast proton transfer follows electron transfer then quenching may be said to involve H-atom transfer.

The dynamical processes of formation and decay of exciplexes are schematically presented in Figure 6.7. An intermediate nonrelaxed electron transfer state $(^1D^+ A^-)$ is produced by electron transfer in the encounter complex which relaxes to the exciplex state after readjustment of coordinates and solvent orientation. Triplet formation and ion-pair

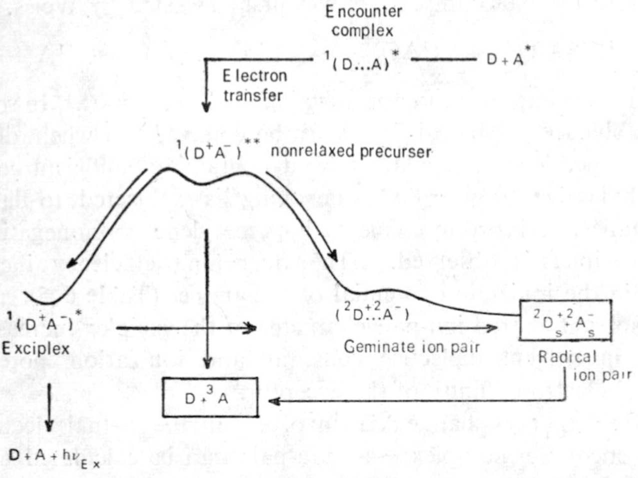

Figure 6.7 The dynamical processes of formation and decay of exciplex by electron donor-acceptor mechanism.

formation may occur in competition to exciplex formation and also as a mode of relaxation of exciplex. The geminate ion-pair $(^2D^+ \ldots {}^2A^-)_s$ lead to radical ion-pair which decays via triplet state.

An example of exciplex emission is the system anthracene or biphenyl and an easily oxidized amine such as NN′-diethylaniline. In nonpolar solvents, fluorescence of the hydrocarbon is replaced by a broad and structureless emission spectrum lying at a longer wavelength. It is associated with the formation of a fluorescent charge transfer complex in the excited state. The absorption spectra are unchanged showing absence of complexing in the ground state.

$$^1A^* + \phi N(C_2H_5)_2 \rightarrow (\phi N(C_2H_5)_2^{+} . A^-)$$
$$\rightarrow A + \phi N(C_2H_5)_2 + h\nu_f$$

In polar solvents, such as acetonitrile, the CT-fluorescence disappears due to solvation of ion-pair to form solvent separated radical ions. In strongly polar solvent, ion-pair formation is very fast and the rate of fluorescence quenching is diffusion controlled:

$$A^* + Q \rightarrow (A^- Q^+) \rightarrow (A^-)_{solv} (Q^+)_{solv}$$

If definite stoichiometry is maintained in the exciplex formation, an *isoemissive point* similar to *isosbestic point* in absorption may be observed. An interesting example of intra-molecular exciplex formation has been reported for 9-methoxy-10-phenanthrenecarboxanil. The aniline group is not necessarily coplanar with the phenanthrene moiety but is oriented perpendicular to it. The *n*-electron located on its N-atom interacts with the excited π-electron system and an intramolecular exciplex with T-bone type structure is formed in rigid glassy medium where rotation is restricted. Temperature dependence of fluorescence of this compound in methylcyclohexane-isopentane (3:1) solvent shows a definite isoemissive point (Figure 6.8). As the solvent melts and movement is restored to the molecule, structured fluorescence reappears.

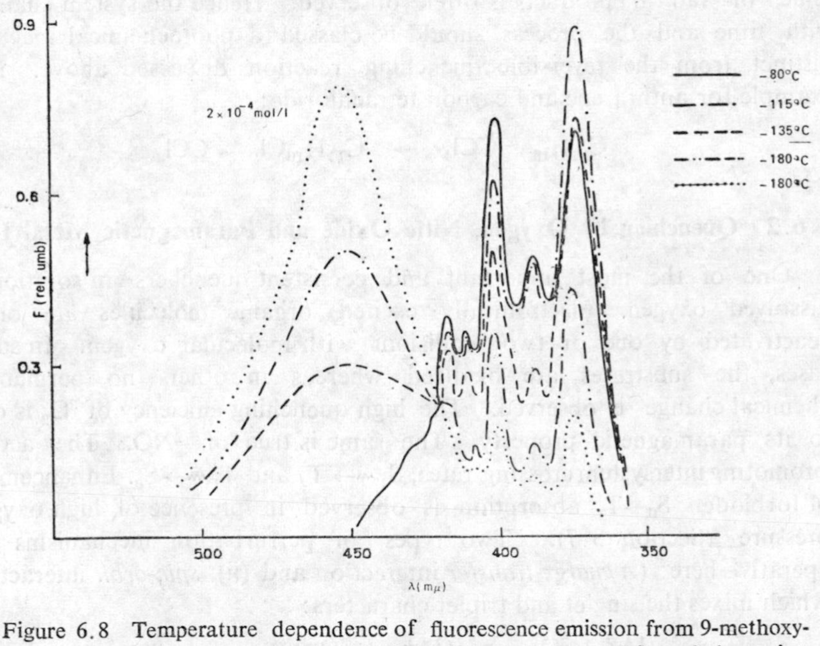

Figure 6.8 Temperature dependence of fluorescence emission from 9-methoxy-10-phenanthrenecarboxanil. Exciplex emission and isoemissive point.

The fluorescence of a number of dyes is quenched by ions in the order:

$$I^- > CNS^- > Br^- > Cl^- > C_2O_4^{=2} > SO_4^{=2} > NO_3^- > F^-$$

This correlates with the increasing ionization potential which shows that the fluorescence quenching efficiency of these ions is related to the ease of charge transfer from the ion.

The quenching effects can be produced by *solvents* in absence of any other added quenchers. Polar solvents may involve chemical interactions of the type

$$A^* + S \rightarrow (A^- S^+)^* \rightarrow A + S$$

where S represents the solvent molecule. Most of the energy degradation is believed to occur via the triplet state during the encounter complex formation. Stronger the interaction greater is the energy loss. In non-polar solvents dispersion force interactions are important. The reverse of this process, that is charge transfer from the fluorescer to the solvent may also occur. It is not a single solvent molecule which receives the electron but the complete solvent shell. The *solvated electrons* can be identified by flash spectroscopic techniques.

Some solvents containing heavy atoms can induce enhancement of phosphorescence at the expense of fluorescence, e.g. ethyl iodide, nitromethane, CS_2 (external heavy atom effect). Irreversible conversion to ionic or radical products is often observed. Hence the system changes with time and the process should be classed a photochemical reaction distinct from the reversible quenching reactions discussed above. For example for anthracene and carbon tetrachloride:

$$C_{14}H_{10} + CCl_4 \rightarrow C_{14}H_{10}Cl. + \cdot CCl_3$$

6.6.2 Quenching by Oxygen, Nitic Oxide and Paramagnetic Metal Ions

One of the most important and persistent quenchers in solution is dissolved oxygen. Electronically excited organic molecules are often deactivated by one or two collisions with molecular oxygen. In some cases, the substrates are oxidized whereas in others no permanent chemical change is observed. The high quenching efficiency of $^3O_2$ is due to its paramagnetic property. The same is true of ^2NO. They act by promoting intersystem crossing rates, $S_1 \rightsquigarrow T_j$ and $T_1 \rightsquigarrow S_0$. Enhancement of forbidden $S_0 \rightarrow T_1$ absorption is observed in presence of high oxygen pressure (Section 3.7). Two types of perturbation mechanisms are operative here: (i) *charge-transfer* interaction and (ii) *spin-orbit* interaction which mixes the singlet and triplet characters:

$$^1A^* + {^3O_2} \rightarrow ({^2A^+} \ldots {^{+2}O_2^-}) \searrow$$
$$\qquad\qquad\qquad\qquad\qquad\qquad {^3A} + {^3O_2}$$
$$^1A^* + {^3O_2} \rightarrow ({^1A^*} \ldots {^3O_2}) \nearrow$$

Because of the spin exchange requirement, a close approach of the fluorescer and the quencher is required. Hence oxygen quenching is diffusion controlled. The increase in the rates of intersystem crossing has been observed for paramagnetic ions such as Ni^{++}, Co^{++} and some rare earth ions.

6.6.3 Quenching of the Triplet Energy States

Normally the quenching of the triplet states by the collisional process

$$\begin{array}{cccc} {}^3A^*(\uparrow\uparrow) + {}^1Q(\uparrow\downarrow) & \rightarrow & A(\uparrow\downarrow) + Q(\uparrow\downarrow) \\ (S=1) \quad (S=0) & & (S=0) \quad (S=0) \end{array}$$

is forbidden by spin selection rules and is therefore inefficient. However, if the quencher is in a spin quantum state greater than zero ($S_Q > 0$), i.e. it is paramagnetic, and fluorescer in spin state ($S_A \geqslant 1$), then the collisional process,

$$\begin{array}{ccccccc} A^* & + & Q & \rightarrow & A & + & Q \\ (S_A = x) & & (S_Q = y) & & (S_A = x-1) & & (S_Q = y) \end{array}$$

is spin allowed. The paramagnetism of the perturbing molecule or ion helps to conserve the total spin, in terms of Wigner's spin conservation rule.

The quenching efficiency by electron transfer appears to decrease with decrease in availability of electrons in the quencher for loose bonding in the complex. Such quenchers can be classified into three groups (Table 6.6).

TABLE 6.6
Types of quenchers and the rate constants for quenching of triplet energy states

	Rate constant for quenching k_q 1 mol^{-1} s^{-1}	Type of quencher
Gr. I	$\simeq 10^{10}$	O_2, NO, aromatic triplets
Gr. II	$\simeq 5 \times 10^7$	Metal ions of transition series
Gr. III	$\simeq 2 \times 10^5$	Ions of lanthanide series

The classification follows the extent of overlap of the electron cloud of the quencher with the fluorescer molecule. The overlap of *p*-orbital of $O_2 >$ *d*-orbitals of solvated ions of transition metals $>$ *f*-orbitals of solvated rare earth ions.

6.6.4 Electronic Energy Transfer Mechanisms

The electronic energy transfer mechanism has become one of the most useful processes in photochemistry. It has wide applications as a mechanistic tool and in photochemical synthesis. It allows photosensitization of physical and chemical changes in the acceptor molecule by the electronically excited donor molecule. The process may be defined by the following two steps:

$$D + h\nu \rightarrow D^* \qquad \text{light absorption by donor}$$

$$D^* + A \rightarrow D + A^* \qquad \text{energy transfer donor} \rightarrow \text{acceptor}$$

The electronically excited D* (donor) is initially formed by direct light absorption. This can transfer the electronic energy to a suitable acceptor molecule A (acceptor) resulting in simultaneous quenching of D* and electronic excitation of A to A*. The transfer occurs before D* is able to radiate and hence is known as *nonradiative transfer of excitational energy*. The A* molecule, thus excited indirectly, can undergo various photophysical and photochemical processes. Such processes are called *photosensitized reactions*. In these reactions, the initial light absorbing species remains unchanged. One of the most important sensitized photochemical reactions is photosynthesis by plants. The green chlorophyll molecules of the leaves are the light absorbers. They initiate reaction between CO_2 and H_2O to produce carbohydrates as the stable products. Examples of many such sensitized photoprocesses are observed in photobiology and photochemistry. Photosensitized oxidation of proteins and nucleic acids is termed *photodynamic action*. Hammond and his coworkers have developed a very useful technique using the principles of energy transfer between molecules for synthetic and mechanistic studies in organic photochemistry.

The *nonradiative energy transfer* must be differentiated from *radiative transfer* which involves the *trivial* process of emission by the donor and subsequent absorption of the emitted photon by the acceptor:

$$D^* \rightarrow D + h\nu$$

$$A + h\nu \rightarrow A^*$$

It is called *trivial* because it does not require any energetic interaction between the donor and the acceptor. It is merely reabsorption of fluorescence radiation in accordance with Beer's Law and shows r^{-2} dependence on donor-acceptor distance. Although called trivial, it causes radiation imprisonment and can be important factor to be considered in fluorescence measurements. It may introduce error and distort emission spectrum by absorbing only that portion which overlaps its absorption spectrum. It is specially troublesome in studies on concentration quenching.

For quenching by energy transfer mechanism, the quencher must have suitable energy levels, singlet or triplet, near or below the energy level of the donor molecule. Such a transfer has the greatest probability if there is an approximate resonance between the donor and the acceptor energy levels.

In general, two different types of mechanisms are postulated for the nonradiative energy transfer phenomenon:

(i) long range transfer by dipole-dipole coupling interactions, and
(ii) short range transfer by exchange interactions.

6.6.5 Donor-Acceptor Interactions in Energy Transfer Processes

Radiationless excitation transfer occurs only when $(D^* + A)$ initial state is in or near resonance with $(D + A^*)$ final state and there is a suitable donor-acceptor interaction between them. The rate of transfer, $k_{D^* \to A}$, is given by the time-dependent perturbation theory,

$$k_{D^* \to A} = \frac{4\pi^2}{h} \rho_E V^2 = \frac{4\pi^2}{h} \rho_E \beta_{el}^2 F \tag{6.30}$$

where the last term F is the familiar Franck-Condon overlap factor.

This expression, it may be recalled, is similar to the one obtained for *intramolecular* radiationless transfer rate for internal conversion and intersystem crossing (Section 5.2.1). For *intermolecular* cases

$$F = \Sigma(\int \chi_{D^*} \chi_D \, d\tau_v \int \chi_{A^*} \chi_A \, d\tau_v)^2 \tag{6.31}$$

where χ_{D^*}, χ_D, χ_{A^*} and χ_A are vibrational wave functions of D^*, D, A^*, A respectively. ρ_E is the density of energy states, and hence, ρ_E F is related to the overlap between absorption and emission spectra. β_{el} is the total interaction energy parameter and includes all the electrostatic interactions of electrons and nuclei of the doner with those of the acceptor plus the electron exchange interactions. Thus, the total interaction can be expressed as a sum of '*Coulomb*' and '*exchange*' terms. The Coulomb term induces long range transfer whereas the exchange term is effective for short range **only.**

If Ψ_i and Ψ_f are the wave functions for the initial and the final states, and $\mathcal{H}$ is the perturbation operator, the electronic interaction energy β_{el} is given by

$$\beta_{el} = \int \Psi_i \mathcal{H} \Psi_f \, d\tau = \int \psi_{D^*} \psi_A \mathcal{H} \psi_D \psi_{A^*} \, d\tau. \tag{6.32}$$

When properly antisymmetrized for two electrons only, the initial and the final states may be expressed as

$$\Psi_i = \frac{1}{\sqrt{2}} \left\{ \psi_{D^*}(1)\, \psi_A(2) - \psi_{D^*}(2)\, \psi_A(1) \right\}$$

$$\Psi_f = \frac{1}{\sqrt{2}} \left\{ \psi_D(1)\, \psi_{A^*}(2) - \psi_D(2)\, \psi_{A^*}(1) \right\}$$

and $\beta_{el} = \int (\psi_{D^*}(1)\, \psi_A(2) - \psi_{D^*}(2)\, \psi_A(1) \mathcal{H} (\psi_D(1)\, \psi_{A^*}(2) - \psi_D(2)\, \psi_{A^*}(1) \, d\tau$

$$= \int \psi_{D^*}(1)\, \psi_A(2) \mathcal{H} \psi_D(1)\, \psi_{A^*}(2) \, d\tau - \int \psi_{D^*}(1)\, \psi_A(2) \mathcal{H} \psi_D(2)\, \psi_{A^*}(1) \, d\tau \tag{6.33}$$

= Coulomb term + exchange term.

6.6.6 Long Range Transfer by Coulombic Interactions

Early observations on long range energy transfer were made by Franck and Cario in a mixture of mercury and thallium vapour. When a mixture of Hg-vapour (0.2 mm) and thallium vapour (0.25 mm) was irradiated by 253.7 nm radiation, absorbed only by Hg atom thereby promoting it to Hg (6^3P_1) energy level, the emitted radiation, in addition to mercury emission lines, consisted of lines due to Tl also. Similar *mercury sensitized emission* of Na-vapour was noted in a mixture of Hg and Na-vapour. The Na-line at 442 nm, which is very weak normally, is much enhanced in presence of Hg-vapour. When a trace of N_2 gas is added, which is known to induce transfer from Hg (6^3P_1) to Hg (6^3P_0), 475 nm line of Na is enhanced. These observations can be rationalized with the help of energy level schemes for Hg and Na atoms. The transition $6^3P_1 \rightarrow 6^1S_0$ in Hg atom requires 469.4 kJ mol^{-1} of energy which is of nearly equal energy to the transition $9^2S_{1/2} \leftarrow 3^2S_{1/2}$ in Na atom. A resonance condition is established as in a coupled system. The excitational energy is easily transferred from Hg to Na atom making up the small energy deficit (162 cm^{-1}) from the surrounding. Once in the 9S state, Na ($9^2S_{1/2} \rightarrow 3^2P_{1/2}$) transition emitting at 442 nm is the most probable one which is then followed by Na ($3^2P_{1/2} \rightarrow 3^2S_{1/2}$) transition at 594.6 nm. In presence of N_2, 6^3P_1 is transferred to 6^3P_0 by $E \rightarrow V$ transfer mechanism. Now the Hg ($6^3P_0 \rightarrow 6^1S_0$) transition is in resonance with Na ($7^2S_{1/2} \leftarrow 3^2S_{1/2}$) transition sensitizing the emission of Na ($7^2S_{1/2} \rightarrow 3^2P_{1/2}$) at 475 nm. The energy deficit 560 cm^{-1} is made up from the relative energies of translation of the colliding partners.

At the pressure prevailing in the mixture, the average donor-acceptor distance may be as large as 100 Å or more. It may appear surprising that such long range transfer is possible at all during the short lifetime of the excited molecules, which is usually of the order of 10^{-8} s. But if we consider that the period of orbital motion is 10^{-15} s, the lifetime of 10^{-8} s is comparatively long. Since some kind of coupling interactions are necessary, the time-dependent perturbation theory predicts that the probability of energy transfer should be porportional to the interaction time t. A comparatively weak interaction between distant atoms or molecules may be sufficient for excitation transfer provided some kind of resonance condition is fulfilled. A rough analogy of such a coupled system can be drawn with two pendulums of similar frequencies, hanging from a horizontal string. The vibration of one pendulum travels to the other via small disturbances of the string (Figure 6.9). Theoretical considerations show that a critical interaction distance $R = \lambda/2\pi \simeq 1000$ Å for the visible radiation may be defined for the energy transfer process in an ideally coupled pendulum. This is a large distance on the atomic scale and corresponds to a concentration of about 0.5×10^{-6} $moles^{-1}$. The idea of resonance transfer between atoms can now be easily extended to molecular systems. Fluorescence sensitization has been observed in gaseous molecular system like mixtures of aniline and indigo.

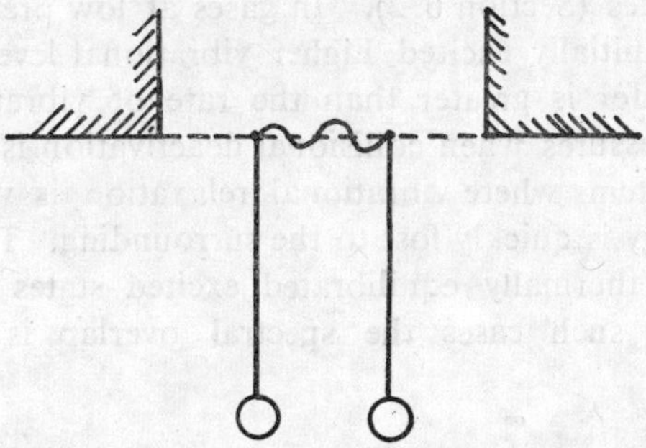

Figure 6.9 A pair of coupled pendulums in resonance.

The concept of long range energy transfer in solution was first proposed by Perrin to explain concentration depolarization of fluorescence. Molecules have characteristic vibrational energy levels also, and hence they can provide a larger number of approximately resonant paths. The greater the number of such resonant paths, the larger will be the transfer probability (Figure 6.10). Therefore, the probability of transfer is a function of the extent of overlap between emission spectrum of donor and absorption spectrum of acceptor, i.e. the density of states ρ_E in the initial

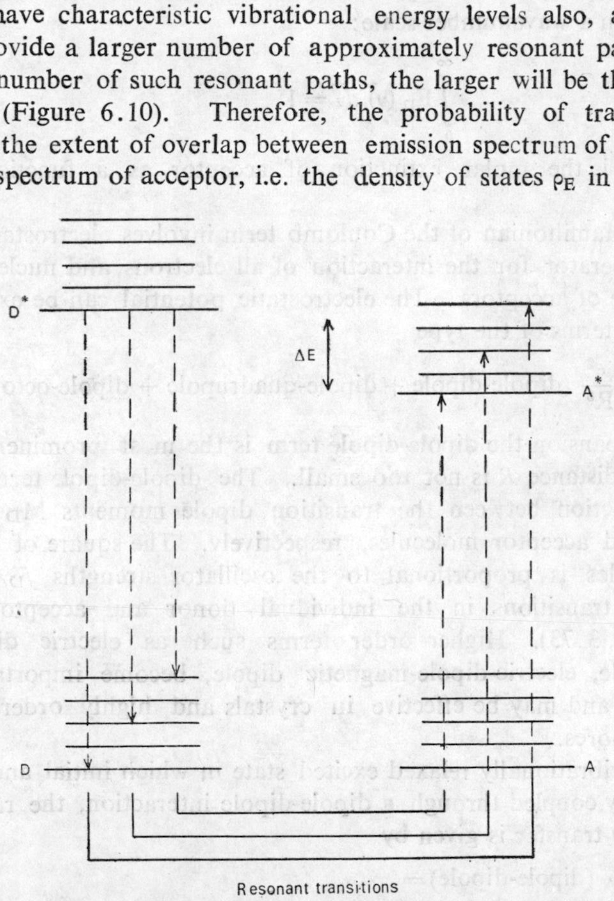

Figure 6.10 Energy level scheme for resonance energy transfer between a donor (D) and an acceptor (A).

and the final states (Section 6.2). In gases at low pressures, transfer can take place from initially excited higher vibrational levels of the donor, if the rate of transfer is greater than the rate of vibrational deactivation. But at higher pressures when collisional deactivation is the faster step and in condensed systems where vibrational relaxation is very efficient, excess vibrational energy is quickly lost to the surrounding. The donor molecules are obtained in thermally equilibrated excited states with a Boltzmann distribution. In such cases the spectral overlap is expressed by the integral,

$$J = \int_0^\infty F_D(\bar{\nu})\, \varepsilon_A(\bar{\nu})\, d\bar{\nu} \tag{6.34}$$

Here $F_D(\bar{\nu})$ is the spectral distribution of donor emission expressed in number of quanta emitted per unit wave number interval and normalized to unity on a wavenumber scale:

$$\int_0^\infty F_D(\bar{\nu})\, d\bar{\nu} = 1$$

and $\varepsilon_A(\bar{\nu})$ is the molar extinction of acceptor as a function of wavenumber.

The Hamiltonian of the Coulomb term involves electrostatic potential energy operator for the interaction of all electrons and nuclei of donors with those of acceptors. The electrostatic potential can be expanded into multipole terms of the type

$$V \simeq \frac{e^2}{R^3} \text{ (dipole-dipole + dipole-quadrupole + dipole-octopole, etc.)}$$

In this expansion the dipole-dipole term is the most prominent if donor-acceptor distance R is not too small. The dipole-dipole term represents the interaction between the transition dipole moments M_D and M_A of donor and acceptor molecules, respectively. The square of these transition dipoles is proportional to the oscillator strengths f_D and f_A for radiative transitions in the individual donor and acceptor molecules (equation 3.73). Higher order terms such as electric dipole-electric quadrupole, electric-dipole-magnetic dipole, become important at close approach and may be effective in crystals and highly ordered array of chromophores.

For vibrationally relaxed excited state in which initial and final states are weakly coupled through a dipole-dipole interaction, the rate constant for energy transfer is given by

$$k_{D^* \rightarrow A} \text{ (dipole-dipole)} =$$

$$k_{ET} = \frac{4\pi^2}{h} C\rho_E \left\{ \frac{M_D . M_A}{R^3}\, \theta(\theta) \right\}^2 \Sigma \left(\int \chi_{D^*} \chi_D\, d\tau \int \chi_{A^*} \chi_A\, d\tau \right)^2 \tag{6.35}$$

where the last term is the Franck-Condon factor and the sum is over all possible final resonant states, C is a constant and $\theta(\theta)$ gives the angular dependence of the dipole-dipole interaction. Appropriate transformation of the equation leads to an expression in which all terms are related to experimentally obtainable parameters. Thus, we have

$$k_{ET} = \frac{4\pi^2}{h} C' \frac{f_D f_A}{R^6 \bar{\nu}^2} \theta(\theta)^2 J \tag{6.36}$$

where C′ is a constant, $\bar{\nu}$ is an 'average' wave number for the transitions, f_D is the oscillator strength of the donor in the emission mode, f_A that of the acceptor in the absorption mode, and J is the spectral overlap integral. In this expression, $f_D \propto \phi_D/\tau_D \bar{\nu}^2$.

The final form of the expression for the rate constant k_{ET}, for energy transfer in solution, expressed in the units of litre mol^{-1} s^{-1} is

$$k_{ET} = \frac{9000 \ln 10}{128\pi^5 N} \frac{\kappa^2 \phi_D}{n^4 \tau_D R^6} \int_0^\infty F_D(\bar{\nu}) \varepsilon_A(\bar{\nu}) \frac{d\bar{\nu}}{\bar{\nu}^4} \tag{6.37}$$

where, N is Avogadro's number, ϕ_D and τ_D are fluorescence efficiency and actual radiative lifetime respectively of the donor, κ is an orientation factor relating the geometry of the donor-acceptor dipoles, n is refractive index of the medium to correct for the field effect. This expression was first derived by Th. Förster in 1946. It applies strictly to those cases where the donors and acceptors are well separated (by at least 20Å) and assumed to be immobile. The orientation factor κ, which is determined by the spatial orientation of the two transition moment vectors, for a fixed geometry is given as

$$\kappa^2 = \cos\theta_{DA} - 3\cos\theta_D \cos\theta_A \tag{6.38}$$

where θ_{DA} is the angle between the two vectors and θ_D and θ_A the angles sustained by the donor and acceptor vectors, respectively, with the line joining the centre of gravity of the two vectors; $\kappa^2 = 2/3$ for random geometry as obtained in solutions and in gaseous systems. In crystalline solids, the factor depends on the orientation of molecules in the crystal lattice.

The expression (6.37) can be written as

$$k_{ET} = \frac{1}{\tau_D}\left(\frac{R_0}{R}\right)^6 \tag{6.39}$$

where R_0 is known as the *critical transfer distance* and is characteristic for a given donor-acceptor pair, being a function of their spectral properties,

$$R_0^6 = \frac{(9000 \ln 10)\, \kappa^2 \theta_D}{128\pi^5 n^4 N} \int F_D(\bar{\nu}) \varepsilon_A(\bar{\nu})\, d\bar{\nu}/\bar{\nu}^4 \tag{6.40}$$

It can be calculated from a knowledge of overlap integral J and quantum

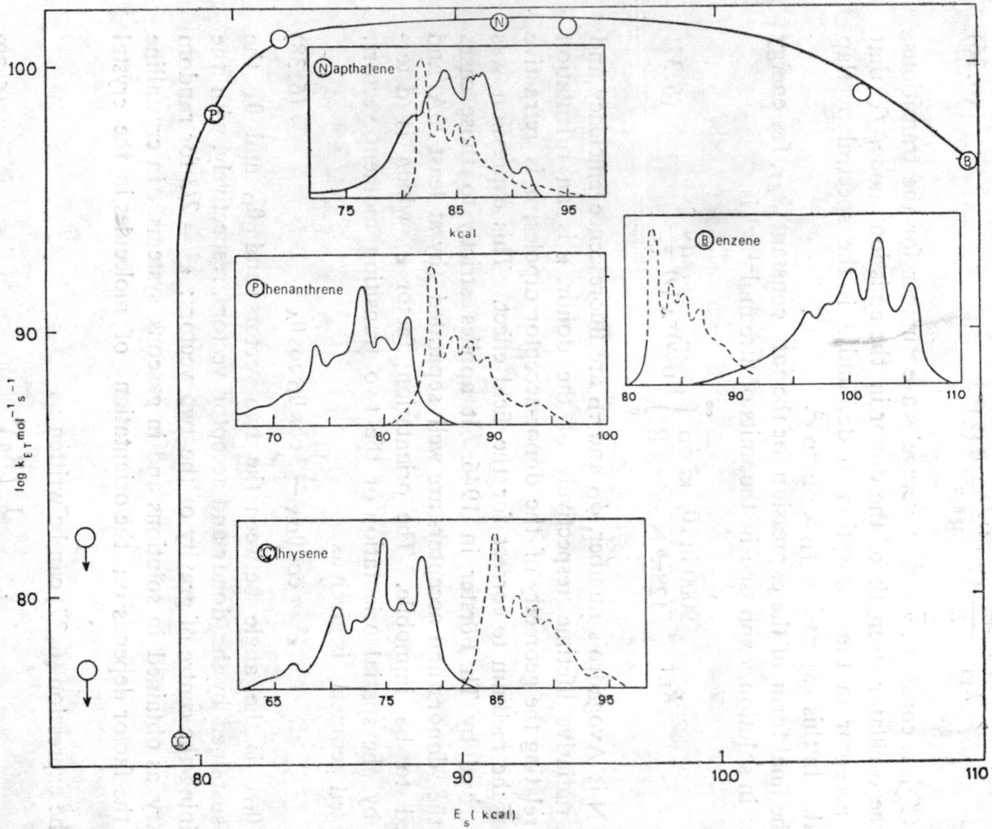

Figure 6.11 The rate of energy transfer k_{ET} as a function of overlap integral between the emission spectra of the donor and the absorption spectra of the acceptor and the rate constant for energy transfer. In each case donor (dotted line) is benzene and acceptor (solid line) are aromatic hydrocarbons of different singlet state energy E_s. The energy scale is expressed in kcal mol^{-1}.
[Ref. P.S. Engel and C. Steel: Acc. Chem. Res. 6 (1973) 275]

yield ϕ_D of donor fluorescence. The greater the spectral overlap, the more efficient is the energy transfer (Figure 6.11). The acceptor absorption should be on the red end of the donor emission. The efficiency of transfer E, is given as

$$E = \frac{k_{ET}}{k_{ET} + k_f + k_i} = \frac{1}{1 + (R/R_0)^6} \tag{6.41}$$

R_0 can also be defined as the distance of separation between donor and acceptor at which the probability of transfer is equal to the probability of decay of the donor :

$$k_{ET}\,[D^*]\,[A]_0 = k_D\,[D^*] = \frac{1}{\tau_D}\,[D^*]$$

and
$$k_{ET}\,\tau_D = \frac{1}{[A]_0} \tag{6.42}$$

The average intermolecular separation depends on the concentration of the solution. The *critical concentration of the acceptor* A_0, is related to R_0 by the expression

$$\frac{1}{[A_0]} = \frac{4}{3}\pi R_0^3 \tag{6.43}$$

where $[A_0]$ is expressed in molecules per cm³. R_0 can be identified with the radius of a sphere with D in the centre and containing one molecule of A only. An experimental value of R_0 (in Å) may be calculated from the half-quenching concentration data. $[A]_{1/2}$, the concentration of the acceptor at which the donor solution is 50% quenched is inversely related to K_{SV} $(= k_{ET}\,\tau_D)$. Expressing the concentrations in moles per litre we have from (6.43)

$$R_0 = \left[\frac{3 \times 1000}{4\pi\, N\, [A]_{1/2}}\right]^{1/3} = \frac{7.35}{[A]_{1/2}^{1/3}}\ \text{Å} \tag{6.44}$$

Because of the Stokes shift for vibrationally relaxed systems (the rate of transfer $<$ the rate of vibrational relaxation), transfer between like molecules is less efficient than that between unlike molecules when acceptor is at a lower energy level (exothermic transfer). No transfer is expected if the acceptor level is higher than the donor level. If (i) the acceptor transition is strong ($\varepsilon_{max} \simeq 10{,}000$), (ii) there is significant spectral overlap, and (iii) the donor emission yields lie within 0.1 — 1.0, then R_0 values of 50-100 Å are predicted.

Selection rules for the electronic energy transfer by dipole-dipole interactions are the same as those for corresponding electric dipole transitions in the isolated molecules. The spin selection rule requires that the total multiplicity of the donor and the acceptor, prior to and after the act of transfer, must be preserved. This implies that $M_{D^*} = M_D$ and $M_A = M_{A^*}$ where M's denote the multiplicity of the states (Section 2.5.1).

The two spin-allowed processes are:

Singlet-singlet transfer :

$$D^* \text{ (singlet)} + A \text{ (singlet)} \rightarrow D \text{ (singlet)} + A^* \text{ (singlet)}$$

Singlet-triplet transfer :

$$D^* \text{ (singlet)} + A \text{ (triplet)} \rightarrow D \text{ (singlet)} + A^{**} \text{ (triplet)}$$

Bowen and his group have unequivocally established the premises of Förster's expression by carefully designed experiments in which all the other modes of transfer such as trivial processes of reabsorption, donor-acceptor complex formation, collisional transfer and excitation migration among donor molecules were avoided (Figure 6.12). Two most important criteria for long range energy transfer processes are that (i) the transfer should occur at distances much greater than the kinetic collision radii, and (ii) the transfer efficiency should be independent of viscosity.

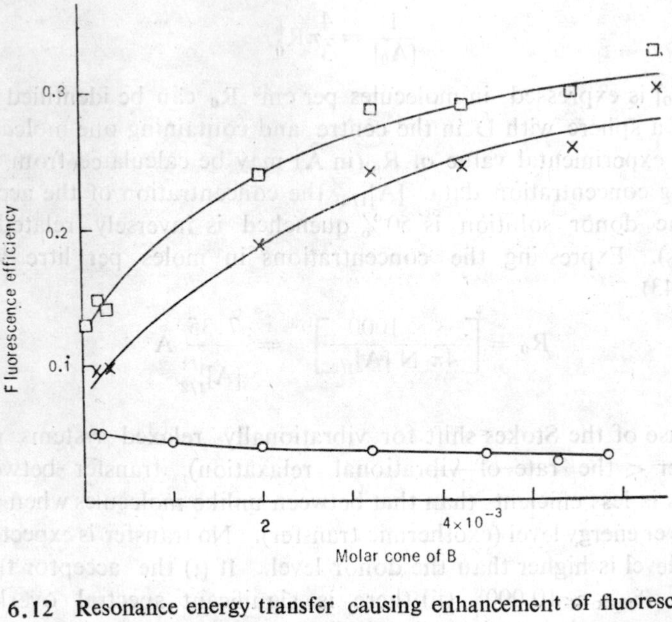

Figure 6.12 Resonance energy transfer causing enhancement of fluorescence in perylene-chloroanthracene system.

The transfer is detected by the sensitization of acceptor fluorescence and the quenching of the donor fluorescence and involves singlet states in both the donor and the acceptor (singlet-singlet transfer) as shown in Table 6.7.

TABLE 6.7

Singlet-singlet transfer by resonance mechanism

Donor	Acceptor	$k_{ET} \times 10^{-11}$	R_0 (Å)
1-Chloroanthracene	Perylene	2	41
1-Chloroanthracene	Rubrene	2	38
1-Cyanoanthracene	Rubrene	3	84

The rate constants k_{ET} are greater than those for diffusion-controlled reactions. They are independent of viscosity for 100-fold change. The critical transfer distances R_0 are very large as calculated from concentration dependence of quenching or sensitization, confirming the predictions of Förster's theory. If the absorption of the acceptor is sufficiently weak such as ($n \rightarrow \pi^*$) transition in biacetyl, R_0 becomes nearly equal to the kinetic collision diameter. The rate of singlet-singlet energy transfer becomes diffusion limited and Förster's expression ceases to be valid. Over such short distances, exchange interactions may become operative.

Long range *triplet-singlet transfer* has been observed to occur effectively although it is a spin-forbidden process :

$$D^* \text{ (triplet)} + A \text{ (singlet)} \rightarrow D \text{ (singlet)} + A^* \text{ (singlet)}$$

This is because the low efficiency of such a transfer is compensated by long lifetime. If other modes of triplet deactivation are less competitive, that is, similarly forbidden, intermolecular transfer may occur with reasonable rates. The slow rate of energy transfer is not incompatible with a large value of R_0 for transfer by a resonance mechanism (Table 6.8).

TABLE 6.8

Critical energy transfer distance R_0 from donor triplet to acceptor singlet (triplet-singlet transfer)

Donor	Acceptor	R_0
Phenanthrene	Fluorescein	35
Triphenylamine	Chlorophyll	54
N, N-Dimethylamine	9-Methylanthracene	24

Such transfers are detected by (i) reduction of the phosphorescence lifetime of donor in presenee of acceptor, (ii) appearance of acceptor fluorescence and (iii) experimental calculation of R_0 and its comparison with the theoretical value obtained from the Förster formulation. In such transfers it is necessary that the phosphorescence spectrum of the donor should overlap with the singlet-singlet absorption spectrum of the acceptor.

The Förster formulation predicts R^{-6} dependence on separation distance between the donor and the acceptor. Therefore the rate of transfer is expected to increase sharply when distance is reduced and vice versa. For example, at a distance $R = R_0/2$, the rate of transfer

$$k_{ET} = \frac{1}{\tau_D}\left(\frac{R_0}{R_0/2}\right)^6 = \frac{1}{\tau_D}(2)^6$$

A 64-times increase in transfer probability!

6.6.7 Short Range Transfer by Exchange Interaction

The exchange interaction is a quantum mechanical effect and arises because of the symmetry requirement of electronic wave functions with respect to exchange of space and spin coordinates of any two electrons in the donor or acceptor complex. From the expression (6.33) for total donor acceptor interactions

$$\beta_{(exchange)} = \int \psi_{D^*}(1)\,\psi_A(2)\frac{e^2}{r_{12}}\,\psi_D(2)\,\psi_{A^*}(1)\,d\tau \qquad (6.45)$$

where r_{12} is the distance between electrons (1) and (2). It is the same kind of interaction which causes singlet-triplet splitting. The characteristic features of the exchange interaction energy are described below:

(i) The electrostatic interaction between the charge clouds will be large only when there is a spatial overlap of the donor and the acceptor wave functions, i.e. when the two actually collide in kinetic sense. Therefore, it can only be a short range phenomenon. No transfer outside the boundaries of the molecule is expected by this mechanism. The transfer occurs at diffusion-controlled rate.

(ii) The magnitude of the exchange interaction is not related to the oscillator strengths of the transitions in the donor and the acceptor. For forbidden transitions like singlet to triplet, exchange interactions will be predominant at short distances. But if the transitions are allowed, dipole-dipole interactions may take over.

An expression for transfer by exchange interaction was derived by Dexter for a vibrationally relaxed system. The rate of transfer is given by

$$k_{ET} = \frac{4\pi^2}{h}\, e^{-2R/L} \int_0^\infty F_D(\bar{\nu})\,\epsilon_A(\bar{\nu})\,d\bar{\nu} \qquad (6.46)$$

where L is an effective average orbital radius for the initial and final electronic states of the donor and the acceptor. Other symbols have their usual meaning. In this expression, like $F_D(\bar{\nu})$, $\epsilon_A(\bar{\nu})$ is also normalized to unity. As a result, the transfer rates are independent of the 'allowedness' or 'forbiddenness' of the transition in the acceptor.

The following *triplet-triplet* transfer,

$$\begin{array}{cccc} D^*(\text{triplet}) + & A(\text{singlet}) & \rightarrow D(\text{singlet}) + & A^*(\text{triplet}) \\ S = 1 & S = 0 & S = 0 & S = 1 \end{array}$$

is doubly forbidden by spin selection rules and cannot, therefore, occur by Coulomb interaction. But it is fully allowed by Wigner's rule since the total spin in the initial and the final states is conserved. According to the spin conservation rule, $S_{D^*} = S_{A^*}$ and $S_D = S_A$, where S is the total spin wave function. For transfer by this mechanism the total spin but not the total multiplicity need be conserved.

Intermolecular energy transfer from the *triplet* of the donor to the *triplet* of the acceptor was first established by Terenin and Ermolaev in 1952. They demonstrated sensitized phosphorescence emission from naphthalene in rigid glassy medium at 77K using 365 nm radiation in presence of a suitable donor. They choose donors whose triplet energy E_{TD} was higher than E_{TA} of the acceptor naphthalene but singlet energy E_{SD} was less than E_{SA}. By such combinations the possibility of direct excitation of the acceptor could be excluded by using a suitable light filter to cutoff radiation absorbed by the acceptor (Figure 6.13).

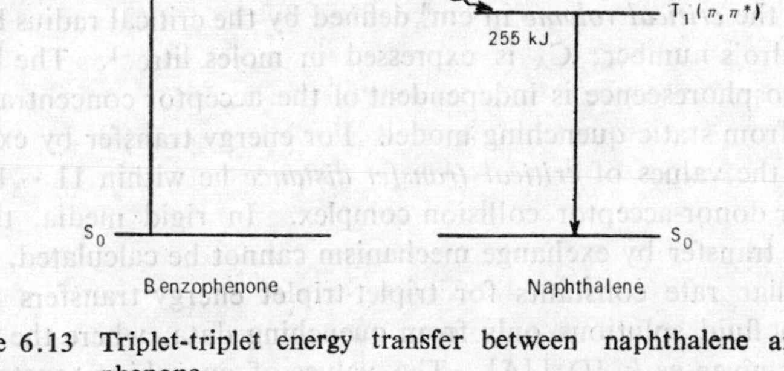

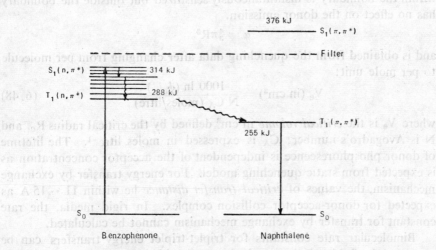

Figure 6.13 Triplet-triplet energy transfer between naphthalene and benzophenone.

Benzophenone was found to possess the criteria for a suitable donor or sensitizer. For this molecule, the intersystem crossing $^1D^* \rightsquigarrow {}^3D$ occurs with unit efficiency ($\phi_{ISC} \simeq 1$). Under the experimental setup, no phosphorescence was observed in absence of benzophenone. After direct excitation of benzophenone by 365 nm radiation the sensitization followed the path:

$$^1D \rightsquigarrow {}^3D \rightsquigarrow {}^3A \rightarrow {}^1A + h\nu_p$$

The critical separation distance calculated from the quenching data was found to be 13 Å which is of the same order as the van der Waals separation. The critical separation distance remained unchanged when halogen substituted naphthalenes were used. The halo-substitution is expected to increase $T_{1A} \rightsquigarrow S_{0A}$ transition probability in naphthalenes. Since oscillator strengths f (naphthalenes) : f(iodonaphthalenes) is as 1 : 1000, no increase in transfer efficiency is clear indication of the lack of dependence on the oscillator strength.

The quenching of donor phosphorescence is a function of the acceptor concentration in rigid media and follows the expression:

$$\frac{\phi_0}{\phi} = e^{\alpha C_A} \tag{6.47}$$

where ϕ_0 and ϕ respectively, are the phosphorescence yields of the excited donor triplet in absence and in presence of the acceptor A at a concentration C_A, and α is a 'sphere of quenching action'. The 'sphere of quenching action' model was first suggested by Perrin for static quenching of fluorescence. It is defined as the sphere of radius R wherein an acceptor within the boundary is instantaneously sensitized but outside the boundary has no effect on the donor emission.

$$\alpha = \tfrac{4}{3}\pi R^3$$

and is obtained from the quenching data after changing from per molecule to per mole unit:

$$V_c \text{ (in cm}^3) = \frac{1000 \ln (\phi_0/\phi)}{N\, C_A \text{ (moles/litre)}} \tag{6.48}$$

where V_e is the *critical volume* in cm³ defined by the critical radius R_c, and N is Avogadro's number; C_A is expressed in moles litre^{-1}. The lifetime of donor phosphorescence is independent of the acceptor concentration as is expected from static quenching model. For energy transfer by exchange mechanism, the values of *critical transfer distance* lie within 11 – 15 Å as expected for donor-acceptor collision complex. In rigid media, the rate constant for transfer by exchange mechanism cannot be calculated.

Bimolecular rate constants for triplet-triplet energy transfers can be measured in fluid solutions only from quenching data, where the rate of quenching is given as k_q [D*] [A]. The values of quenching constants for biacetyl as donor fall into three groups depending on the position of triplet energy E_{TD} of donor relative to E_{TA} of acceptor (Figure 6.14):

(i) When $E_{TD} > E_{TA}$, i.e., transfer is exothermic by 3–4 kcal/mole, transfer occurs on nearly every collision, i.e. the rate of transfer is diffusion-controlled.

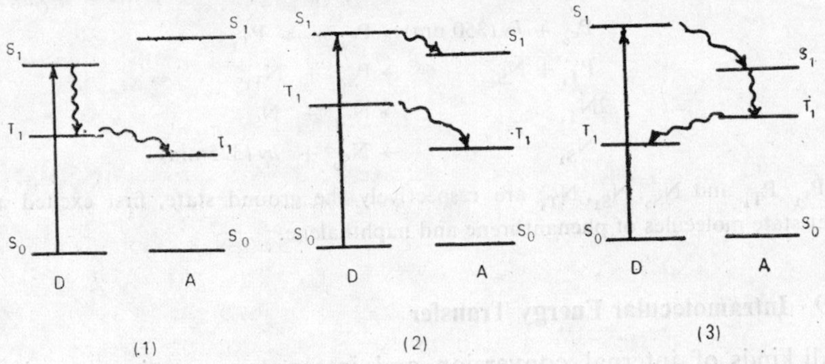

Figure 6.14 Three donor-acceptor energy level schemes for quenching by energy transfer. 1. Benzophenone (D)+naphthalene (A); 2. Naphthalene (D)+biacetyl (A); and 3. Couramin (D)+ benzophenone (A)

(ii) When $E_{TD} \simeq E_{TA}$, the quenching rate drops suddenly, probably due to the possibility of back transfer. In this range energy transfer may show temperature dependence ($\Delta E \approx kT$).

(iii) When $E_{TD} < E_{TA}$ by 3–4 kcal or more, the quenching rate becomes at least a million times slower than the diffusion-controlled rate.

The effectiveness of an energy transfer quencher is determined mainly by the position of its lower triplet level and not by its molecular structure. The transfer rate is diffusion-controlled even when $E_{TD} \gg E_{TA}$, probably because higher triplet states get involved. In practical application, both singlet-singlet and triplet-triplet transfers can be arranged in suitable donor-acceptor systems so chosen as to provide conveniently located energy levels. Three different quenching schemes are possible as shown in Figure 6.14.

Aromatic carbonyl compounds are good sensitizers of triplet state because of small singlet-triplet splitting. On the other hand, aromatic hydrocarbons and olefines have convenient singlet-triplet levels to act as good triplet quenchers or acceptors.

6.6.8 Sensitized Delayed Emission

An interesting variation of energy transfer between two triplet molecules of the same compound under conditions such that triplets can be generated in large numbers has been observed in some aromatic hydrocarbons. The two long-lived triplets can collide with each other in solution, whereby tripleis are 'annihilated' giving rise to one molecule in the excited singlet state and the other in the gound state (Section 5.9).

This phenomenon is possible in molecules like naphthalene, anthracene and pyrene, each of which happens to have the lowest singlet energy level about twice that of the respective triplet. For the generation of these triplets in high concentrations (T – T) type energy transfer from a suitable donor is necessary. Delayed fluorescence in naphthalene has been sensitized by phenanthrene, according to the following scheme:

$$P_{S_0} + h\nu\,(350\text{ nm}) \rightarrow P_{S_1} \rightsquigarrow P_{T_1}$$
$$P_{T_1} + N_{S_0} \rightarrow P_{S_0} + N_{T_1}$$
$$2N_{T_1} \rightarrow N_{S_1} + N_{S_0}$$
$$N_{S_1} \rightarrow N_{S_0} + h\nu\,(325\text{ nm})$$

P_{S_0}, P_{S_1}, P_{T_1} and N_{S_0}, N_{S_1}, N_{T_1} are respectively the ground state, first excited and triplet state molecules of phenanthrene and naphthalene.

6.6.9 Intramolecular Energy Transfer

All kinds of internal conversion and intersystem crossings are intramolecular energy transfer within a chromophoric group. Excitation energy transfer within a molecule from one chromophoric group to another has been observed to occur even when the chromophores are separated by insulating groups. The absorption spectra of these compounds are sum of the absorption spectra of individual chromophores, which indicates that there is little, if any, interaction in the ground state. For example, In 4-benzoylbiphenyl (I) the absorption is very much like benzophenone but phosphorescence emission is characteristic of biphenyl. Similar is the case for naphthyl ketones (II). These compounds present energy level

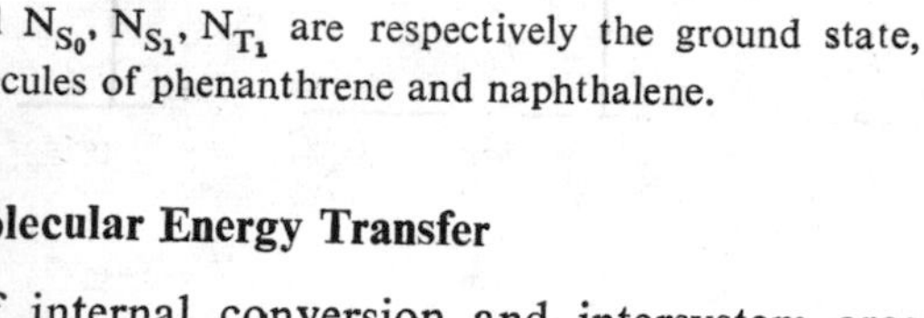

I II

scheme (i) as given in Section 6.6.7 (Figure 6.14). The lowest excited state is S (n, π^*) and the excitation resides on the carbonyl moiety. But the $T(\pi, \pi^*)$ of biphenyl is lower than the $T_1(n, \pi^*)$ state of benzophenone. The excitation energy is transferred by exchange mechanism (T — T type) after intersystem crossing in the carbonyl moiety. Very fast intersystem crossing in benzophenone-like molecules precludes the possibility of transfer between $S(n, \pi^*)$ of C=O to $T(\pi, \pi^*)$ of biphenyl. The sequence of events is:

$$S_0 + h\nu_a \rightarrow S_1(n, \pi^*) \rightsquigarrow T_0(n, \pi^*) \rightsquigarrow T(\pi, \pi^*) \rightarrow S_0 + h\nu_e$$
$$T(\pi,\pi^*) \rightsquigarrow S_0^{\,v} \rightsquigarrow S_0$$

In compounds like:

$h\nu_a \rightarrow$ 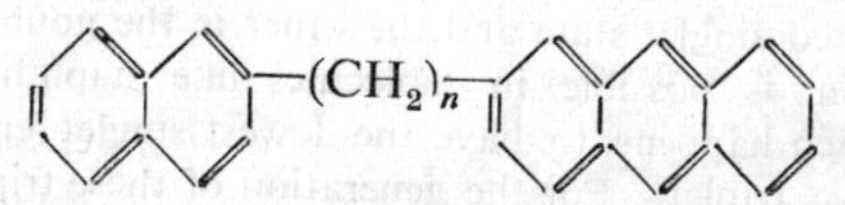 $\xrightarrow{h\nu_e}$

efficient singlet-singlet (S — S) type energy transfer occurs when only naphthalene is excited by light absorption. Although insulating CH_2-groups preclude direct conjugation between naphthyl and anthryl moieteis, the emission is exclusively from anthracene. Changing the distance of separation by increasing the number of CH_2-groups does not affect the transfer efficiency. If the naphthyl group is similarly attached to benzophenone moiety, both (S – S) and (T – T) types of transfers are possible. The energy scheme for two individual moieties is as given in Figure 6.4, for type (iii) system coumarin and benzophenone. The path of energy transfer can be traced as follows:

$$N_{S_0} \xrightarrow{h\nu} \underbrace{N_{S_1}(\pi, \pi^*) \rightsquigarrow B_{S_1}(n, \pi^*)}_{\substack{\text{intramolecular} \\ (S-S)\ \text{transfer}}}$$

$$\downarrow$$

$$\underbrace{B_{T_1}(n, \pi^*) \rightsquigarrow N_{T_1}(\pi, \pi^*)}_{\substack{\text{intramolecular} \\ (T-T)\ \text{transfer}}} \rightarrow N_{S_0} + h\nu_p$$

where N = naphthalene and B = benzophenone.

The validity of the orientation term κ^2 has been established by synthesizing rigid molecules with suitable spatial spacer and orientation. In those systems where internal rotation is completely restricted and the absorbing and emitting chromophore angles can be established unequivocally, the Förster formulation with $\kappa^2 = \cos^2 \theta_{DA} - 3 \cos \theta_D \cos \theta_A$ agrees with experiment (Figure 6.15). The compounds given in the figure are: I. k-anthryl-l′-naphtyl altqanes. II. Nicotinamide-adenine-dinucleotide (reduced) NADH. III. Oligomers of poly-L-proline with α-naphthyl group as energy donor and dansyl group as energy acceptor. III. p-methoxy benzene and flourene chromophere in norbornadiene-spirocylopropane frame. V and VI. Anthrone and naphthalene chromphore in spiro-compounds.

Most interesting applications of intramolecular energy transfer between nonconjugated chromophores are found in the conformational studies of biomolecules like nucleic acids and proteins. The experiments on rotational depolarization of emission from intrinsic fluorescent groups on externally attached fluorescent probes, have resulted in a vast store of knowledge which has helped to enrich the subject of photobiology.

6.6.10 Energy Transfer in Rare Earth Chelates

Rare earth ions, especially Eu^{+3}, Sm^{+3}, Gd^{+3}, Tb^{+3}, and Dy^{+3} with electronic configuration (f^5 through f^9), emit characteristic line spectra from their $4f$ energy states. The energies of the emitting levels are fairly low so that the rare earth ions can be very useful species as acceptors to detect triplet states in solution by energy transfer. Such quenching processes

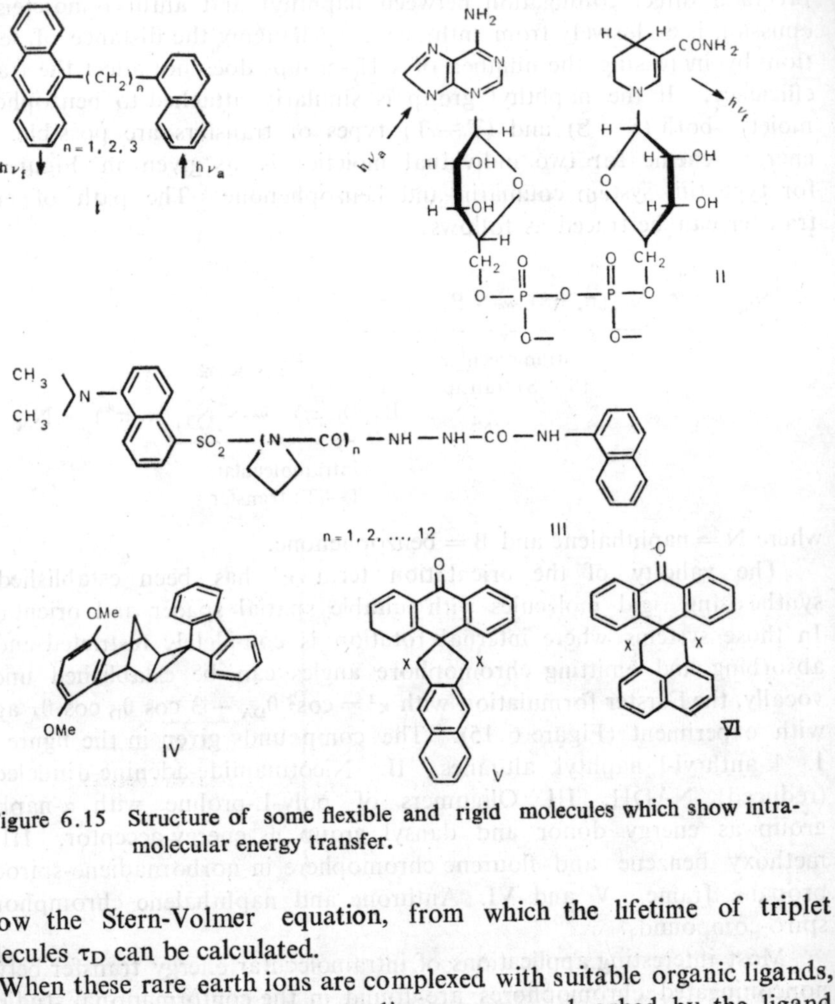

Figure 6.15 **Structure of some flexible and rigid molecules which show intramolecular energy transfer.**

follow the Stern-Volmer equation, from which the lifetime of triplet molecules τ_D can be calculated.

When these rare earth ions are complexed with suitable organic ligands, such as the diketone-1, 3-propanedione, energy absorbed by the ligand appears as the line emission of the central metal ions, *if* the triplet state of the ligand lies above the emitting energy level of the ion. The most probable path of energy transfer within the chelate is:

(i) L_{S_1} (ligand singlet) ⇝ L_{T_1} (ligand triplet) ⇝ M^* (metal excited) → M_0 (metal ground)

If the ligand triplet lies more than 2–3 kcal below the emitting level of the ions, energy is not transferred and broad molecular phosphorescence characteristic of the ligand is observed. Few other schemes are also visualized depending on the relative energy levels of ligand and the metal ions, such as:

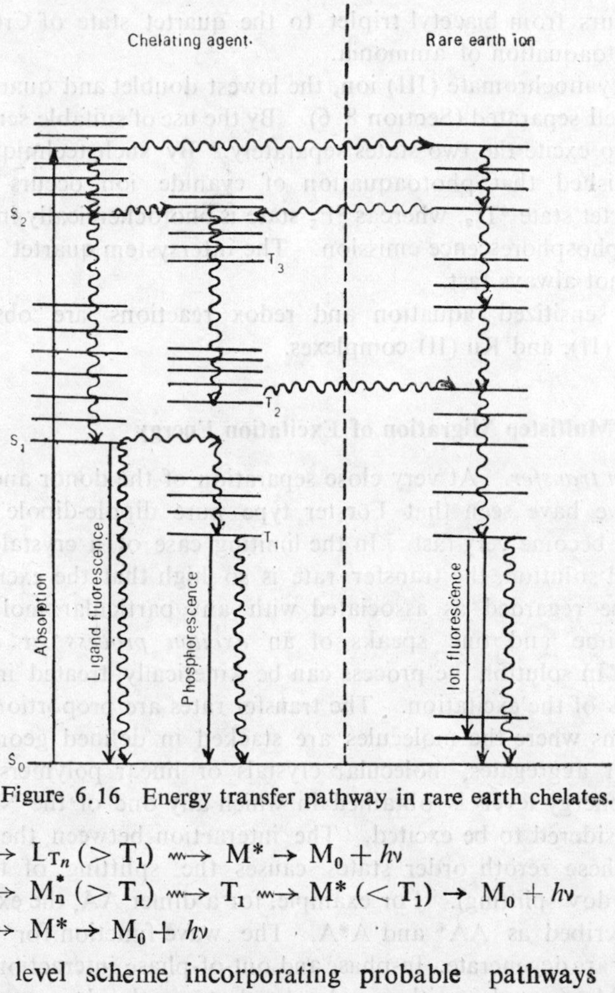

Figure 6.16 Energy transfer pathway in rare earth chelates.

(i) $L_{S_1} \rightsquigarrow L_{T_n} (> T_1) \rightsquigarrow M^* \rightarrow M_0 + h\nu$

(ii) $L_{S_1} \rightsquigarrow M_n^* (> T_1) \rightsquigarrow T_1 \rightsquigarrow M^* (< T_1) \rightarrow M_0 + h\nu$

(iii) $L_{S_1} \rightsquigarrow M^* \rightarrow M_0 + h\nu$

Energy level scheme incorporating probable pathways for energy transfer within the chelate is given in Figure 6.16. In these chelates, transfer occurs by exchange mechanism. If suitable coupling interaction exists, transfer may occur from the singlet state of the ligand to the higher energy state of the metal ion. The rate of energy transfer $k_{ET} \simeq 10^{10} - 10^{12}\ s^{-1}$ is predicted.

6.6.11 Energy Transfer Processes Involving Coordination Compounds

Energy transfer from singlet or triplet states of suitable organic molecules can cause excitation of the central metal ion in a coordination compound. Photosensitization of Cr(III) complexes by biacetyl leads to aquation reaction of $Cr(NH_3)_5(NCS)^{2+}$ ion. The aquation of (NH_3) is hundred times more than that of (NCS) for the sensitized reaction, whereas it is only 66 times on direct excitation of Cr(III). This shows that energy transfer occurs from biacetyl triplet to the quartet state of Cr(III) which favours photoaquation of ammonia.

In hexacyanochromate (III) ion, the lowest doublet and quartet excited states are well separated (Section 8.6). By the use of suitable sensitizers, it is possible to excite the two states separately. By such techniques, it has been established that photoaquation of cyanide ion occurs from the excited quartet state $^4T_{2g}$, whereas ^2E_g state is photochemically inactive and the seat of phosphorescence emission. The intersystem quartet to doublet crossing is not always fast.

Similar sensitized aquation and redox reactions are observed for Co (III), Pt (II), and Ru (II) complexes.

6.6.12 Fast Multistep Migration of Excitation Energy

(*a*) *Exciton transfer.* At very close separation of the donor and acceptor molecules we have seen that Förster type pure dipole-dipole resonance transfer can become very fast. In the limiting case of a crystal or a very concentrated solution, the transfer rate is so high that the excitation can no longer be regarded as associated with any particular molecule at a particular time and one speaks of an *exciton process* or *delocalized excitation.* In solution the process can be kinetically treated in terms of random walk of the excitation. The transfer rates are proportional to R^{-3}.

In systems where the molecules are stacked in defined geometry such as molecular aggregates, molecular crystals or linear polymers, a set of degenerate energy levels is obtained in which any one of the N molecules may be considered to be excited. The interaction between the transition dipoles of these zeroth order states causes the splitting of the energy levels (Davydov splitting). For example, for a dimer AA, the excited state can be described as AA* and A*A. The wave function for these two descriptions are degenerate. In-phase and out-of-phase interaction of transition dipole M, created in either molecule, destroys the degeneracy. A set of two energy levels results. The interaction is highly geometry dependent and the energy difference ΔE between the two exciton levels is given as

$$\Delta E = \frac{2|M|^2}{R^3} G$$

where R is the distance between the centres of the two molecules and G defines the geometry of stacking (Figure 6.17). The interaction energy varies between 2000 cm^{-1} for a strong coupling case and 10 cm^{-1} for a weak coupling case.

Although all the exciton levels are not equally allowed, they are important for energy migration or exciton delocalization. Using time dependent perturbation equation, time period for transfer τ_{tr} is given as

$$\tau_{tr} = \frac{h}{2\Delta E} \simeq 10^{14}\ s^{-1}\ (\text{for } \Delta E \simeq 1500\ cm^{-1}) \qquad (6.49)$$

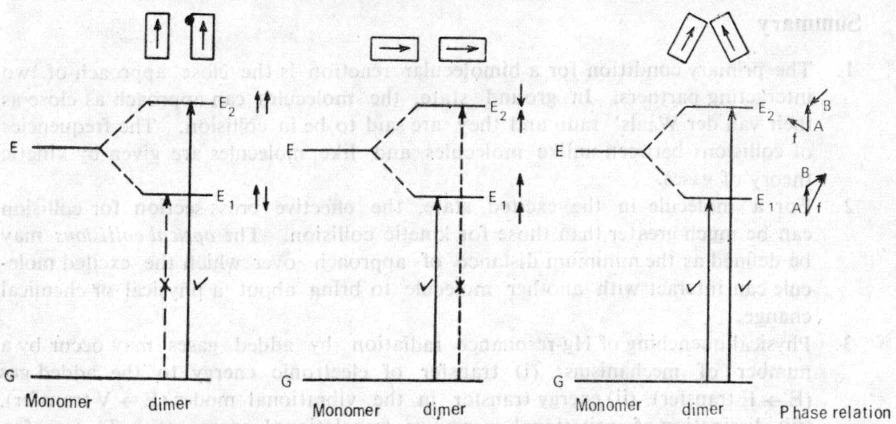

Figure 6.17 Exciton splitting and geometry of ordered array of planar molecules.

the fastest rate predicted by any mechanism. Even when the splitting is about 10 cm^{-1} the rate of exciton migration is fairly high and much greater than that calculated for diffusion-controlled rate. A well known example of such fast exciton migration is a mixed crystal of anthracene contianing very small concentration of naphthacene $\simeq 10^{-4}$ ppm as trace impurity. When excited in anthracene absorption band, emission from naphthacene is observed with high efficiency. The host-sensitized emission from guest molecules suggests that the excitation energy explores about 10^6 lattice sites during the lifetime of the excited state. There is no transfer if these crystals are dissolved in fluid or rigid solutions. Such exciton transfer modes are said to work in photosynthetic units in plants and many other biological systems.

(*b*) *Liquid scintillators.* If a very dilute solution of anthracene or *p*-terphenyl is excited by x-rays or γ-rays, emission from the solute present in trace amount is observed. In such systems excitation energy supplied by the ionising radiation preferentially interacts with the solvent and finally resides in the low lying excited singlet state of the solvent molecule. Because of the close proximity of like molecules excitation energy migrates from solvent to solvent until it is captured by a solute molecule. This molecule then becomes the centre of emission with high efficiency. Such solutes are called liquid scintillators and are widely used for detection of ionizing radiations. Sometimes they are embedded in plastics or other polymeric materials. The important scintillator solutes are oxazole and oxadiazole derivatives. The scintillator diphenyloxazole is commonly called PPO. For liquid scintillators, the nature of solvent is also important. Good scintillation solvents are generally unsaturated molecules such as benzene, toluene and xylene.

Summary

1. The primary condition for a bimolecular reaction is the 'close' approach of two interacting partners. In ground state, the molecules can approach as close as their van der Waals' radii and they are said to be in collision. The frequencies of collisions between unlike molecules and like molecules are given by kinetic theory of gases.
2. For a molecule in the excited state, the effective cross-section for collision can be much greater than those for kinetic collision. The *optical collisions* may be defined as the minimum distance of approach over which the excited molecule can interact with another molecule to bring about a physical or chemical change.
3. Physical quenching of Hg-resonance radiation by added gases may occur by a number of mechanisms: (i) transfer of electronic energy to the added gas ($E \rightarrow E$ transfer), (ii) energy transfer in the vibrational modes ($E \rightarrow V$ transfer), (iii) dissipation of excitational energy as translational energy ($E \rightarrow T$) transfer, (iv) by complex formation in the excited state leading to chemical reactions or radiative return to the ground state.
4. Collisions in solution are diffusion-controlled and hence depend on the viscosity of the solvent. Due to Franck-Rabinowitch cage effect they occur in sets called '*encounters*'.
5. The rate constant k_2 for bimolecular reactions becomes encounter limited if energy of activation ΔE is negligible and steric effect is absent. For such favourable cases k_2 is expressed by the equation

$$k_2 = 4\pi N\, R_{AB}\, (D_A + D_B) \times 10^{-3}\ l\ mol^{-1}\ s^{-1}$$

where D_A and D_B are diffusion coefficients of the two colliding partners and R_{AB} is the encounter radius. A special case of this equation is when Stokes radii become equal to interaction radii ($r_A = r_B = a$). The rate constant becomes independent of molecular dimensions,

(i) $k_2 = \dfrac{8\,RT}{3000\eta}\ l\ mol^{-1}\ s^{-1}$ (sliding friction $\neq 0$)

(ii) $k_2 = \dfrac{8\,RT}{2000\eta}\ l\ mol^{-1}\ s^{-1}$ (sliding friction $= 0$)

6. The quenching of fluorescence by added substance Q at concentrated [Q] is expressed by the Stern-Volmer equation,

$$\frac{\phi_f^0}{\phi_f} = 1 + k_q\,\tau[Q] = 1 + K_{SV}\,[Q]$$

where ϕ_f^0 and ϕ_f are fluorescence quantum yields in absence and in presence of quencher respectively and K_{SV} is the Stern-Volmer quenching constant. This constant is the product of bimolecular quenching rate constant k_q and the lifetime of the fluorescer τ in absence of the quencher. This must be differentiated from *static* quenching, where τ is unaffected.
7. For dilute solutions, the fluorecsence intensity F is linearly related to the concentration of the fluorescer as

$$F = \phi_f\, 2.3\, I_0\, \epsilon\, Cl$$

where the symbols have their usual meaning. If I_0 is kept constant, then F as a function of wavelength will reproduce the absorption spectrum assuming ϕ_f to be wavelength independent. This is known as an *excitation spectrum*.

8. In concentrated solutions, concentration quenching may occur. The concentration quenching is said to proceed via intermediate formation of *excimers* by interaction between an excited and a ground state molecule. Excimers are stable only in the excited state and thus differ from the excited state of a dimer.
9. Excimer may relax (i) by emission of characteristic structureless band shifted to about 6000 cm^{-1} to the red of the normal fluorescence, (ii) dissociate nonradiatively into original molecules, (iii) form a photodimer. Those systems which give rise to photodimers may not decay by excimer emission. The binding energy for excimer formation is provided by interaction between charge transfer (CT) state $A^+A^- \rightleftharpoons A^-A^+$ and charge resonance state $AA^* \rightleftharpoons A^*A$.
10. Quenching by added substances generally occurs through two broad-based mechanisms: (a) charge transfer mechanism and (b) energy transfer mechanism.
11. If added substances absorbing at higher frequencies are present the quenching mechanism is visualized to proceed through transient complex formation in the excited state. They are known as *exciplexes* and have pronounced charge transfer character. Exciplexes may decay radiatively, nonradiatively or lead to reaction products.
12. When the quencher contains heavy atoms nonradiative relaxation of the exciplex occurs via the triplet state (heavy atom perturbation). A second mode of exciplex dissociation is through electron transfer between the excited molecule and the quencher. Ionization potential of the donor, electron affinity of the acceptor and solvent dielectric constant are important parameters in such cases.
13. Paramagnetic molecules $^3O_2$ and ^2NO are very efficient quenchers of singlet and triplet states. Oxygen may form CT complex and lead to peroxide formation in the triplet state. Paramagnetic ions of the transition series and lanthanides also quench the triplet states.
14. Second important mechanism of quenching is nonradiative transfer of electronic energy from the fluorescer to the quencher. For such transfers the energy level of the acceptor should be equal to or lower than that of the donor and some donor-acceptor coupling interactions should be present. There are two types of interactions : (i) dipole-dipole coupling interactions leading to long range transfer, and (ii) exchange interaction responsible for short range transfer.
15. The expression for rate constant for transfer is similar to the one derived for nonradiative processes within a molecule

$$k_{D^* \to A} = \frac{4\pi^2}{h} \rho_E \beta_{el}^2 F$$

where F, the Franck Condon factor $= \Sigma < \chi_{D^*} \mid \chi_D > < \chi_{A^*} \mid \chi_A > \mid^2$

$$= \Sigma \left(\int \chi_D \chi_{D^*} d\tau_D \int \chi_{A^*} \chi_A d_A \right)^2$$

and β_{el} is the electronic interaction energy parameter which can be broken up into (i) Coulomb term and (ii) exchange term.
16. The rate constant for long range transfer by dipole-dipole mechanism is given by

$$k_{ET} = \frac{9000 \ln 10}{128 \pi^5 N} \frac{\kappa^2 \phi_D}{n^4 \, {}_D R^6} \int F_D(\bar{\nu}) \, \epsilon_A(\bar{\nu}) \frac{d\bar{\nu}}{\bar{\nu}^4}$$

where ϕ_D and τ_D are respectively fluorescence efficiency and actual radiative lifetime of the donor, κ is an orientation factor and the integral defines the overlap between the fluorescence spectrum of the donor and the absorption spectrum of the acceptor. The transfer probability is usually expressed in terms of critical transfer distance R_0. Spin selection rule applies to such transfers. In an ideal situation R_0 value of 5 nm to 10 nm are predicted. These are much greater than kinetic encounter radii.

17. Short range transfer by exchange mechanism occurs when donor and acceptor electronic wavefunctions spatially overlap. The rate follows diffusion-controlled kinetics if donor and acceptor energy levels are in near resonance. Transitions forbidden by dipole-dipole mechanism may occur by exchange mechanism, e.g.

$$D^* \text{ (triplet)} + A \text{ (singlet)} \rightarrow D \text{ (singlet)} + A^* \text{ (triplet)}$$

Such triplet-triplet transfers are very important in photochemistry because they provide means for populating the long-lived triplet state to which transitions are forbidden by direct absorption of radiation.

18. Intramolecular transfers between two chromphores separated by insulating groups can lead to absorption by one chromophore and emission from the other. Complexes of rare earths specially of Eu^{+3}, Sm^{+3}. Gd^{+3} and Dy^{+3} emit line spectrum characteristic of the central metal ion when absorption takes place in the ligand moeity.

19. Multistep fast transfer of excitation energy may occur in ordered array of molecules such as molecular aggregates, crystals and polymers. The phenomenon is known as delocalized excitation or *exciton migration.* In liquid scintillators, excitation energy, after decaying down to the lowest excited singlet state of the solvent, migrates from solvent to solvent until it is captured by the solute molecule from which emission may take place.

SEVEN

Photochemical Primary Processes

7.1 CLASSIFICATION OF PHOTOCHEMICAL REACTIONS

A photochemical reaction may be classified as adiabatic or diabatic, depending on its course along the potential energy surface as a function of reaction coordinates. If the chemical change occurs on the same continuous potential energy surface the reaction is said to be *adiabatic* : if crossing of potential surfaces is involved, it is classified as *diabatic*. According to this criterion, in an adiabatic photochemical reaction the reactants and products must correlate with each other and with the transition state. The products will be in electronically excited state and may be detected by their luminescence and/or photochemical properties. The photodissociation of small molecules in the vapour state like $I_2 + h\nu \rightarrow I^* + I$, and proton transfer in the excited state may be classified as adiabatic reactions. But the majority of photochemical reactions in condensed phase produce product molecules in the ground state indicating a radiationless transition from the upper to a lower potential energy surface of the system before the chemical reaction is completed, i.e. before the product configuration is achieved. There might be intermediate cases also as illustrated in Figure 7.1. In the intermediate case certain fractions of the reacting species may escape deactivation long enough to attain the product configuration. The

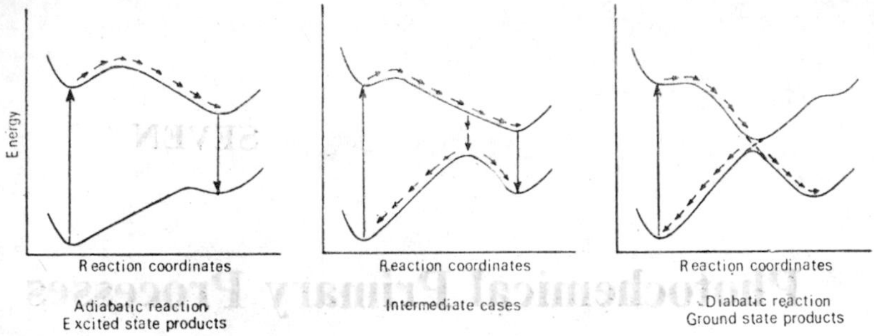

Figure 7.1 Classification of photochemical reactions according to the nature of potential energy surfaces.
(a) Adiabatic reaction—excited state products.
(b) Intermediate case.
(c) Diabatic reaction—ground state products.

crossing process is governed by the laws for the conservation of energy, momentum and symmetry and the selection rules for radiationless processes are applicable. These rules may be influenced by local forces at the point of intersection, which include (i) electronic configuration interaction, (ii) vibronic interaction or Franck-Condon factor, and (iii) magnetic interactions such as spin-orbit coupling and hyperfine coupling.

7.2 RATE CONSTANTS AND LIFETIMES OF REACTIVE ENERGY STATES

For a photoexcited molecule, the time allowed for a reaction to occur is of the order of the lifetime of the particular excited state, or less when the reaction step must compete with other photophysical processes. The photoreaction can be *unimolecular* such as photodissociation and photoisomerization or may need another molecule, usually unexcited, of the same or different kind and hence *bimolecular*. If the primary processes generate free radicals, they may lead to *secondary* processes in the dark.

	Rate	
$A^* \xrightarrow{k_r} .C + D$	$k_r[A^*]$	(unimolecular)
$A^* + B \xrightarrow{k_r} P$	$k_r[A^*][B]$	(bimolecular)

The seat of photoreaction may not be the initially excited state. In general, the photochemistry is observed to occur from lowest excited singlet and triplet states. The singlet excited state has more energy but short lifetime ($10^{-9} - 10^{-7}$ s), whereas triplet state has less energy but more time ($10^{-6} - 1$ s) for the reaction and greater reactivity due to their unpaired spin. An important aspect of photochemical studies is to establish the energy state involved in a given reaction.

The efficiency of a photochemical reaction is defined (Section 1.2) in terms of quantum yield of reaction ϕ_R,

$$\phi_R = \frac{\text{rate of reaction}}{\text{rate of absorption of radiation}}$$

$$= \frac{+\frac{d[\mathrm{P}]}{dt} \text{ or } -\frac{dx}{dt}}{\mathrm{I}_0 \times \text{fractional absorption s}^{-1}\,\text{cm}^{-3}}$$

$$= \frac{\text{moles of product formed or reactant consumed s}^{-1}\,\text{cm}^{-3}}{\text{einstein absorbed s}^{-1}\,\text{cm}^{-3}}$$

The quantum yield of a reaction can be related to reactivity of a given state only after the rates of other competing processes are identified and measured. Consider the reaction

$$\mathrm{A}_i^* \rightarrow \mathrm{P} \text{ (product)}$$

where A_i^* is not the initially excited state but reaches subsequently by photophysical processes. If A_1^* is the initially excited molecule, then the following general scheme can be drawn up for the reaction to occur from A_i^* (Figure 7.2).

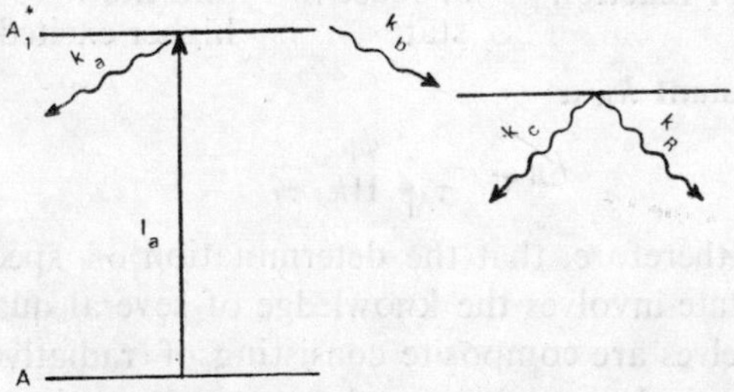

Figure 7.2 General scheme for a photochemical reaction which does not occur from initially excited state.

k_a = sum of rate constants for deactivation of A_1^* except by transfer to A_i^*.
k_b = rate constant for crossing over to A_i^*.
k_c = sum of rate constants for deactivation of A_i^*, except by photoreaction.
k_R = rate constant for formation of the product.
I_a = rate of absorption = rate of formation of A_1^*.
Since the lifetimes of excited states are small, by applying the steady state approximation, respective concentrations are obtained,

$$[\mathrm{A}_1^*] = \frac{\mathrm{I}_a}{k_a + k_b} \qquad (7.1)$$

$$[\mathrm{A}_i^*] = \frac{k_b\,[\mathrm{A}_1^*]}{k_c + k_R} = \frac{k_b\,\mathrm{I}_a}{(k_c + k_R)\,(k_a + k_b)} \qquad (7.2)$$

Since the rates of formation of product are given by:

$$\frac{d[\mathrm{P}]}{dt} = k_{\mathrm{R}}\,[\mathrm{A}_i^*] \quad \text{unimolecular reaction} \tag{7.3}$$

$$\frac{d[\mathrm{P}]}{dt} = k_{\mathrm{R}}\,[\mathrm{A}_i^*]\,[\mathrm{B}] \quad \text{bimolecular reaction} \tag{7.4}$$

we have, for unimolecular reaction

$$\text{Rate} = k_{\mathrm{R}}\,[\mathrm{A}_i^*] = k_{\mathrm{R}}\,\frac{k_b}{k_c + k_{\mathrm{R}}}\,\frac{\mathrm{I}_a}{k_a + k_b} \tag{7.5}$$

and

$$\phi_{\mathrm{R}} = \frac{k_r\,[\mathrm{A}_i^*]}{\mathrm{I}_a} = \frac{k_{\mathrm{R}} k_b}{(k_c + k_r)\,(k_a + k_b)} \tag{7.6}$$

$$= (k_{\mathrm{R}} \tau_{\mathrm{A}_i^*})\,(k_b\,\tau_{\mathrm{A}_1^*}) \tag{7.7}$$

where $\tau_{\mathrm{A}_i^*}$ and $\tau_{\mathrm{A}_1^*}$, respectively are the lifetimes of reactive state and the initially excited state under the actual experimental conditions. In general, if there are a number of states through which the initially excited molecule has tumbled through to come to the reactive state, then reactivity as defined by quantum yield ϕ_{R}, is expressed as

$$\phi_{\mathrm{R}} = k_{\mathrm{R}} \qquad \tau_{\mathrm{A}_i^*} \qquad \Pi k_j\,\tau_j \tag{7.8}$$

= rate constant for reaction × lifetime of reactive state × product of rate constant and lifetimes for all the higher excited states.

and the rate constant k_{R} as

$$k_{\mathrm{R}} = \frac{\phi_{\mathrm{R}}}{\tau_{\mathrm{A}_i^*}\,\Pi k_j\,\tau_j} \tag{7.9}$$

It is evident, therefore, that the determination of specific reactivity of a given excited state involves the knowledge of several quantities k_a, k_b, k_c etc. which themselves are composite consisting of radiative and radiationless processes, physical quenching and chemical reactions other than that of direct interest. The following specific cases can be identified.

1. If the reaction occurs entirely from the state reached by absorption,

$$\mathrm{A}_i^* = \mathrm{A}_1^*$$

$$k_{\mathrm{R}} = \frac{\phi_{\mathrm{R}}}{\tau_{\mathrm{A}_i^*}} \tag{7.10}$$

2. If the quantum yield of reaction is unity and the reaction is simple,

$$k_{\mathrm{R}} = \frac{1}{\tau_{\mathrm{A}_i^*}} \tag{7.11}$$

3. In the general case of a first order reaction which does not occur from the primary excited state and whose quantum yield is not unity, experimental determination of the rates of radiationless processes and of the quantum yield of formation of reacting electronic state becomes necessary.

4. For second order reactions, a knowledge of the concentration of B is required.

$$\phi_R = \frac{k_R [A^*][B]}{I_a}$$

$$= \frac{k_R k_b [B]}{(k_c + k_R)(k_a + k_b)} \quad (7.12)$$

The reactivity of the given excited state can be obtained by simply measuring the lifetimes of that state at two different concentrations of [B]:

$$k_R = \left(\frac{1}{\tau_2} - \frac{1}{\tau_1}\right) \Big/ \left([B_2] - [B_1]\right) \quad (7.13)$$

Therefore, *the limiting parameter in all reactivity studies is the lifetime of the particular state.* A short lived state of high reactivity may be less efficient in product formation than a long lived state of lower reactivity. Let us consider a general scheme in which a reaction may occur from either the initially excited singlet state or the subsequently populated triplet state. Assuming no emission from the singlet or the triplet states, we have

	Rate
$A + h\nu \rightarrow {}^1A^*$	I_a
${}^1A^* \rightarrow A$	$k_{IC}[{}^1A^*]$
${}^1A^* \rightarrow {}^3A$	$k_{ISC}[{}^1A^*]$
${}^1A^* + C \rightarrow C + D$	${}^1k_r [{}^1A^*][B]$
${}^3A \rightarrow A$	$k_{ISC}[{}^3A]$
${}^3A + B \rightarrow C + D$	${}^3k_r [{}^3A][B]$

For singlet state reaction, the quantum yield ${}^1\phi_R$ is given by

$${}^1\phi_R = \frac{{}^1k_R [B]}{k_{IC} + k_{ISC} + {}^1k_R [B]} = {}^1k_R \, {}^1\tau \, [B] \quad (7.14)$$

where ${}^1\tau$ is the actual life of the singlet state. Similarly, for triplet state reaction, the quantum yield ${}^3\phi_R$ is given by

$${}^3\phi_R = \frac{{}^3k_R [B]}{k_{ISC}^T + {}^3k_R [B]} \, \frac{k_{ISC}}{k_{ISC} + k_{IC}}$$

$$= {}^3k_R [B] \, {}^3\tau \, \phi_{ISC} \quad (7.15)$$

For a reaction in solution occurring at a diffusion-controlled rate, ${}^1k_R \simeq 10^9$ l mol^{-1} s^{-1}, and if k_{IC} is negligible and $k_{ISC} \simeq 9 \times 10^8$ s^{-1}, then for $[B] = 0.1$ M, the efficiency of singlet state reaction is (equation 7.14)

$${}^1\phi_R = \frac{10^9 (0.1)}{9 \times 10^8 + 10^9 (0.1)} = 10^{-1}$$

If the same reaction occurs in the triplet state with reactivity about

10^4 times less than the singlet state, $^3k_R \simeq 10^5$ l mol^{-1} s^{-1}, and assuming $\phi_{ISC} \simeq 1$ and $k^T_{ISC} \simeq 9 \times 10^4$ s^{-1}, a commonly observed value, then again for [B] = 0.1 M, (equation 7.15)

$$^3\phi_R = \frac{10^5\,(0.1)}{9 \times 10^4 + 10^5\,(0.1)} = 10^{-1}$$

In this hypothetical example, although the triplet state is less reactive by four orders of magnitude, the net efficiency of reaction is the same. The actual lifetime of the singlet state $^1\tau$, works out to be $\simeq 10^{-9}$ s and that of triplet state to be $\simeq 10^{-5}$ s. The short lifetime does not allow enough time for reaction even though the reactivity of the state is high.

The quantum yield of photochemical processes can vary from a low fractional value to over a million (Section 1.2). High quantum yields are due to secondary processes. An initially excited molecule may start a chain reaction and give rise to a great number of product molecules before the chain is finally terminated. For nonchain reactions, the quantum yields for various competitive photophysical and photochemical processes must add up to unity for a monophotonic process if the reaction occurs from the singlet state only:

$$\phi_R + \phi_f + \phi_{IC} + \phi_{ISC} \simeq 1.0$$

The quantum yield for the primary photochemical process differs from that of the end product when secondary reactions occur. Transient species produced as intermediates can only be studied by special techniques such as flash photolysis, rotating sector devices, use of scavengers, etc. Suitable spectroscopic techniques can be utilized for their observations (UV, IR, NMR, ESR, etc.). A low quantum yield for reaction in solutions may sometimes be caused by recombination of the products due to *solvent cage effect*.

7.2.1 Determination of Rate Constants of Reactions

Consider the reaction scheme as given in Section 7.2 for a bimolecular reaction. Assume that the reaction occurs from the triplet states only and step (4) is replaced by fluorescence emission step. The rate of reaction is expressed in terms of quantum yield of disappearance of the reactant or the appearance of the product

$$\phi^0_R = -\frac{1}{I_a}\frac{d[x]}{dt} = \frac{k_{ISC}}{k_{IC} + k_{ISC}}\,\frac{k_R[B]}{k^T_{ISC} + k_R[B]}$$

$$= \phi_{ISC}\,\frac{k_R[B]}{k^T_{ISC} + k_R[B]}$$

$$\frac{1}{\phi^0_R} = \frac{1}{\phi_{ISC}} + \frac{1}{\phi_{ISC}}\,\frac{k^T_{ISC}}{k_R[B]} \qquad (7.16)$$

In this simplified scheme, a plot of $[\phi_R^0]^{-1}$ vs $[B]^{-1}$ should be linear. From the intercept ϕ_{ISC} can be obtained and the slope/intercept ratio is k_{ISC}^T/k_R. The rate constant for the reaction k_R can be estimated by the method of competition. The reaction is carried out in presence of a third substance which quenches the reaction physically or chemically by competing for the reaction intermediate. In this case the triplet state of the photoexcited reactant is the intermediate and if the quenching step is

$$^3A + Q \rightarrow A + Q \qquad k_Q[^3A][Q]$$

The expression for the quantum yield of reaction becomes

$$^3\phi_R = \phi_{ISC} \frac{k_R[B]}{k_{ISC}^T + k_R[B] + k_Q[Q]}$$

The ratio of quantum yields in presence and in absence of quencher is then given as

$$\frac{\phi_R^0}{\phi_R} = \frac{k_{ISC}^T + k_R[B] + k_Q[Q]}{k_{ISC}^T + k_R[B]}$$

$$= 1 + \frac{k_Q[Q]}{k_{ISC}^T + k_R[B]} = 1 + K_Q[Q] \qquad (7.17)$$

A Stern-Volmer type expression is obtained. At constant [B], a plot of ϕ_R^0/ϕ_R *vs* [Q] will be linear and the slope is equal to K_Q.

If the experiment is repeated for a set of reactant concentrations [B], a set of straight lines of unit intercept will be obtained. The slope K_Q will be a function of [B]

$$K_Q = \frac{k_Q}{k_{ISC}^T + k_R[B]} = k_Q\tau$$

where τ is the lifetime of ^3A in absence of quencher. If k_Q is considered diffusion-controlled, then τ can be estimated and since

$$\frac{1}{K_Q} = \frac{k_{ISC}^T}{k_Q} + \frac{k_R[B]}{k_Q}, \qquad (7.18)$$

k_{ISC}^T and k_R can be calculated from the slope and intercept of the linear plot between $1/K_Q$ *vs* [B].

7.3 EFFECT OF LIGHT INTENSITY ON THE RATE OF PHOTOCHEMICAL REACTIONS

Monophotonic photochemical reactions are those where each absorbed quantum excites one molecule which then reacts. Rates are usually directly proportional to the light intensity. Where secondary reactions are set up, however, the proportionality changes, depending on the chain termination processes. If chain intermediates are terminated by unimolecular reactions, as with vessel walls or with other molecules to give relatively stable products, rates are proportional to I, but if atoms or radicals recombine bimolecularly, rates vary as $\sqrt{I}$. In this latter case lifetime of intermediates may be obtained from the variation of rates when the incident light is interrupted by a sector disc rotating at different speeds (Section 10.3.1).

At very high intensities *biphotonic* reactions are encountered; either two quanta are simultaneously absorbed by one molecule, or, more commonly, reaction occurs by the interactions of *two excited* molecules (Section 5.9). Rates here become proportional to I^2. Because of their long life, triplet states are involved in the second of these processes, and provide a mechanism whereby high energy bond cleavage may be effected with lower energy photons.

7.4 TYPES OF PHOTOCHEMICAL REACTIONS

The chemical behaviour of molecules generally depends on the most weakly bound electrons. A molecule in the excited state differs from the ground state molecule with respect to both energy and electron wavefunction, and therefore differs in its chemistry. Irradiation to produce an excited electronic state alters the reactivity of molecules in a number of ways which decide the nature of any photochemical reaction:

(1) Since nuclei are often more weakly bound in the excited state than in the ground state, the molecule may be more easily dissociated. If excited to the repulsive state dissociation may occur with unit efficiency (photodissociation).

(2) Because of the Franck-Condon principle, different vibrational and rotational modes may be excited which can bring about reactions, normally not possible in the ground state (valence isomerization).

(3) The excited electron is usually in a more weakly bound orbit, often extending over a larger region of space, and, therefore, is more likely to be removed by an electrophilic reagent (oxidation).

(4) The excited electron may interact with any other odd or unpaired electron attacking agent and form a bond, as soon as the orbit begins to overlap with that of the excited molecule due to increase in electron affinity.

(5) In inorganic compounds or complexes with variable valence system, a redox reaction may be set up by intramolecular or intermolecular electron transfer processes (redox reactions).

7.4.1 Photofragmentation or Photodissociation

The direct dissociation of a molecule on absorption of a quantum of radiational energy becomes very probable when the energy absorbed is equal to or more than the bond dissociation energy. Such photodissociative mechanisms are best appreciated with the help of potential energy curves for diatomic molecules as given in Figure 7.4 for Cl_2 and HI molecules. A molecule excited into the continuum region of the absorption spectrum dissociates immediately with unit quantum efficiency.

When the molecule dissociates from an excited state it is called *photolysis.* In such a case it is possible that at least one of the atoms is in its excited state. Any excess energy will appear as the kinetic energy of the partners which fly apart if in the gaseous state. The excess potential and kinetic energies may impart increased chemical activity to the photofragmented particles. The symmetry correlation rules are helpful in

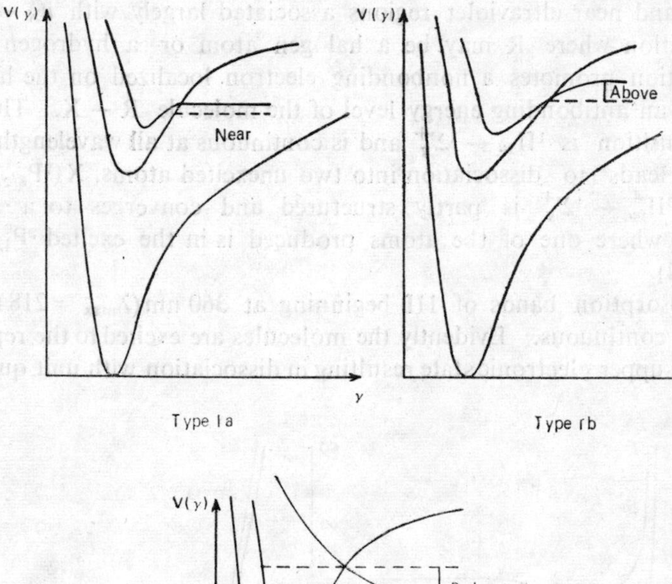

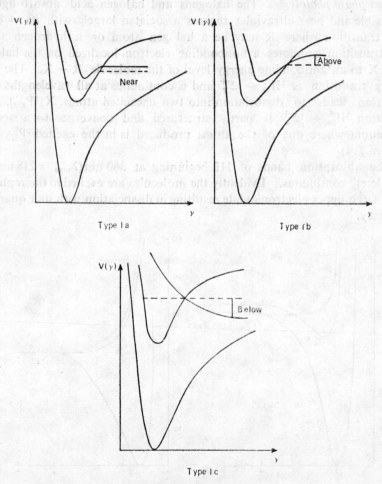

Figure 7.3 Situations for predissociation of a diatomic molecule on photoexcitation.

predicting the energy states of the product particles. These rules are: (i) the symmetry of the products must correlate with that of the reactants, (ii) in setting up the reactant-product symmetry correlation rules it is not possible to leave a lower energy state of given symmetry uncorrelated, and (iii) the energy states of the same symmetry do not cross (noncrossing rule). Thus reactants and products must lie on the same potential energy surface (adiabatic reaction).

Photodissociation may occur even on excitation to below a spectroscopic continuum region if a higher energy state intersects the curve at a suitable point creating a dissociative situation (Figure 7.3). The spectra of halogens and halogen acids are instructive in this respect.

A. *Gas phase photolysis.* The halogens and halogen acids absorb light in the visible and near ultraviolet regions associated largely with $n_X \rightarrow \sigma^*_{R-X}$ type transition where R may be a halogen atom or a hydrogen atom. The transition promotes a nonbonding electron localized on the halogen atom X to an antibonding energy level of the molecule R — X. The high energy transition is $^1\Pi_{1u} \leftarrow {}^1\Sigma_g^+$ and is continuous at all wavelengths. The transition leads to dissociation into two unexcited atoms, $X(^2P_{3/2})$. The transition $^3\Pi_{0u}^+ \leftarrow {}^1\Sigma_g^+$ is partly structured and converges to a second continuum where one of the atoms produced is in the excited $^2P_{1/2}$ state (Figure 7.4).

The absorption bands of HI beginning at 360 nm ($\lambda_{max} = 218$ nm) is completely continuous. Evidently the molecules are excited to the repulsive part of the upper electronic state resulting in dissociation with unit quantum efficiency.

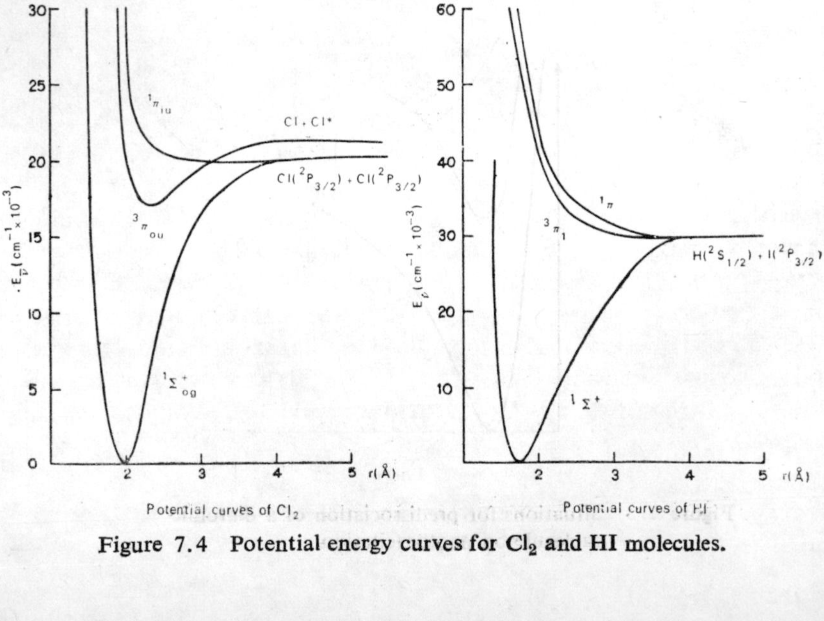

Figure 7.4 Potential energy curves for Cl_2 and HI molecules.

In gaseous mixtures of H_2 and Cl_2, Cl atoms generated by photodissociation initiate a chain reaction by abstraction of H from H_2:

$$\text{Initiation:} \quad Cl_2 + h\nu \xrightarrow{\lambda\, 478\ \text{nm}} 2Cl$$

$$\text{Propagation:} \quad Cl + H_2 \rightarrow HCl + .H \qquad \Delta H = 4.2\ \text{kJ}$$

$$.H + Cl_2 \rightarrow HCl^* + Cl \qquad \Delta H = -\ 189\ \text{kJ}$$

$$.Cl + H_2 \rightarrow HCl + .H$$

Termination of the chains of this reaction is very sensitive to experimental conditions. If atoms are removed by wall effects or by inhibiting molecules such as O_2 present in the system, rates vary directly as the intensity of radiation I. In highly purified gases and large vessels the reaction chains may be as long as 10^5, $\phi_{HCl} \simeq 10^5$. The chain termination by bimolecular atomic recombination is associated with a third molecule to remove excess energy. Here the rate varies as $\sqrt{I}$.

$$\text{Termination:} \quad .Cl + Cl + M \rightarrow Cl_2 + M$$

$$.Cl + .H + M \rightarrow HCl + M$$

The reaction between Br_2 and H_2 needs a somewhat elevated temperature because of the weaker reactivity of Br atoms. Chain termination is mostly by Br atom recombination, giving a $\sqrt{I}$ rate dependence. The expression for quantum yield of HBr formation ϕ_{HBr} can be set up from steady state approximation:

$$\phi_{HBr} = \frac{d\,HBr}{dt} \Big/ \text{rate of light absorption}$$

$$= \frac{k\,[H_2]}{\sqrt{I_{ab}}\left\{1 + \dfrac{k'\,[HBr]}{k''\,[Br_2]}\right\}}$$

No photoreaction occurs between I_2 and H_2 because of the poor reactivity of I atoms. These results emphasize the importance of energy changes involved in bond breaking and bond making in these reactions. The abstraction of H from H_2 by .Cl is thermoneutral ($\Delta H \simeq 4.2$ kJ) but the next step is highly exothermic ($\Delta H = -\ 189$ kJ). Thus the radicals are produced with excess energy. They are called hot *radicals* and have enough energy to propagate the chain. The exothermicity of reaction between Br + H_2 is less and therefore smaller chains are produced whereas I_2 + H reaction is endothermic ($\Delta H \simeq +\ 38$ kcal) and no chain reaction is possible. On the other hand, the quantum yield for photodecomposition of HI, $\phi_{HI} = 2$.

$$HI + h\nu \rightarrow .H + .I$$

$$.H + HI \rightarrow H_2 + .I$$

$$.I + .I \rightarrow I_2$$

The reaction

$$.F + H_2 \rightarrow HF^{*} + .H \qquad \Delta H = -139.9 \text{ kJ}$$

is also exothermic and can produce energy rich HF^{*} molecules. The heat of chemical reaction is distributed in various vibrational-rotational modes to give vibrationally excited HF^{*} or HCl^{*} in large numbers. Emission from these hot molecules can be observed in the infrared region at λ 3.7 μm. The reaction system in which partial liberation of the heat of reaction can generate excited atoms or molecules is capable of laser action (Section 3.2.1). They are known as *chemical lasers.* The laser is chemically pumped, without any external source of radiation. The active molecule is born in the excited state. Laser action in these systems was first observed by Pimental and Kasper in 1965. They had termed such a system as *photoexplosion laser.*

Another system in which laser action was observed is the photolysis of CF_3I emitting at 1.315μ*m.*

$$CF_3I \xrightarrow{h\nu} .CF_3 + I^*$$

The laser action originates from electronically excited I* atoms. This type of laser is termed a *photodissociation laser.* Since there are no vibrational and rotational modes in the I atom, the efficiency of I* production may be 100%. These systems emit in the infrared region.

The photodissociation of I_2 molecules by suitable wavelengths generates one I atom in the ground state I $(5^2P_{3/2})$ and one in the excited state I $(5^2P_{1/2})$.

$$I_2 \xrightarrow{\lambda < 499 \text{ nm}} I + I^*$$

The excited I $(^2P_{1/2})$ has enough energy to abstract H atom from a hydrocarbon in gas phase reactions

$$I(^2P_{1/2}) + CH_3CH_2CH_3 \rightarrow HI + .CH_2CH_2CH_3$$

$$\rightarrow HI + \dot{C}H_3CH\,CH_3$$

In presence of unsaturated compounds, chain reactions can be generated by halogen atoms. For photohalogenation reactions

$$X_2 \xrightarrow{h\nu} 2X\cdot$$

$$\cdot X + H_2\dot{C} = CR_2 \rightarrow R_2CXCR_2$$

$$R_2CX\dot{C}R_2 + X_2 \rightarrow R_2CXCXR_2 + X\cdot$$

The reactivity of halogens in such addition reactions decreases in the order chlorine > bromine > iodine. Usually chloro- and bromo-compounds are produced by this method :

$$CH_2 = CHCl + Cl_2 \xrightarrow{h\nu} CH_2Cl\,CHCl_2$$

$$(CH_3)_3C.C \equiv CH + Br_2 \xrightarrow{h\nu} (CH_3)_3\,CCBr = CHBr$$

$$\phi CH = CHCOOH + Br_2 \xrightarrow{h\nu} \phi CHBr\,CHBr\,COOH$$

The photoaddition of hydrogen bromide to olefines and acetylenes is used to synthesize alkyl and alkenyl bromides

$$CH_3C=CH_2 + HBr \xrightarrow{h\nu} CH_3CH_2CH_2Br$$

$$CH=CH + HBr \xrightarrow{h\nu} CH_2=CHBr$$

In the vapour phase photolysis of hydrocarbons in the *vacuum UV region* (120–200 nm), the main fragmentation process results in the elimination of hydrogen.

$$RCH_2R \xrightarrow{h\nu} R\ddot{C}R + H_2$$

$$\begin{aligned} (H)(R)C=C(H)(H) &\xrightarrow{h\nu} RC\equiv CH + H_2 \\ &\rightarrow RCH=C: + H_2 \\ &\rightarrow R\dot{C}=CH_2 + H. \\ &\rightarrow RCH=CH\ + H^{\cdot} \\ &\rightarrow CH=CH_2 + R^{\cdot} \end{aligned}$$

Azoalkanes when photolyzed in the vapour phase are readily fragmented into alkyl radicals and nitrogen

$$CH_3N=NCH_3 \xrightarrow{h\nu} 2CH_3 + N_2$$

Amides and amines give a variety of products:

$$RCONH_2 \xrightarrow{h\nu} RCO + NH_2; \quad RCH_2NH_2 \xrightarrow{h\nu} RCH_2NH + H^{\cdot}$$

$$\longrightarrow R^{\cdot} + {}^{\cdot}CONH_2 \qquad \rightarrow R\dot{C}HNH_2 + {}^{\cdot}H$$

$$\rightarrow R^{\cdot} + {}^{\cdot}CH_2NH_2$$

B. *Atomspheric photochemistry.* The photodissociation of oxygen in sunlight is the major photochemical process occurring in earth's atmosphere. The first intense allowed transition in O_2 is $B\,^3\Sigma_a^- \leftarrow X\,^3\Sigma_g^-$ which occurs at 202.6 nm and is called the Schumann-Runge band system (Section 2.8). It merges into a continuum beyond 175.9 nm and correlates with one oxygen atom O (2^3P) in the ground state and one in the excited state O (2^1D)

$$O_2(X\,^3\Sigma_g^-) \xrightarrow{h\nu < 175.9} O(2^3P) + O(2^1D)$$

Excited $O(2^1D)$ atoms are metastable and have very long life in the thin upper atmosphere. The atoms are deactivated by energy transfer on collision with oxygen molecules generating singlet molecular oxygen,

$$O(2^1D) + O_2(X\,^3\Sigma_g^-) \rightarrow O(2^3P) + O_2(b\,^1\Sigma_g^+)$$

The ground state oxygen atoms react with oxygen molecules to give ozone in a vibrationally excited state which must be stabilized by a third body M in a number of steps:

$$O(2^3P) + O_2(X^3\Sigma_g^-) \rightarrow O_3^v$$

$$O_3^v + M \rightarrow \rightarrow \rightarrow O_3 + M$$

Two oxygen atoms can also recombine to form O_2 molecules:

$$2\,O(2^3P) + M \rightarrow O_2 + M$$

The emission from electronically excited atomic and molecular oxygen is the cause of the *night airglow.* The atmospheric glow originates in the upper atmosphere where the pressure is so low that the radiative decay can compete favourably with the collisional deactivations.

$$O(^1D) \rightarrow O(^3P) + h\nu(\lambda = 630 \text{ nm}, 636.4 \text{ nm})$$

$$O(^1S) \rightarrow O(^3P) + h\nu(\lambda = \text{green line})$$

$$O_2\,{}^1\Delta_g \rightarrow O_2(^3\Sigma_g^-) + h\nu(\lambda = 1270 \text{ nm})$$

$$O_2\,{}^1\Sigma_g^+ \rightarrow O_2(^3\Sigma_g^-) + h\nu(\lambda = 762.0 \text{ nm})$$

There is sharp variation in night and daytime intensities of these bands suggesting photochemical control of the concentrations of these species. Formation and dissociation of ozone is one of the important mechanisms. The steady state concentration of ozone in the upper atmosphere is very small but it absorbs so strongly in the near ultraviolet that it screens the earth's surface from the light of wavelength $\leqslant 290$ nm which is deleterious to life. Without this shield, the life would not have emerged on earth from under water where it is thought to have originated. Ozone has weak and diffuse absorption throughout the visible region and even in the near infrared. Absorption in the visible region decomposes ozone into oxygen atoms and molecules, both in the ground state, but at $\lambda = 264$ nm, excited O-atoms are produced:

$$O_3 + h\nu \rightarrow O_2(^3\Sigma_g^-) + O(2^1S)$$

$$O_3 + h\nu \rightarrow O_2(^1\Sigma_g^+) + O(2^1D)$$

$$O(2^1D) + O_2(^3\Sigma_g^-) \rightarrow O(2^3P) + O_2(^1\Delta_g)$$

The maximum quantum yield for decomposition of ozone at low pressures is 4.

Though most of the oxygen in the atmosphere has been formed by photosynthesis in plants, some is produced by photolysis of water vapour in the vacuum ultraviolet region $\lambda < 200$ nm. Photolysis of N_2, NO, NO_2, NH_3, CO, CO_2 and small aliphatic hydrocarbons (alkanes) set up complex reactions in the upper atmosphere.

C. *Photochemical Formation of Smog.* In cities with many industries and automobile traffic, the atmosphere is polluted by the smoke coming out from the chimneys of factories and the exhaust of cars. The main components of the smoke are unsaturated hydrocarbons, NO and some sulphur compounds. In the early morning, atmospheric NO concentration is high but after sunrise, NO decreases and NO_2 appears. NO_2 is formed by photochemical reaction between NO and O_3. The ozone is depleted by this reaction and is regenerated by reaction with O-atoms.

$$NO + O_3 \xrightarrow{h\nu} NO_2 + O_2$$

$$NO_2 \xrightarrow{h\nu} NO + O$$

$$O + O_2 + M \rightarrow O_3 + M$$

Three specific eye irritants have been identified in photochemical smog: formaldehyde, acrolein and peroxyactyl nitrate (PAN). The possible reaction sequences are:

$$O_2(^1\Delta_g) + -C=C-\underset{H}{C} \rightarrow -\underset{OOH}{\underset{|}{C}}-C=C-$$

$$\downarrow \text{thermal}$$

$$\text{radicals, .RCO, etc}$$

$$\cdot RCO + O_2\,(^3\Sigma_g^-) \rightarrow \cdot RCO_3$$

$$\cdot RCO_3 + NO \rightarrow \cdot RCO_2 + NO_2$$

$$\cdot RCO_2 + NO \rightarrow .RCO + NO_2$$

$$RCO_3 + NO_2 \rightarrow RCO_3NO_2 \text{ (PAN)}$$

$$R'CH_2 + O_2 \rightarrow R'CH_2O_2$$

$$R'CH_2O_2 + NO \rightarrow R'CH_2O + NO_2 \rightarrow R'CH_2ONO_2$$

(alkyl nitrate)

Cities liable to meterological conditions of 'atmospheric inversion' are specially subject to smog trouble.

D. *Mercury photosensitized reations.* Mercury atoms are frequently used as photosensitizers in vapour phase reactions. The mechanisms involved are

$$Hg\,(6^1S_0) \xrightarrow[h\nu]{253.7\text{ nm}} Hg\,(6^3P_1)$$

$$Hg\,(6^3P_1) \rightarrow Hg\,(6^1S_0) + h\nu \quad \text{(a) resonance radiation}$$

$$\xrightarrow{+A} Hg\,(6^1S_0) + A^* \quad \text{(b) energy transfer}$$

$$\xrightarrow{+M} Hg\,(6^1S_0) + M \quad \text{(c) collisional deactivation}$$

$$\xrightarrow{+M} Hg\,(6^3P_0) + M \quad \text{(d) collisional deactivation to } 6^3P_0 \text{ state}$$

$$\xrightarrow{RH} Hg\,(S_0) + R + H^\cdot \quad \text{(e) radical formation}$$

Hg (6^3P_0) cannot return to the ground state Hg (6^1S_0) by emission since the transition is doubly forbidden by the selection rules $J = 0 \not\to J = 0$ and $\Delta S \neq 1$. Thus Hg (6^3P_0) has a very long radiative lifetime and can be deactivated mainly by collisions.

The mercury sensitized emission, (process (b)) from Tl and Na vapour has been discussed in Section 6.6.6. The quenching to 6^3P_0 (process (d)) by CO, N_2 and N_2O, NH_3 are presented in Section 6.2. In presence of H_2, the possible photochemical reactions are

$$\begin{aligned} Hg\,(6^3P_1) + H_2 &\rightarrow Hg\,(6^3P_0) + H_2 \\ &\rightarrow Hg\,(6^1S_0) + H_2 \\ &\left.\begin{aligned}&\rightarrow Hg\,(6^1S_0) + 2H \\ &\rightarrow HgH + H\end{aligned}\right\} \phi = 1 \end{aligned}$$

The first two are of minor importance. With olefins;

$$Hg\,(6^3P_1) + \text{olefin} \rightarrow Hg\,(^1S_0) + {}^3\text{olefin}\ (v > 0\text{:hot olefin triplets})$$

The relative reactivities of various substances towards mercury triplets are expressed as quenching cross-sections for mercury phosphorescence at 253.7 nm. The quenching crosssections are large for compounds having low lying triplet states. The large amount of vibrational energy accompanying the transfer from such energetic species yields 'hot' triplets which possess many interesting modes of decomposition normally not observed for relaxed triplets:

$$\begin{aligned} Hg^* + CH_2 = CH_2 \rightarrow CH_2 = CH_2^* &\rightarrow HC \equiv CH + H_2 \\ &\rightarrow CH_2 = CH + H^* \\ &\rightarrow CH_3CH: \\ &\xrightarrow{C_2H_4} \square\ (\text{dimer}) \end{aligned}$$

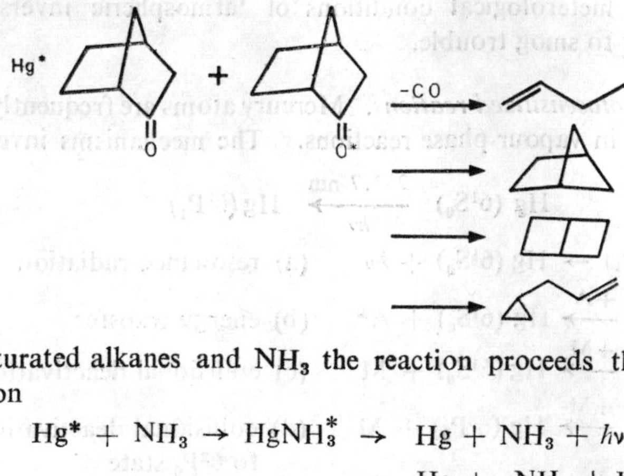

With saturated alkanes and NH_3 the reaction proceeds through complex formation

$$\begin{aligned} Hg^* + NH_3 \rightarrow HgNH_3^* &\rightarrow Hg + NH_3 + h\nu \\ &\rightarrow Hg + .NH_2 + H\cdot \\ Hg^* + (CH_3)_2CH_2 &\rightarrow (CH_3)_2 - CH_2 \ldots Hg^* \end{aligned}$$

E. *Photofragmentation in the liquid phase.* Photodissociation reactions in liquid phase occur at much reduced quantum yields because of the possibility of recombination within the *solvent cage*. Furthermore the product formation and distribution also differ because of collisional deactivation of initially produced vibrationally excited hot molecules. Where CT absorption produces radical ions, the solvent may react with the ionic species.

When azobenzene vapour is photolyzed, fragmentation occurs producing molecular nitrogen and alkyl radicals, but in liquid phase the main reaction is isomerization:

$$CH_3{\diagdown}N=N{\diagdown}CH_3 \quad \overset{vapour}{\longrightarrow} \quad 2 \cdot CH_3 + N_2$$

$$\underset{solution}{\longrightarrow} \quad CH_3{\diagdown}N=N{\diagup}CH_3$$

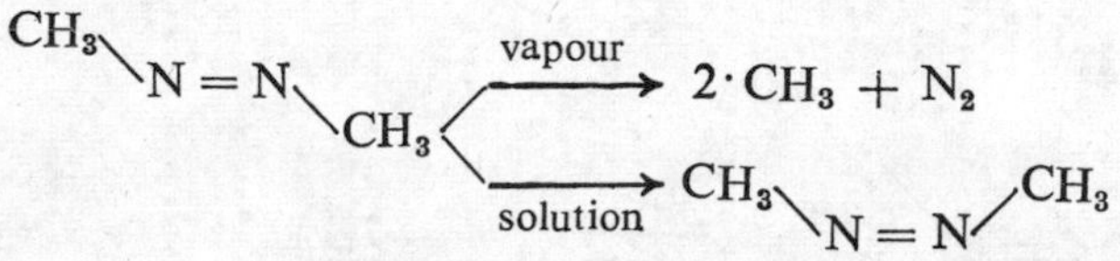

F. *Photodegradation of polymers*: Photodegradation of polymers assumes importance in two different contexts: (i) ultraviolet and visible radiations are harmful to biopolymers like DNA, polysaccharides, proteins, etc. and an understanding of their mode of photolysis is important in life processes and (ii) more and more use of plastic materials in everyday life has created a problem of disposal.

The degradation can be photochemically induced (a) homolytic or (b) heterolytic cleavage at the weaker bonds. The photolysis of the type (a) may lead to elimination reactions and the type (b) may lead to free radical formation. The point of bond cleavage may not be the seat for light absorption. The energy can migrate from unit to unit until it finds itself at the seat of reaction.

In presence of O_2 and a photosensitizing dye, cotton fibres undergo photodegradation by rather complex reactions. The singlet oxygen formed as an intermediate is the reactive species.

7.4.2 Isomerization and Other Rearrangement Reactions

A. *Cis-trans isomerization.* The simplest molecules which can undergo cis-trans isomerization are derived from ethylene $H_2C = CH_2$. The lowest excited state of ethylene results from excitation of a π-electron into an antibonding π^*-orbital. In this state, the bonding effect of π orbital is cancelled by antibonding effect of π^* orbital and the resulting $C - C$ bond is effectively a single bond. Consequently, free rotation around $C - C$ bond becomes possible. The planar molecule obtained from Franck-Condon excitation, relaxes to a configuration in which two $-CH_2$ groups are perpendicular to each other giving rise to a diradical (Figure 7.5).

The possibility of free rotation facilitates cis-trans isomerization and can be demonstrated in substituted ethylenes. Starting from any given

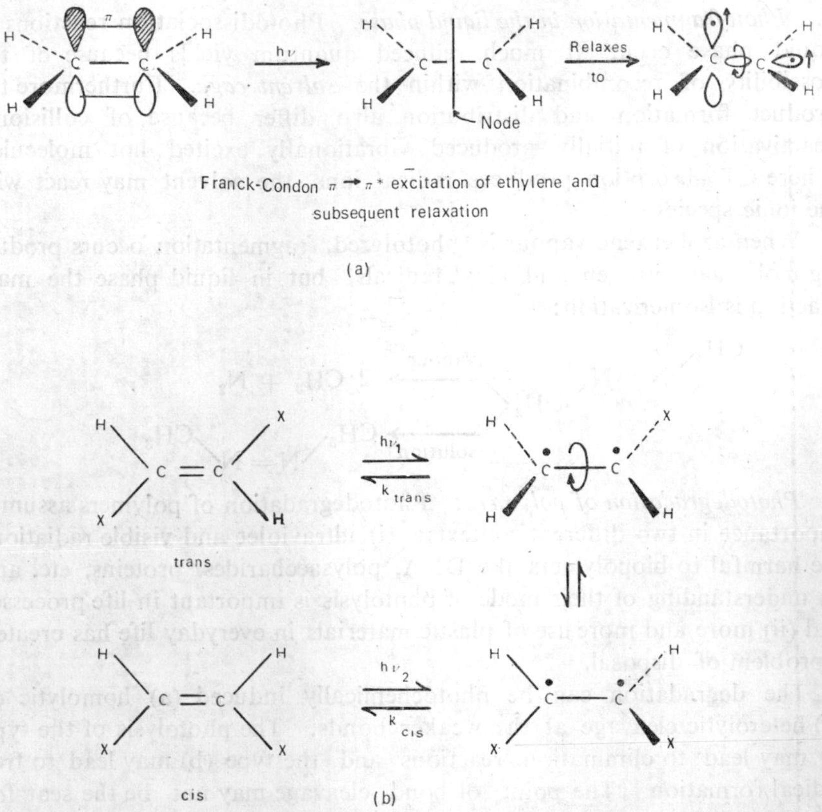

Figure 7.5 (a) Franck-Condon $\pi \rightarrow \pi^*$ excitation of ethylene and subsequent relaxation; (b) trans-cis isomerization in the excited state.

configuration, the corresponding rotational isomer can be produced on photoexcitation by suitable wavelength. A photostationary equilibrium is established, differing from the thermodynamic equilibrium of the ground states.

For the next higher polymer, 1, 4-butadiene, rotation is considerably restricted in the (π, π^*) state because of migration of the double bond in the centre in the diradical configuration but the ground state rotamers are in rapid equilibrium.

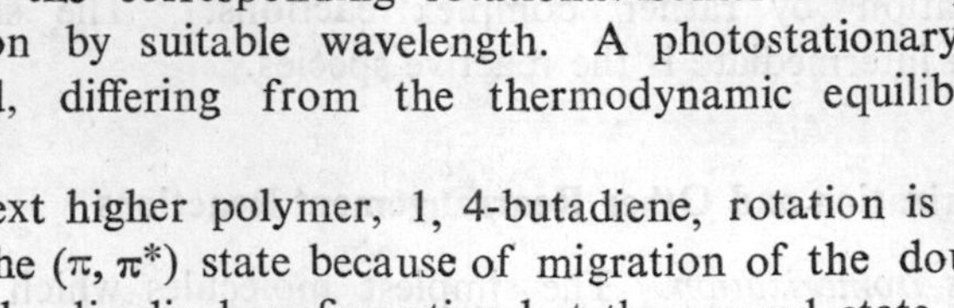

In the next higher vinylog, *cis-trans* isomerization in the excited state again becomes feasible around each single bond, increasing the number of intermediate configuration.

Substituted 1,3 butadienes are well known to undergo such *cis-trans* isomerization as primary photochemical processes.

In phenyl substituted ethylene, stilbene $\phi CH{=}CH\phi$, the central π-system can conjugate with the phenyl π-orbitals. The ground state (I) and the lowest excited state (II) MO's in transtilbene are indicated as HOMO (highest occupied molecular orbital) and LUMO (lowest unoccupied molecular orbital) in Figure 7.6.

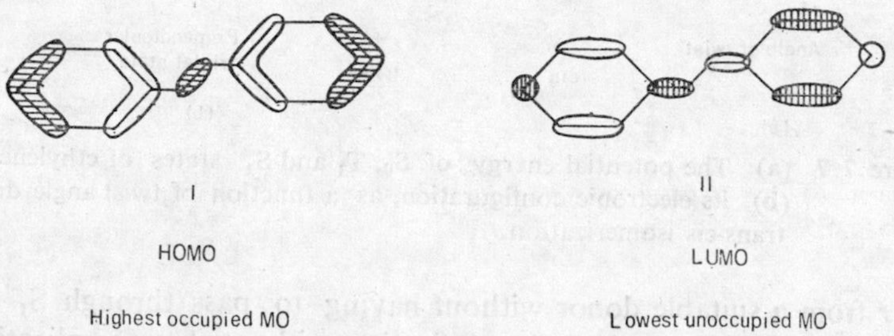

Figure 7.6 Nodal properties of HOMO and LUMO of stilbene.

A photostationary state *cis-trans* equilibrium is established on excitation of either cis-or trans form of stilbene on promotion to the S_1 state.

The photosensitized isomerization of ethylenes and stilbenes has been studied in detail by Hammond and his group and it has been found that in each case both the *cis-trans* and the *trans-cis* isomerizations occur with equal efficiency. A common (π, π^*) triplet state is assumed to be the intermediate in each case. This triplet state is referred to as the *perpendicular triplet state* and was termed a *phantom* triplet state by Hammond. The energy of the perpendicular triplet state is expected to be low since in this configuration the overlap between the π and π^* orbitals is minimized. The potential energy of S_0, T_1 and S_1 states as a function of twist angle is given in Figure 7.7.

The perpendicular triplet state T can even be lower than the ground singlet state in 90° configuration. Deactivation of molecules from this state can give either the ground state of *cis*-isomer or the ground state of *trans*-isomer depending on the mode of rotation. In either case molecules regain their planar geometry.

The perpendicular triplet state can be directly populated by *energy*

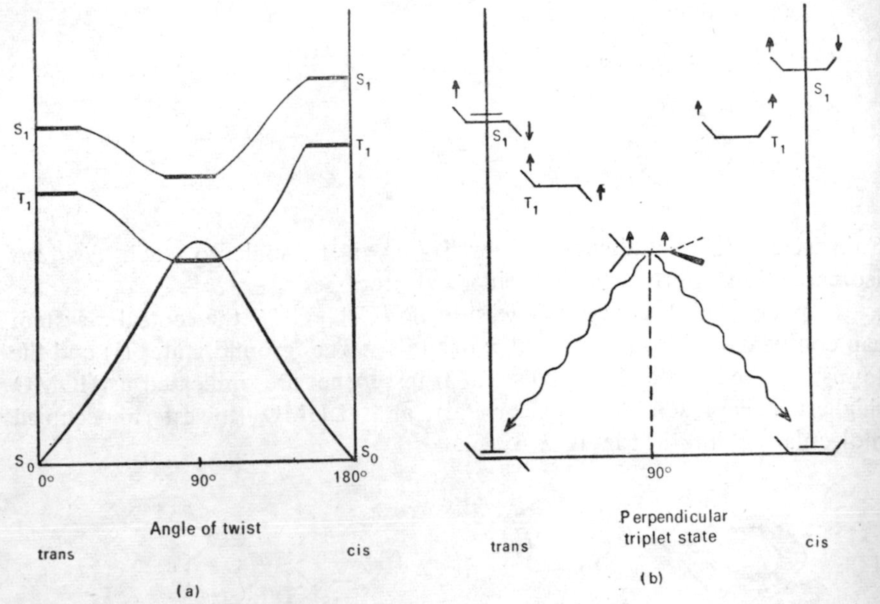

Figure 7.7 (a) The potential energy of S_0, T_1 and S_1 states of ethylene and (b) its electronic configuration, as a function of twist angle during trans-cis isomerization.

transfer from a suitable donor without having to pass through S_1 and T states. This leads to clean isomerization without the complications o photoproducts from these excited states. Extensive studies of sensitize *cis-trans* isomerization has confirmed its existence. Sensitizers wit energies *less* than the T_1 states of either *cis*-or *trans* forms are observed t induce isomerization. Such endothermic transfer is not expected fron simple theories of energy transfer and also does not follow the Franck Condon principle of vertical excitation. Therefore, transition t perpendicular triplets is termed a *nonvertical transition.*

The dienes, specially 1,3-pentadiene (piperylene) and hexadiene quench the triplets of suitable sensitizers by energy transfer with uni efficiency. Hence, they are used widely in mechanistic studies of photo chemical reactions, either to count the triplets or to establish the triple energy of a sensitizer whose E_T is not determinable from spectroscopi data (chemical spectroscopy).

B. *Valence isomerization.* The conjugated polyenes in their excite biradical state can lead to intramolecular cyclization in a number c ways giving rise to a number of products. In such cyclization, valenc bonds are reorganized without migration of atoms or groups but b migration of σ or π electrons only. For example, on excitation c pentadiene, the following products are observed:

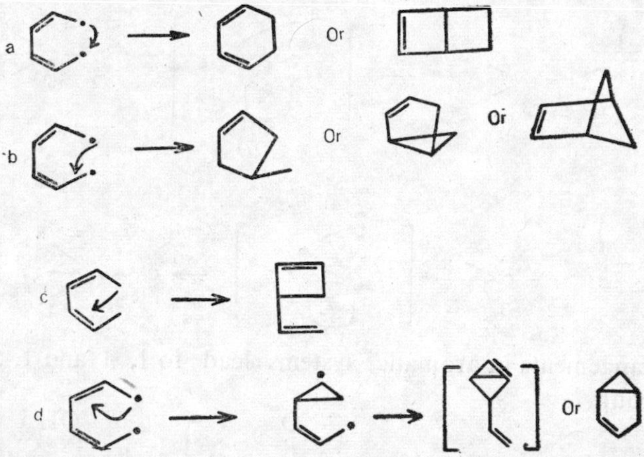

For butadiene, two photochemical cyclization products are cyclobutene or bicyclobutane.

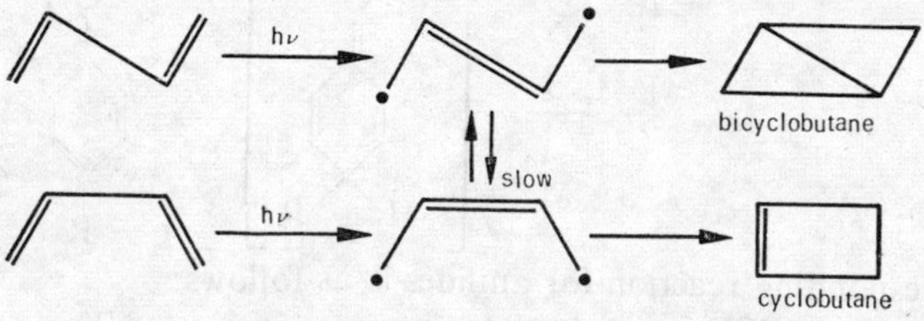

In some cases the nature of the products resulting from photolysis of dienes and trienes depends on whether the reaction occurs from the excited singlet state or the triplet state. The triplet states can be populated exclusively by suitable sensitizers. For example, 3-methylene-1, 5-hexadiene gives a cyclobutene derivative (A) from the singlet state and tricyclo ring compound (B) from triplet state:

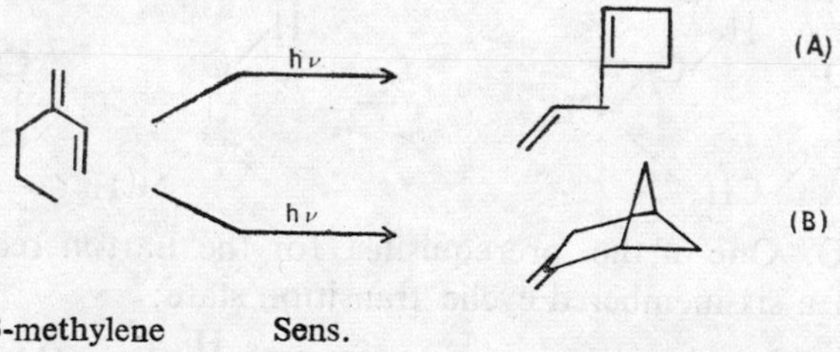

3-methylene
1, 5-hexadiene Sens.

C. *Photo-Fries rearrangement*: Photo-Fries rearrangement involve migration of a group across a double bond (1, 3 migration) in structures of the following type:

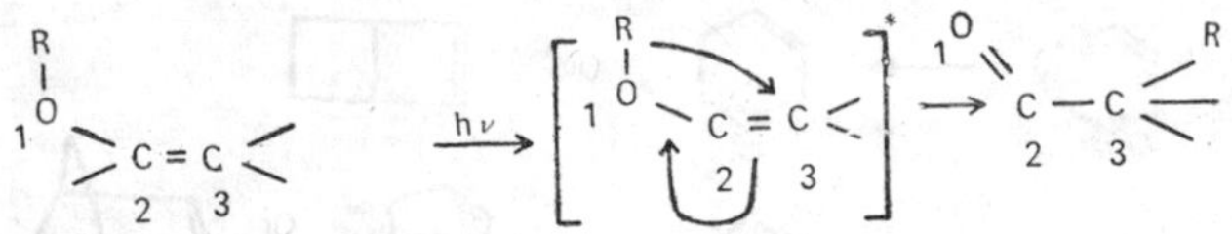

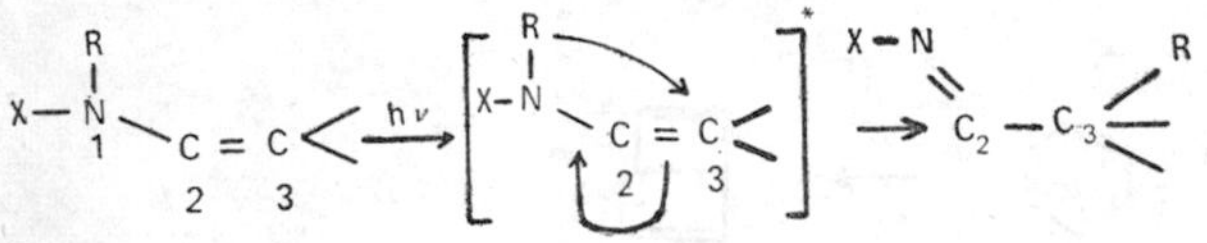

Such rearrangements in aromatic systems lead to 1, 3 and 1, 5 migration of an R group

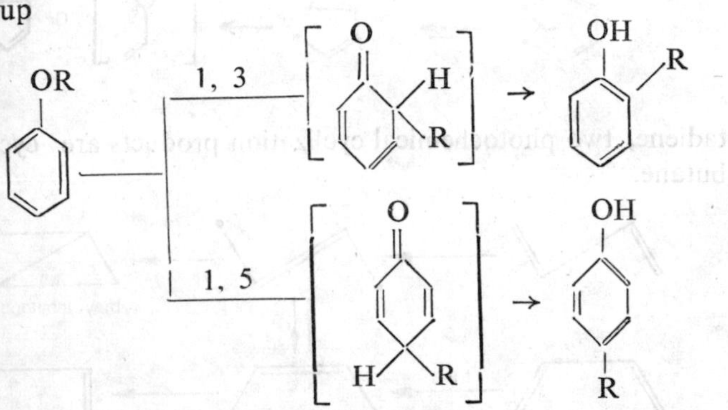

The corresponding reaction for anilides is as follows:

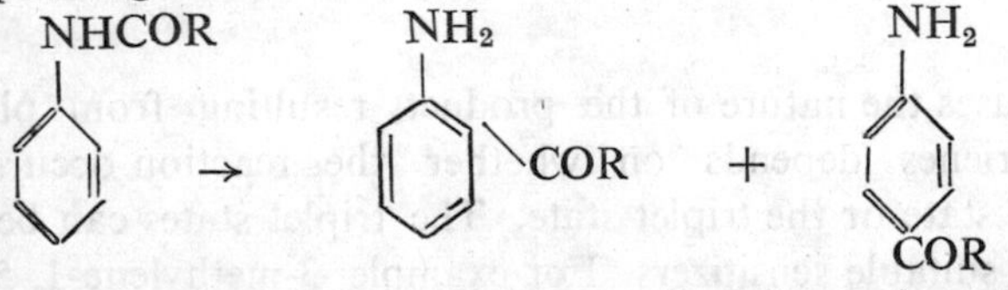

These rearrangements occur through intermediate formation of free radicals.

D. *Barton reaction*

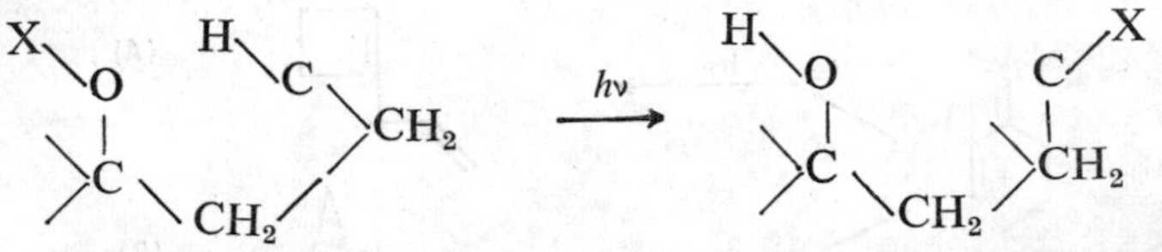

where X = NO. One of the prerequisites for the Barton reaction is the availability of a six-membered cyclic transition state:

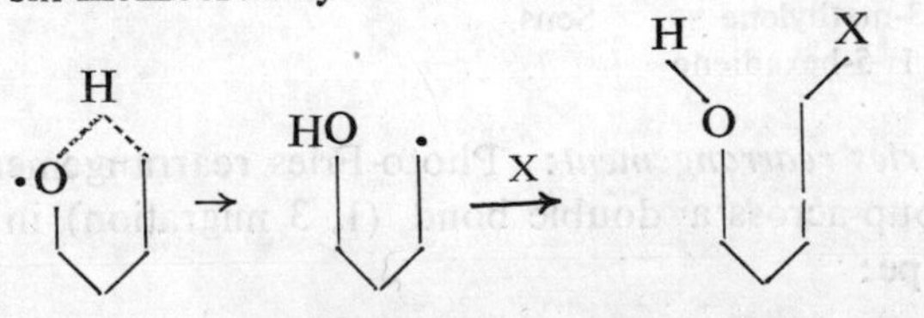

The oxygen atom of the activated alkoxy radical and the hydrogen to be abstracted and subsequently replaced by X, form two adjacent corners of the six-membered transition state. The major application of the Barton reaction have been in the synthesis of steroids particularly with compounds involving functionalization of C_{18} and C_{19} which are difficult to prepare in other ways.

E. *Photoisomerization of benzene.* Photolysis of liquid benzene by excitation to its (π, π*) states brings about interesting photochemical transformations from vibronically excited singlet as well as triplet states (Figure 7.8).

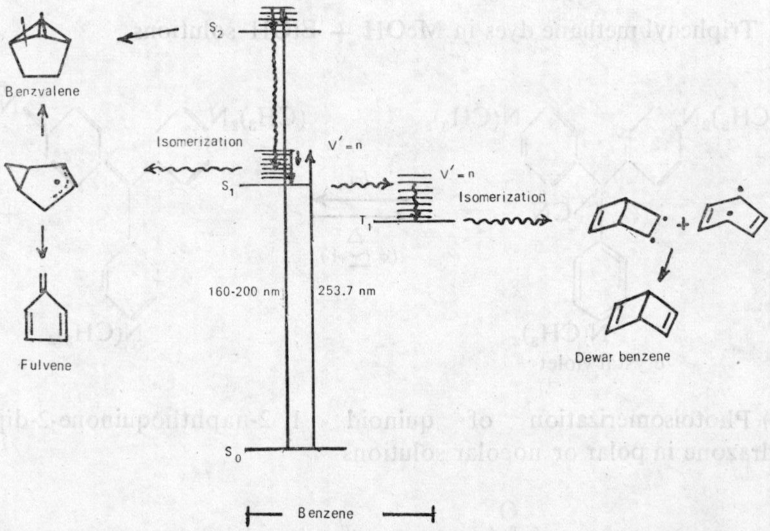

Figure 7.8 Photolysis products of liquid benzene under different excitation conditions (1) excitation λ 253.7 nm → fulvene + benzvalene (2) excitation λ 160 – 200 nm → benzvalene + fulvene + Dewar benzene.

F. *Photochromism or phototropism.* In many compounds, light induced reversible colour change is observed. Such light induced reversible transitions lead to states with quantum-mechanical stability but thermodynamic instability

$$D \overset{h\nu}{\rightleftharpoons} {}^1D \rightleftharpoons {}^3D \rightleftharpoons \underset{\text{photochromic species}}{X}$$

Photochromism in solution and in the solid state can be the result of intramolecular changes, e.g. tautomerism, ring opening, cis-trans isomerization, free radical formation, stereoisomeric transition, formation of dimers and such similar reversible reactions. Some examples are given below:

(i) Free radical formation in cooled solid solution of tetrachloro-1-oxo-dihydronaphthalene

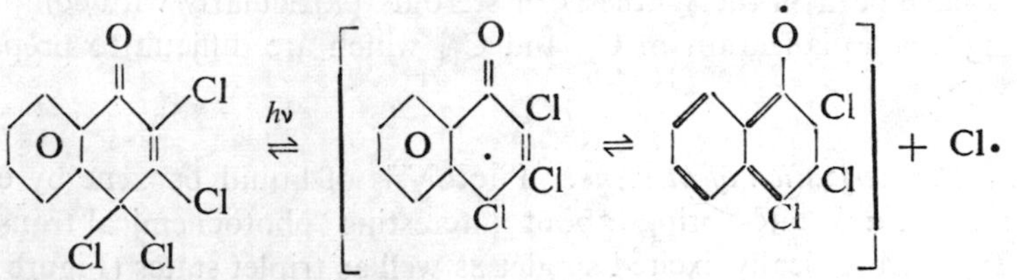

tetrachloro-1-oxo-dihydronaphthalene

(ii) Triphenyl methane dyes in MeOH + EtOH solutions

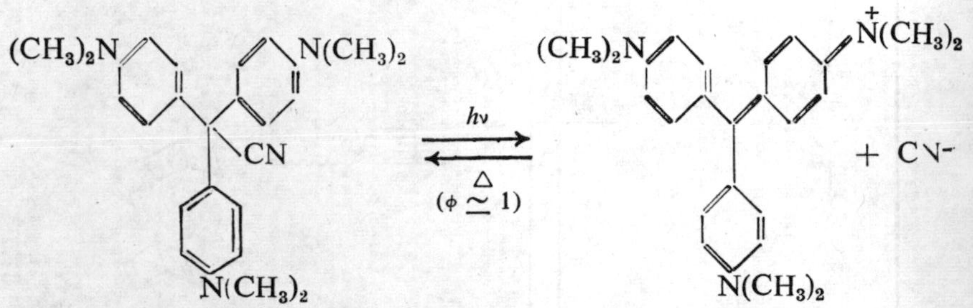

crystal violet

(iii) Photoisomerization of quinoid 1, 2-naphthoquinone-2-diphenyl hydrazone in polar or nopolar solutions

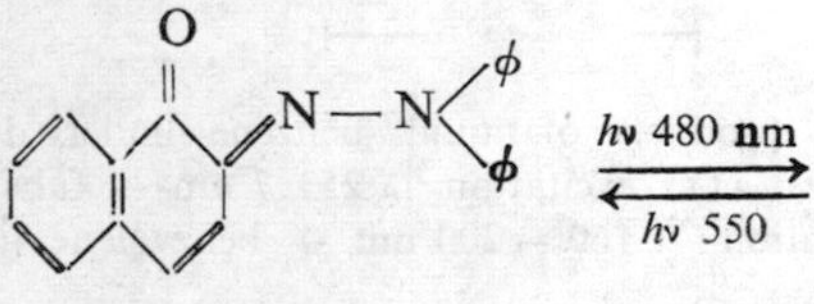

Photochromic or phototropic dye stuffs are used as the basis of photochemical high speed memory with an erasable image. They can also be used as automatic variable density filters for example as Q-switches in high intensity lasers. For thermochromic substances colour change is observed on change of temperature (spirans, bianthrones).

EIGHT

Some Aspects of Organic and Inorganic Photochemistry

8.1 PHOTOREDUCTION AND RELATED REACTIONS

The prototype of photoreduction reactions is hydrogen abstraction by carbonyl compounds in presence of suitable H-donors. Such H-atom transfer may be visualized to occur first by transfer of an electron followed by proton transfer. An electron deficient centre is the seat of reaction and the efficiency of the reaction depends on the nucleophilic nature of the donor.

The MO description of simple aldehydes and ketones assigns a pair of nonbonding electrons to the carbonyl oxygen which are directed in the plane of the molecule but perpendicular to the C=O bond axis. (Section 2.9, Figure 2.16). The lowest energy transitions are $n \rightarrow \pi^*$ and $\pi \rightarrow \pi^*$ in which either an n or a π-electron is transferred to an antibonding orbital. For simple aliphatic ketones the lowest singlet state is S_1 (n, π^*) and the corresponding triplet T_1 (n, π^*). In aromatic ketones, singlet may still be S_1 (n, π^*) but the lowering of (π, π^*) state may bring the $T(\pi, \pi^*)$ in the proximity of $T(n, \pi^*)$ or even below the latter. Therefore in substituted aromatic ketones, the nature of substituent decides the lowest triplet state. The low energy triplet state may be a mixture of $T(\pi, \pi^*)$ and $T(n, \pi^*)$ characters: $T_1 = aT(n, \pi^*) + bT(\pi, \pi^*)$ (Figure 8.1). Energies, E_T, of $T_1 \rightarrow S_0$ transitions as obtained from phosphorescence spectra and the corresponding configurations of the

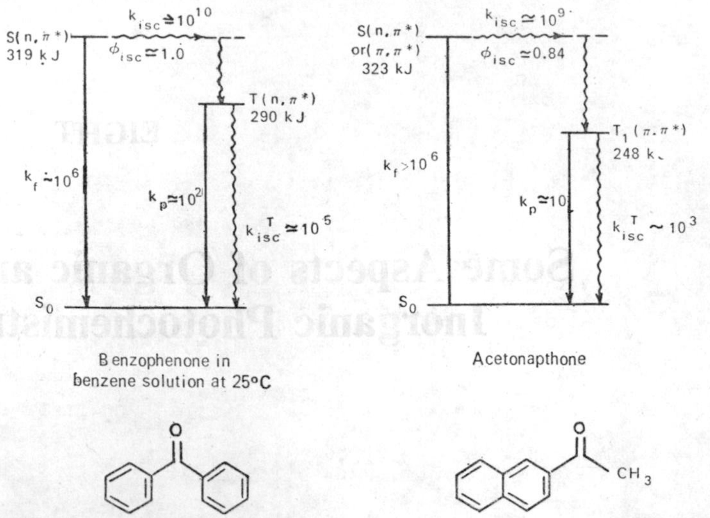

Figure 8.1 Energy state diagram for benzophenone and acetonaphthone.

lowest triplet state of some aldehydes are given in the Table 8.1.

TABLE 8.1

Energies and electron configuration of the lowest triplet state of some aryl aldehydes

Compound	E_T	T_1
Benzaldehyde	72	(n, π^*)
2-hydroxybenzaldehyde	71	(n, π^*)
2-naphthaldehyde	59.5	(π, π^*)
1-naphthaldehyde	56	(π, π^*)
9-anthraldehyde	40	(π, π^*)

A third possibility arises when an electron releasing substituent is suitably attached to an aromatic ring of a ketone. The charge may eventually migrate to the carbonyl oxygen producing a charge transfer state (CT). All the three types of excited states $(n, \pi*)$, (π, π^*) and (CT) have different electron distribution, hence exhibit differences in their photochemical reactivity. The major contributing forms for these three electronic states are given below :

(1) For (n, π^*) state in benzophenone,

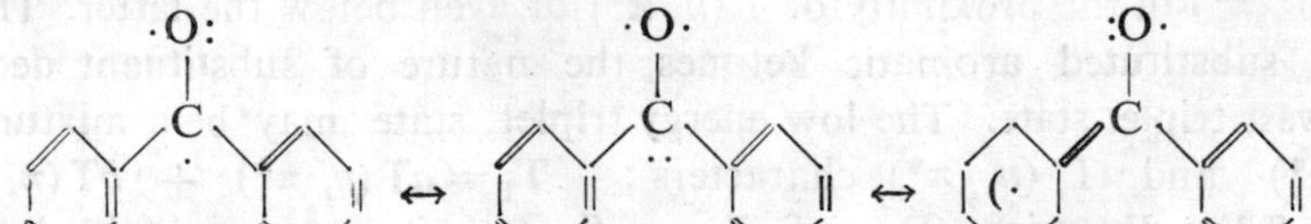

major contributing form

(2) For (π, π^*) state in acetophenone,

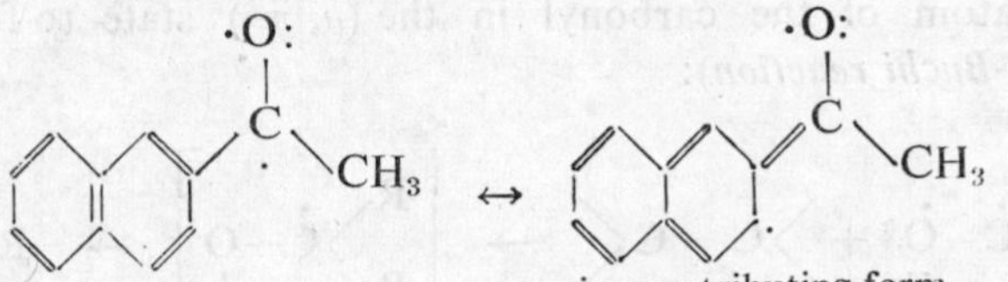

major contributing form

(3) For (CT) state in p-aminobenzophenone,

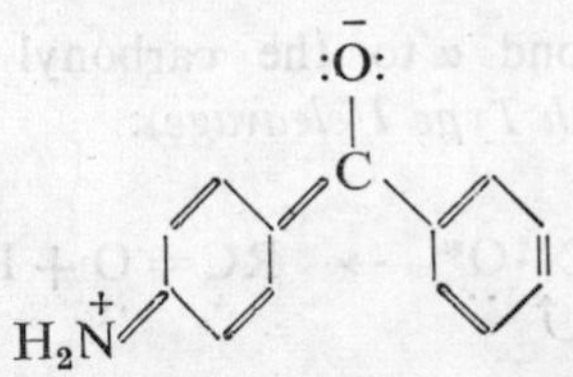

In most simple aldehydes and ketnoes, including benzophenone, the longest wavelength absorption is a low intensity $n \rightarrow \pi^*$ transition. The promotion of a n-electron, localized on O-atom to a π-orbital, leaves behind a 'positive hole' on this atom. The charge density on C-atom is increased creating a 'bipolar' state. The dipole moment of $\rangle C=O$ bond is reduced. Three primary processes are commonly encountered for this electrophilic centre :

(a) If a suitable H-donor is present in the vicinity, the H-atom is extracted by the excited carbonyl compound, producing a ketyl radical, which can subsequently dimerize to pinacols. For benzophenone in 2-propanol as solvent, benzpinacol is formed with unit quantum yield.

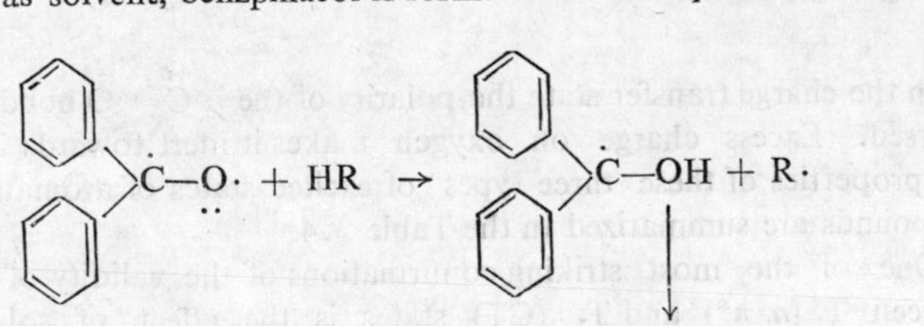

benzpinacol

At high alcohol concentration or high light intensity ϕ_R tends to 2.

$$(C_6H_5)_2C=O^* + (CH_3)_2CHOH \rightarrow C_6H_5\dot{C}OH + (CH_3)_2\dot{C}OH$$

$$(C_6H_5)_2C=O + (CH_3)_2\dot{C}OH \rightarrow (C_6H_5)_2\dot{C}OH + (CH_3)_2C=O$$

$$2\,(C_6H_5)_2\dot{C}OH \rightarrow (C_6H_5)_2COH.COH\,(C_6H_5)_2$$

(b) A compound with unsaturated linkage can undergo addition with the oxygen atom of the carbonyl in the (n, π^*) state to form oxetane (*Paterno-Buchi reaction*):

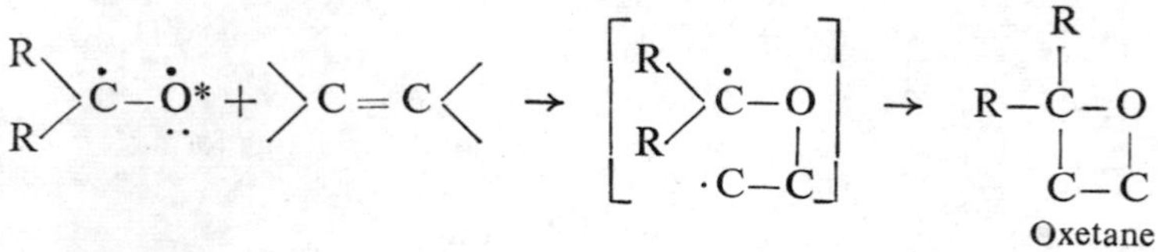

Oxetane

(c) Cleavage of the bond α to the carbonyl group produces radical intermediates (*Norrish Type I cleavage*):

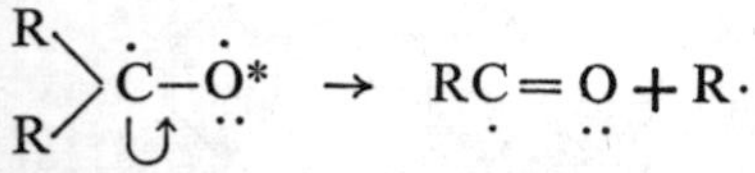

For a $\pi \rightarrow \pi^*$ transition, the charge density on C=O oxygen increases due to the antibonding character of π^* orbital. The electrophilic character of the oxygen centre is decreased considerably leading to negligible reactivity for H-abstraction from 2-propanol. The hydrogen abstraction may still occur if a weakly bonded hydrogen donor such as tributylstannate is available.

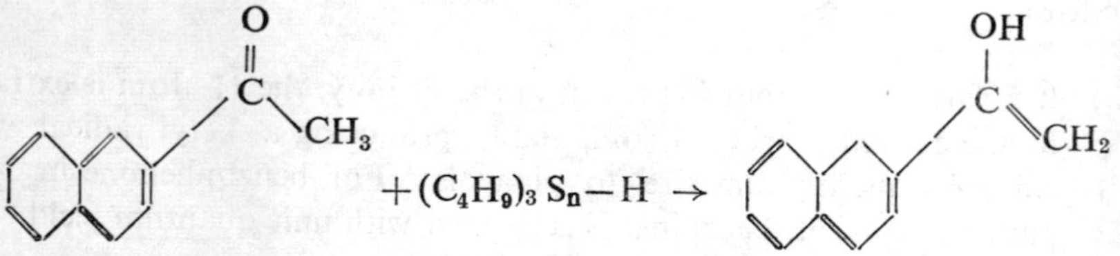

In the charge transfer state the polarity of the $>\overset{+\delta}{C}=\overset{-\delta}{O}$ bond is actually reversed. Excess charge on oxygen makes it inert towards 2-propanol. The properties of these three types of excited states of aromatic carbonyl compounds are summarized in the Table 3.4.

One of the most striking confirmations of the validity of distinction between T_1 (n, π^*) and T_1 (CT) states is the effect of solvent on the reactivity of some of these compounds such as p-aminobenzophenone. From a survey of the energy states in 2-propanol and in cyclohexane for this compound, it is observed that because of the highly polar nature of the (CT) state, which lies very close to (n, π^*) state, there is a large solvent shift in the polar solvent 2-propanol. This brings about a lowering of the (CT) state with respect to (n, π^*) state. A switching of state energies in polar solvents occurs which imparts exceptional reactivity to this ketone. It can abstract H-atom from cyclohexane solvent but not from 2-propanol.

In acidic media p-aminobenzophenone undergoes ready photoreaction with 2-propanol since no charge transfer is possible in the protonated ketone.

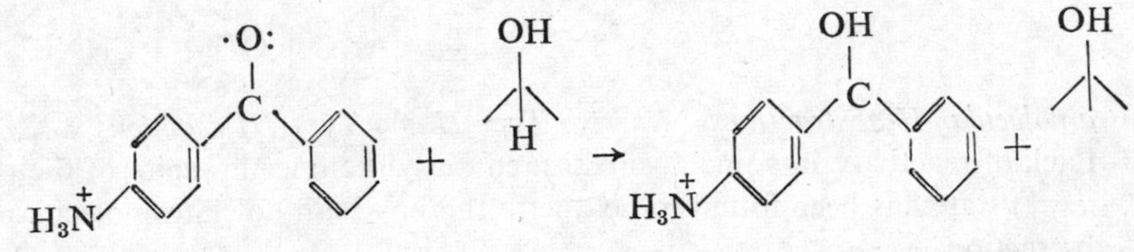

A variety of H-donors have been used for the photoreduction of carbonyl compounds. They include amines, alcohols, hydrocarbons, phenols and amides. The presence of compounds with double bonds lead to *oxetane* formation.

(a) $(C_6H_5)_2 C = O + RCH_2NHR' \xrightarrow{h\nu}$
$(C_6H_5)_2 \underset{\underset{OH}{|}}{C} - \underset{\underset{OH}{|}}{C} (C_6H_5)_2 + RCH = NR'$

(b) $(C_6H_5)_2 C = O + CH_3CON (CH_3)_2 \xrightarrow{h\nu}$
$(C_6H_5)_2 \underset{\underset{OH}{|}}{C} - CH_2CO N (CH_3)_2$

(c) $C_6H_5CHO \xrightarrow{h\nu} \underset{15\%}{\phi \, CHOHC_6H_5} + \underset{\underset{11\%}{\overset{OH \quad OH}{\underset{|}{|} \quad \underset{|}{|}}}}{\phi \cdot CH - CH \cdot \phi}$

$\downarrow \begin{array}{c} h\nu \\ + 2\text{-methyl-} \\ 2\text{-butene} \end{array}$

40%   25%

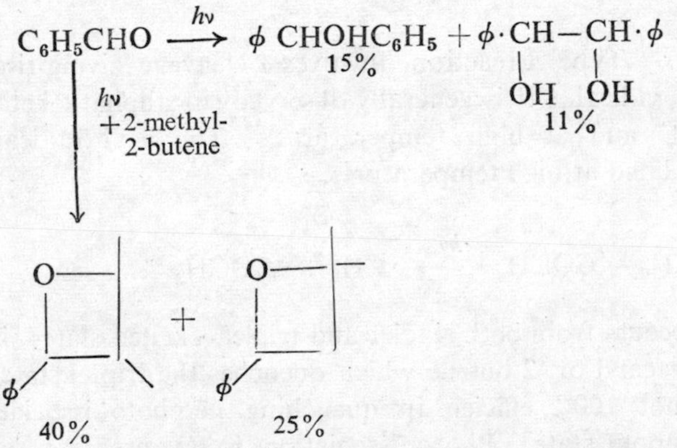

The photoreduction of aryl ketones by amines generally occur via a charge transfer interaction between the triplet state of ketone and the amines, as shown in the following scheme. A ketone radical anion and an amine radical cation are formed in the intermediate stage.

$$Ar_2C{=}O\,(T_1) + RCH_2NR_2' \rightarrow [Ar_2\dot{C}-\bar{\dot{O}}\ldots RCH_2-{}^{+}\dot{N}R'_2]$$

$$\downarrow$$

$$Ar_2\dot{C}-OH + RCHNR'_2$$

Intramolecular H-abstraction: *Norrish Type II and Type III reactions*

Lack of reactivity in some *O*-substituted diaryl ketones in spite of their T_1 (n, π^*) state has been found to be due to the possibility of intramolecular H-abstraction

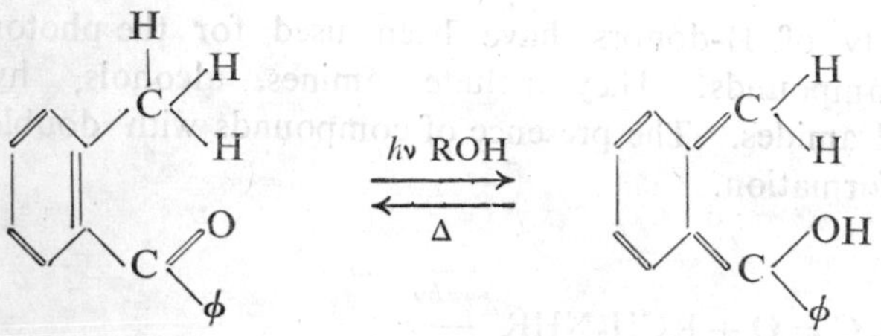

(A)

These compounds are often used as photostabilizers in plastic materials. 2-methylbenzophenone (A) appears to undergo intramolecular H-abstraction in 2-propanol in the excited state. The unstable enol reverts to the ketone in the ground state. The possibility of such a six-membered transition state does not exist for 2-t-butyl benzophenone, and it is as reactive towards H-abstraction as benzophenone itself.

Intramolecular interactions have been classified as Norrish Type II and Type III reactions.

Norrish type reactions. Type I reaction involves α-cleavage giving rise to an acyl and an alkyl radical. It is generally observed in aliphatic ketones in the vapour state and at high temperatures. The acyl radical is essentially decarbonylated at high temperatures.

$$CH_3-COCH_3 \xrightarrow{h\nu} \cdot CH_3 + \cdot COCH_3$$

Type I cleavage occurs from both singlet and triplet excited states since high pressures of biacetyl or 2-butene which quenches the triplet state by energy transfer are not 100% efficient in quenching the photodissociation of acetone in the vapour state. Photodissociation is favoured if a relatively stable alkyl radical is ejected. The yields of Type I cleavage are usually lower in inert solvents than in vapour phase due to radical recombination by solvent cage effect. In solvents containing C–H bonds, fragmentation reaction has to compete with H-abstraction reaction. Dibenzyl ketones fragment very readily in solution to give carbon monoxide

xide and benzyl radical. Tetramethyl oxetanone undergoes fragmentation with a quantum yield close to unity from the singlet state.

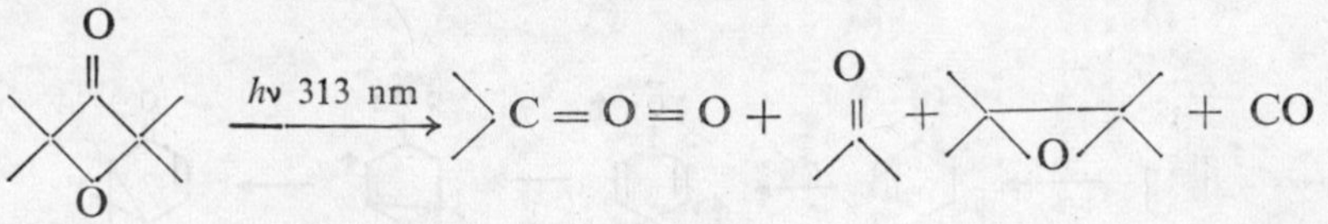

Type II reaction occurs in carbonyl compounds which possess γ-hydrogen atoms. The reaction proceeds by a shift of a γ-hydrogen to oxygen with subsequent cleavage to an olefine and an enol.

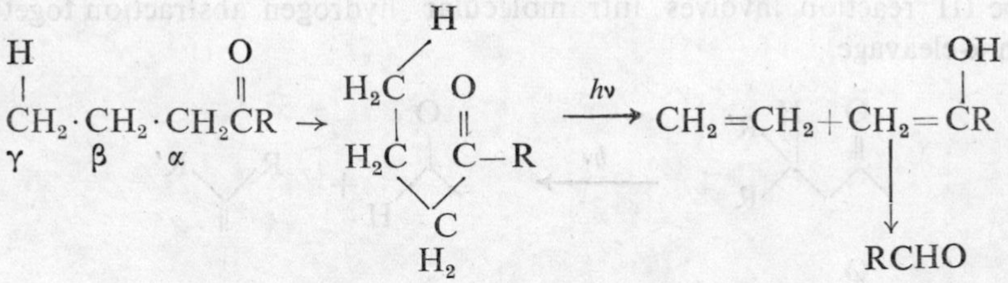

In effect it is intramolecular H-abstraction followed by β-cleavage and is also called Type II photoelimination.

Sometimes ring closure to form cyclobutanols is favoured. For 6-hepten-2-one (B) in pentane solution the total photoreaction is:

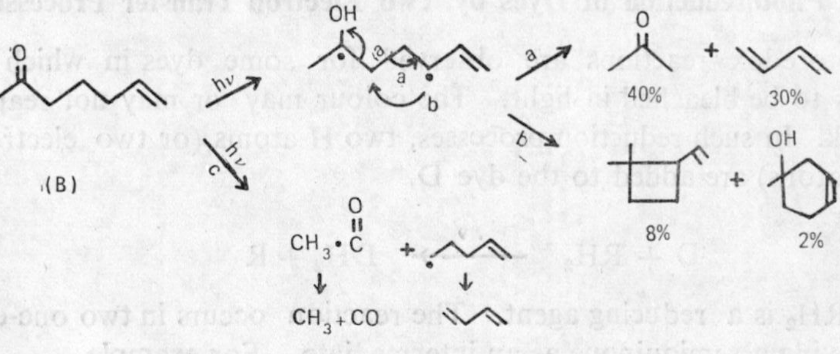

The reaction can occur from the excited singlet and the triplet state as evidenced from the quenching of reaction in 2-pentanone and 2-hexanone with piperylene. Piperylene can quench the reaction only partially, suggesting singlet state reaction mechanism for the unquenched fraction. Photoenolization resembles Type II process in that a γ-hydrogen migrates to the carbonyl oxygen.

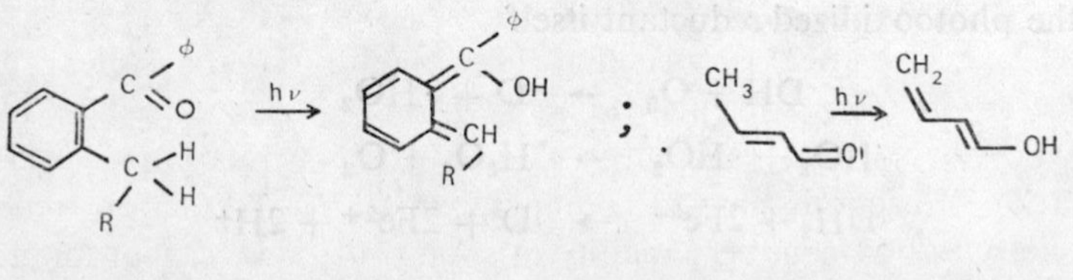

Several photoisomerization reactions also proceed by intramolecular hydrogen abstraction in a triplet state

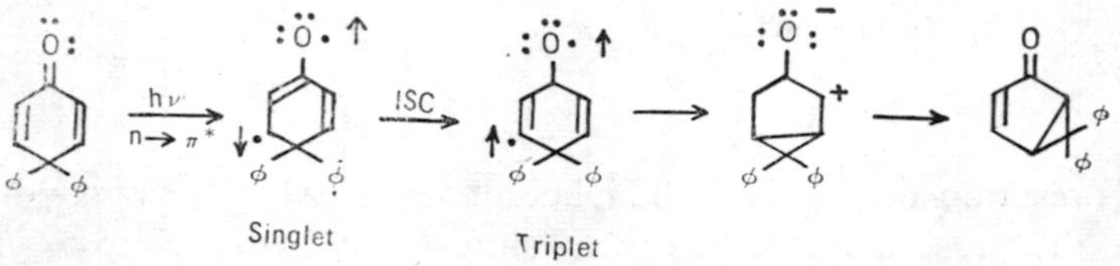

Type III reaction involves intramolecular hydrogen abstraction together with α-cleavage.

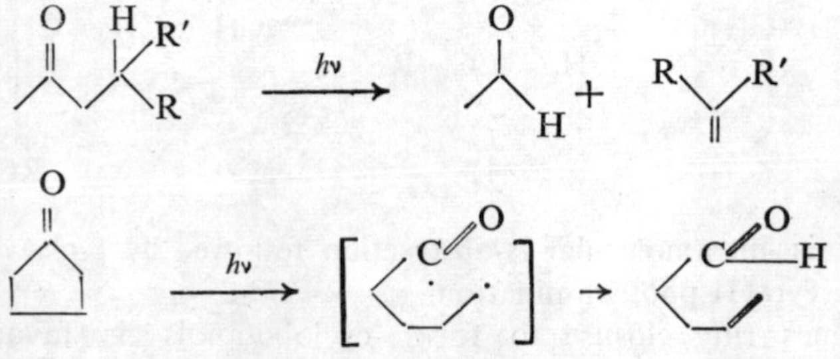

8.1.1 Photoreduction of Dyes by Two Electron Transfer Processes

Photoredox reactions are observed for some dyes in which the dye appears to be bleached in light. The colour may or may not reappear in the dark. In such reduction processes, two H-atoms (or two electrons and two protons) are added to the dye D,

$$D + RH_2 \xrightarrow{h\nu} DH_2 + R$$

where RH_2 is a reducing agent. The reaction occurs in two one-electron steps giving a semiquinone as an intermediate. For example

$$D + Fe^{2+} \xrightarrow{h\nu} \underset{\text{semiquinone}}{DH} + Fe^{3+}$$

$$DH + DH \rightarrow D + \underset{\text{leucodye}}{DH_2}$$

In some cases the reaction may be reversed by H_2O_2 or atmosphere O_2 or by the photooxidized reductant itself

$$DH + O_2 \rightarrow D + \cdot HO_2$$

$$\cdot HO_2 + \cdot HO_2 \rightarrow H_2O_2 + O_2$$

$$DH_2 + 2Fe^{3+} \rightarrow D + 2Fe^{2+} + 2H^+$$

From thermodynamic considerations, the course of such redox reactions can be expressed in terms of the redox potentials of the two couples, E_D and E_R

$$\begin{array}{lcll} D + 2H^+ + 2e & \rightarrow & DH_2 & E^\circ_D \\ R + 2H^+ + 2e & \rightarrow & RH_2 & E^\circ_R \\ \hline D + RH_2 & \rightarrow & R + DH_2 & E^\circ_D - E^\circ_R \end{array}$$

In terms of these redox potentials the free energy change for the net two electron transfer reaction is given as

$$\Delta G = -2 \times F(E_D - E_R)$$

where ΔG is expressed in Joule and F is the Faraday of electric charge. Two situations can arise leading to two types of reactions:

Type I: When $E_D < E_R$ and $\Delta G > 0$. These reactions are not thermodynamically feasible in the dark but may occur on photoexcitation of the dye if the free energy differences become favourable (Section 4.9). The photoreaction goes against the thermochemical gradient and reverses spontaneously in the dark. When the reaction is carried out in an electrochemical cell, a potential is developed at suitable electrodes. This is known as the *photogalvanic effect*. The thionine-ferrous system falls in this category. The most important example of such endergonic reactions is the photosynthesis in plants. The light absorbed by chlorophyll molecules pushes an electron energetically uphill, thus providing the driving force for the primary processes. Duplicating such a system in the laboratory could provide a method for photochemical conversion and storage of solar energy, if the problem of spacially separating the products could be solved (Section 9.5).

Type II: When $E_D > E_R$ and $\Delta G < 0$, the reaction should be possible thermodynamically but may not occur because of a high energy of activation $\Delta G^{\neq}$. The role of photoexcitation in this case is to provide an easier route for the reaction. These are called *photocatalyzed reactions*. The photoreduction of methylene blue by EDTA or other electron donors like stannous chloride fall in this category. The reaction is pH and concentration dependent and the reductant is consumed during the reaction. Dyes in the reduced state can act as very powerful reducing agents.

8.2 PHOTOOXIDATION AND PHOTOOXYGENATION

8.2.1 Photooxidation

Photooxidation reactions in absence of molecular O_2 or in which oxygen does not participate are better described as photochemical oxidoreduction reactions in which an electronically excited donor molecule (D) transfers an electron to a suitable acceptor molecule (A), i.e. an oxidizing agent.

$$D + h\nu = D^*$$

$$D^* + A = D_{ox} + A_{red}$$

For example:

$$(\text{eosin})^* + Fe(CN)_6^{3-} \rightarrow \underset{\text{semioxidized eosin}}{[\text{eosin}]^+} + FeCN_6^{4-}$$

Some examples of photooxidation reaction in inorganic complexes have been discussed in Section 8.6.

Photoredox reactions can be mediated by an intermediate which is called a *sensitizer*. The photosensitized oxidation is illustrated below by energy level schemes for donor D, sensitizer S and acceptor A. Three steps can be identified for the final generation of oxidized donor D^+ and reduced acceptor A^- (Figure 8.2). (a) excitation of an electron in S leaving a (+) hole in the ground state, (b) transfer of an electron from the lowest filled energy level of D to neutralize this hole and (c) transfer of excited electron from S to the unfilled energy level of A to produce A^-. The sensitizer S finally emerges unreacted. It thus acts as a *photocatalyst* in the general sense of the term.

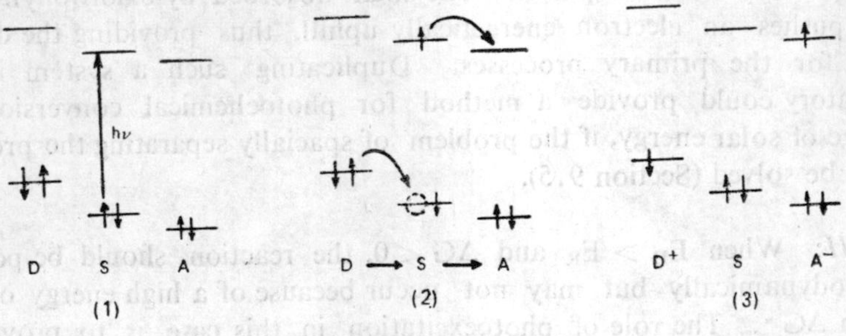

Figure 8.2 Sensitized electron transfer reaction obeying spin conservation rule.

8.2.2 Photooxygenation

In *air-saturated* or *oxygen-saturated solution*, molecular O_2 participates as an oxidizing agent in the photosensitized oxidation of organic substrates. When an addition product AO_2, a *peroxide*, is formed:

$$A + O_2 \rightarrow AO_2$$

the reaction is called *photooxygenation* or *photoperoxidation*, to distinguish it from such photooxidative processes as electron transfer or H-atom transfer. For the addition of molecular oxygen at the double bond aromatic hydrocarbons and olefins are good substrates. The unsaturated cyclic compounds give endoperoxides.

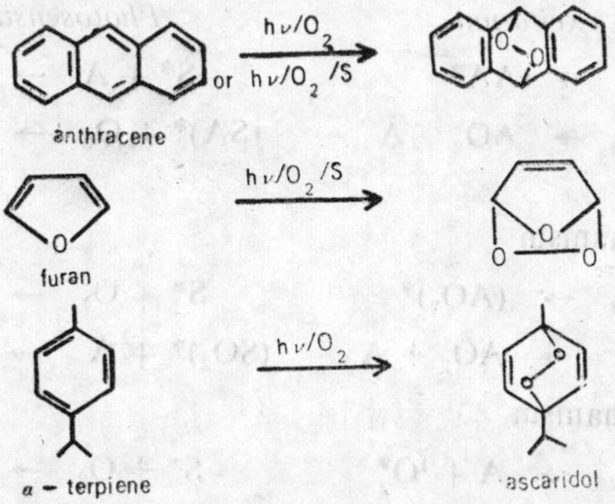

The olefins give hydroperoxy compounds and the reaction is called an *'ene' reaction*:

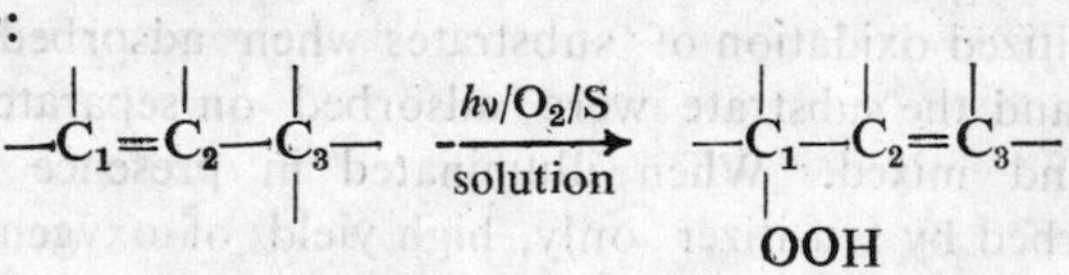

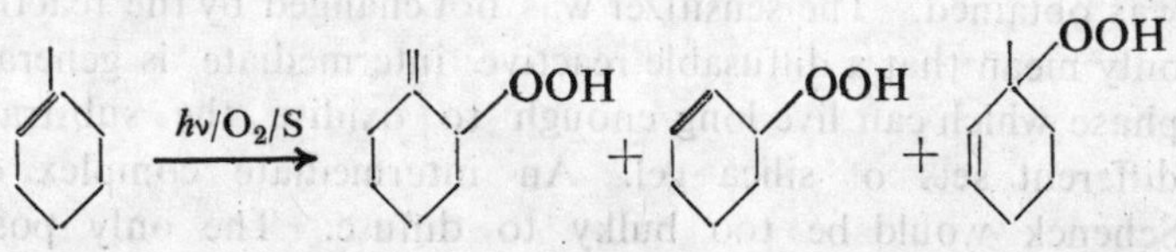

Photooxygenation may take place either as (a) a *direct process,* where light is absorbed by A designated as the substrate or as (b) an *indirect process* or *photosensitized process* where light is absorbed by a molecule other than that which reacts and is called a sensitizer. The reaction is said to occur from the triplet state of the excited molecule.

Photooxygenation reactions are conveniently divided into (i) *Type I processes* in which free radicals and electronically excited molecules are involved as intermediates, and (ii) *Type II processes,* in which only electronically excited molecules are intermediates. The Type II mechanism is further subdivided into (*D-D*) and (*D-O*) *mechanisms.* In (*D-D*) *mechanism,* the acceptor is activated in the intermediate step, which finally reacts with O_2 to form AO_2. In (*D-O*) *mechanism,* O_2 is activated in the intermediate step and the final peroxidation step involves active O_2 and the ground state substrate. Two types of intermediates have been proposed for *D-O mechanism*: (a) formation of moloxide—*Schenck mechanism,* (b) formation of singlet oxygen generated by energy transfer from excited sensitizer—*Kautsky mechanism.*

All these types of reactions can occur either by direct excitation or by photosensitized pathways:

D-D mechanism

Direct excitation	*Photosensitized*
$A^* + A \rightarrow AA^*$	$S^* + A \rightarrow (SA)^*$
$AA^* + O_2 \rightarrow AO_2 + A$	$(SA)^* + O_2 \rightarrow AO_2 + S$

D-O mechanism

(a) Schenck mechanism

$A^* + O_2 \rightarrow (AO_2)^*$	$S^* + O_2 \rightarrow (SO_2)^*$
$(AO_2)^* + A \rightarrow AO_2 + A$	$(SO_2)^* + A \rightarrow AO_2 + S$

(b) Kautsky mechanism

$A^* + O_2 \rightarrow A + {}^1O_2^*$	$S^* + O_2 \rightarrow S + {}^1O_2^*$
${}^1O_2^* + A \rightarrow AO_2$	${}^1O_2^* + A \rightarrow AO_2$

The singlet oxygen mechanism was first proposed by Kautsky in 1930 to explain sensitized oxidation of substrates when adsorbed on silica gel. The sensitizer and the substrate were adsorbed on separate sets of silica gel particles and mixed. When illuminated in presence of oxygen by radiation absorbed by sensitizer only, high yield of oxygenated substrate was obtained. The sensitizer was not changed by the reaction. This could only mean that a diffusable reactive intermediate is generated in the gas phase which can live long enough to oxidise the substrate adsorbed on different sets of silica gel. An intermediate complex or moloxide of Schenck would be too bulky to diffuse. The only possible candidate appears to be an active ${}^1O_2^*$ generated by energy transfer process:

$$ {}^1S \rightarrow {}^3S $$

$$ {}^3S + {}^3O_2 \rightarrow S_0 + {}^1O_2^* $$

$$ A + {}^1O_2^* \rightarrow AO_2 $$

The Kautsky mechanism was not accepted for a long time because not much was known about the reactivity of ${}^1O_2^*$ molecule at that time. The interest in this mechanism was revived when in 1964, C. S. Foote and his collaborators demonstrated that the product distribution in photosensitized oxygenation of a large number of substrates was the same when the oxidation was carried out in the dark using H_2O_2 + NaOCl mixture as the oxidant. The decomposition of hydrogen peroxide in presence of sodium hypochlorite is known to generate *singlet oxygen.* A red glow (chemiluminescence) emitted from the reaction mixture is due to emission from pairs of singlet oxygen molecules produced in the course of reaction. In most photochemical reaction mixtures, O_2 merely quenches the triplet excitation of the sensitizer and thus inhibits other photochemical events. But in some favourable substrates, oxygenation occurs with high quantum yields if the concentration of O_2 is greater than 10^{-5} M approximately. At ordinary temperatures, in air-saturated aqueous solutions $[O_2] \simeq 6\times10^{-4}$ M.

The high specificity of organic photooxygenations can be explained by three major requirements for highly probable reaction: (1) the reaction must be exothermic, this means that O_2 must enter the reaction in an excited state, (2) the total spin of the products must equal that of the reactants (rule for conservation of spin) and (3) the symmetry of the system must not change during the reaction (rule for conservation of parity). The reaction of ground state $O_2\,(^3\Sigma_g)$ with ground state singlet organic molecule S is not very probable since the complex SO_2 will be generated in the triplet state from spin considerations:

$$\underset{\text{triplet}}{O_2\,(\uparrow\uparrow)} + \underset{\text{singlet}}{S\,(\uparrow\downarrow)} \rightarrow \underset{\text{triplet}}{SO_2\,(\uparrow\uparrow\uparrow\downarrow)}$$

The product in the triplet state implies it to be in a higher excited state. Therefore, this reaction should be strongly endothermic and hence not very probable. On the other hand, if O_2 reacts in its singlet state, we have the product in the singlet state which may be its ground state:

$$\underset{\text{singlet}}{^1O_2\,(\uparrow\downarrow)} + \underset{\text{singlet}}{S\,(\uparrow\downarrow)} \rightarrow \underset{\text{singlet}}{SO_2\,(\uparrow\downarrow\uparrow\downarrow)}$$

A. *Nature and importance of singlet oxygen*: The electronic energy states of molecular oxygen have been discussed in Section 2.7 while the potential energy curves have been given in Figure 2.11. The ground state of O_2 is a triplet $^3\Sigma_g^-$ and the next higher energy states of the same electronic configuration are doubly degenerate $^1\Delta_g$ (22.3 kcal or 93.21 kJ) and $^1\Sigma_g^+$ (38 kcal or 158.8 kJ) above, which emit at 1270 and 760 nm, respectively. Besides these two lines, emission spectrum of chemiluminescence from $H_2O_2 + OCl^-$ reaction consists of lines at 480 and 634 nm. These lines are observed in absorption spectrum of liquid oxygen which has a blue appearance. These higher energy states of O_2 are explained to be generated by the simultaneous transitions in a pair of oxygen molecules with a single photon. They are called *dimol absorption or emission.* Thus 634 nm state has the energy of two $^1\Delta_g$ molecules

$$2\,O_2\,(^1\Delta_g)\ (\lambda = 1260\text{ nm},\ \bar{\nu} = 7882\text{ cm}^{-1}) \rightarrow {}^1[O_2{}^1\Delta_g + O_2{}^1\Delta_g]\text{ pair state}$$
$$(\lambda = 634\text{ nm},\ \bar{\nu} = 15765\text{ cm}^{-1})$$

and the energy of the 476 nm state is just the sum of the energies of $^1\Delta_g$ and $^1\Sigma_g^+$ states:

$$O_2\,{}^1\Delta_g\ (\bar{\nu} = 7882\text{ cm}^{-1}) + O_2\,{}^1\Sigma_g^+\ (\bar{\nu} = 13{,}120\text{ cm}^{-1})$$
$$\rightarrow {}^1[O_2\,(^1\Delta_g) + O_2(^1\Sigma_g^+)]\text{ pair state}$$
$$(\lambda = 476\text{ nm},\ \bar{\nu} = 21{,}003\text{ cm}^{-1})$$

At still higher energy, absorption at 381 nm corresponds to summation of two $^1\Sigma_g^+$ energy states.

$$2\,O_2\,(^1\Sigma_g^+)\,(\bar{\nu} = 13120\ \text{cm}^{-1}) \rightarrow {}^1[O_2\,(^1\Sigma_g^+) + O_2\,(^1\Sigma_g^+)]\ \text{pair state}$$
$$(\lambda = 381\ \text{nm};\ \bar{\nu} = 26242\ \text{cm}^{-1})$$

The intensity of these collision induced transitions is borrowed from the dipole-allowed transition $B^3\Sigma_u^- \leftarrow X^3\Sigma_g^-$, commonly known as Schumnan-Runge system (Figure 2.12). These cooperative double molecule and single molecule transitions in O_2 have profound influence in environmental chemistry and existence of life itself on our planet earth. The $^1\Sigma_g^+$ state of O_2 has a much shorter radiative lifetime than $^1\Delta_g$: $\tau(^1\Sigma_g^+) = 7s$ and $\tau(^1\Delta_g) = 45$ min. In condensed systems the actual lifetimes are: $\tau(^1\Sigma_g^+) \simeq 10^{-10}$ s and $\tau(^1\Delta_g) = 10^{-6}$ s. The $^1\Sigma_g^+$ state is very sensitive to collision and quickly deactivates to $^1\Delta_g$. The lifetime of $^1\Delta_g$ is solvent dependent and has a high value in non-hydrogen containing solvent such as CCl_4. In water the lifetime is 2×10^{-6} s. From these considerations, $^1\Delta_g$ has been suggested to be the reactive state of singlet oxygen. Furthermore, $^1\Delta_g$ state has the proper symmetry to act as a *dienophile* in concerted addition reactions to double bonds.

The range of phenomena, reactions and systems in which singlet oxygen is now believed to be directly or indirectly involved is indeed remarkable. Singlet oxygen is clearly involved in numerous dye-sensitized photooxygenation reactions of olefins, dienes and aromatic hydrocarbons and in the quenching of the excited singlet and triplet states of molecules There is reason to suspect that singlet oxygen may also be involved in chemiluminescent phenomena, photodynamic action, photocarcinogenicity, and perhaps even in metal catalyzed oxygenation reactions.

B. *Quenching of fluorescence by* $^3O_2$. The oxygen quenching of fluorescence of aromatic hydrocarbons both in solution and in vapour phase is in general diffusion-controlled. The rate constants for the O_2-quenching of excited singlet and triplet states are given in Table 8.2

TABLE 8.2

Rate constants for O_2-quenching of excited singlet and triplet state molecules.

	$E\,(S_1 - T_1)$ cm^{-1}	Singlet quenching constant $k_q \times 10^{10}$ l mol^{-1} s^{-1}	Triplet quenching constant $k_q \times 10^{9}$ l mol^{-1} s^{-1}
Benzene	8700	16.0	12.0
Naphthalene	10,500	2.8	2.1
Anthracene	12,000	2.6	2.1

The triplet quenching constants are one-ninth of the singlet quenching constant because of spin statistical factors. A collision complex between a substrate in the singlet state ^1A and a ground state oxygen $^3O_2$ will be in the triplet state. But the complex between ^3A and $^3O_2$ can produce quintet, triplet and singlet states according to Wigner's rule: $S_1 = 1$, $S_2 = 1$, $S = 2, 1, 0$ and multiplicity $2S + 1 = 5, 3, 1$. Thus a total of nine states are obtained and the singlet state is only one of these nine states.

In aromatic hydrocarbons, the oxygen-quenching of singlet states does not involve energy transfer but is entirely due to enhanced intersystem crossing, which may proceed via a CT-complex state (Section 6.6.2).

$$^1A + {}^3O_2 \rightarrow (A^+ \ldots O_2^-) \rightarrow {}^3A + {}^3O_2$$

The quenching of triplet state can occur by energy transfer through exchange mechanism to generate singlet oxygen if the sensitizer triplet state energy E_T lies above the singlet states of oxygen molecule. No change in the total spin orientation of the electrons takes place during the process (Figure 8.3):

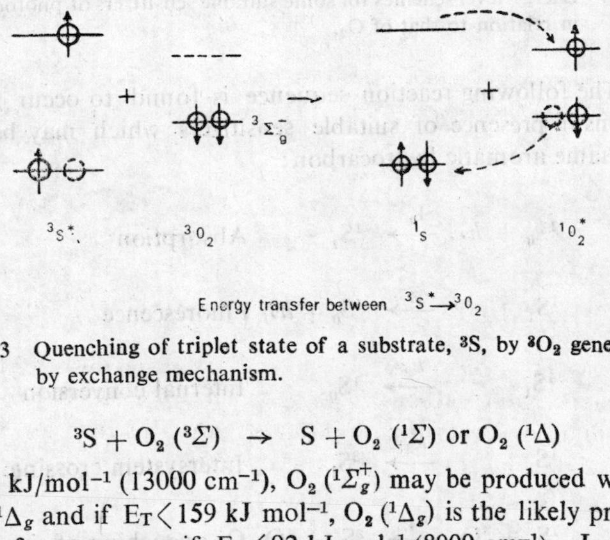

Figure 8.3 Quenching of triplet state of a substrate, ^3S, by $^3O_2$ generating $^1O_2$ by exchange mechanism.

$$^3S + O_2\,(^3\Sigma) \rightarrow S + O_2\,(^1\Sigma) \text{ or } O_2\,(^1\Delta)$$

If $E_T > 159$ kJ/mol^{-1} (13000 cm^{-1}), $O_2\,(^1\Sigma_g^+)$ may be produced which decays rapidly to $^1\Delta_g$ and if $E_T < 159$ kJ mol^{-1}, $O_2\,(^1\Delta_g)$ is the likely product. No energy transfer can occur if $E_T < 92$ kJ mol^{-1} (8000 cm^{-1}). In such cases quenching involves enhanced intersystem crossing to the ground state. Otherwise energy transfer is the favoured mechanism being nearly *1000 times* faster than the intersystem crossing rate. The energy level schemes for some suitable sensitizers of photooxygenation in relation to that of O_2 are given in Figure 8.4.

C. *Kinetics of photoperoxidation reaction.* In the sensitized photooxygenation by Type IIb mechanism the $O_2\,(^1\Delta_g)$ acts as dienophile and adds to give the endoperoxide in concerted addition reaction similar to Diels-Alder

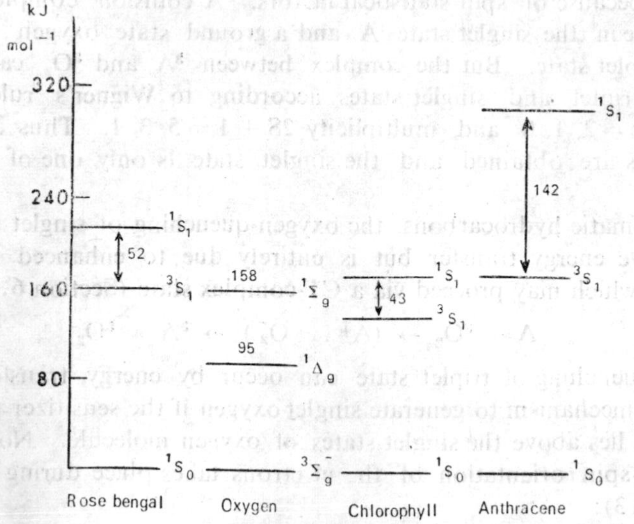

Figure 8.4 Energy level schemes for some suitable sensitizers of photooxygenation in relation to that of O_2.

adducts. The following reaction sequence is found to occur in aromatic hydrocarbons in presence of suitable sensitizers which may be a dye or another or same aromatic hydrocarbon:

$${}^1S_0 + h\nu \xrightarrow{I_a} {}^1S_1 \quad \text{Absorption}$$

$${}^1S_1 \xrightarrow{k_1} {}^1S_0 + h\nu_f \quad \text{Fluorescence}$$

$${}^1S_1 \xrightarrow{k_2} {}^1S_0 \quad \text{Internal conversion}$$

$${}^1S_1 \xrightarrow{k_3} {}^3S_1 \quad \text{Intersystem crossing}$$

$${}^1S_1 + {}^3O_2 \xrightarrow{k_4} {}^3S_1 + {}^3O_2 \quad O_2\text{-quenching of singlet state}$$

$${}^3S_1 \xrightarrow{k_5} {}^1S_0 \quad \text{Reverse intersystem crossing}$$

$${}^3S_1 + {}^3O_2 \xrightarrow{k_6} {}^1S_0 + {}^1O_2 \quad \text{Energy transfer to generate } {}^1O_2$$

$${}^1O_2 \xrightarrow{k_7} {}^3O_2 \quad \text{Decay of singlet oxygen}$$

$${}^1O_2 + A \xrightarrow{k_8} AO_2 \quad \text{Photooxygenation}$$

For direct oxygenation S = A, (S = sensitizer, A = substrate). The quantum yield of AO_2 formation, ϕ_{AO_2}, is given by the product of quantum yields of (i) formation of sensitizer triplet state ϕ_T, (ii) formation of singlet oxygen $\phi_{^1O_2}$ and (iii) of oxygenation step ϕ_R

$$\phi_{AO_2} = \phi_T \cdot \phi_{^1O_2} \cdot \phi_R$$
$$= \frac{k_3 + k_4\,[O_2]}{k_1 + k_2 + k_3 + k_4[O_2]} \, \frac{k_6\,[O_2]}{k_5 + k_6\,[O_2]} \, \frac{k_8\,[A]}{k_7 + k_8\,[A]} \quad (8.1)$$

The ratio $k_7/k_8 = \beta$ is known as the substrate reactivity parameter. It can be obtained as slope/intercept ratio from a plot of $(\phi_{AO_2})^{-1}$ vs $[A]^{-1}$ at constant $[O_2]$, when the photoreaction is studied as a function of substrate concentration.

The β values for photoperoxidation vary with substrates over five orders of magnitude: β = 55 M for relatively unreactive cyclohexene and $\beta = 3 \times 10^{-4}$ M for reactive dimethylanthracene and rubrene. β values are also sensitive to the nature of solvent. For autoperoxidation of anthracene in solvents benzene, bromobenzene and CS_2, respective values are 0.42, 0.12, 0.002.

D. *Mechanism of 'ene' reaction.* In the 'ene' reaction, the double bond is shifted to the allyl position with respect to the starting material according to the scheme proposed by Schenck. This involves (i) attachment of the oxygen molecule to one of the carbon atoms of the double bond (e.g. C_1), (ii) shift of the double bond to allyl position (e.g. to $C_2 \rightarrow C_3$) and (iii) migration of the allyl hydrogen atom to the terminus of the peroxy group.

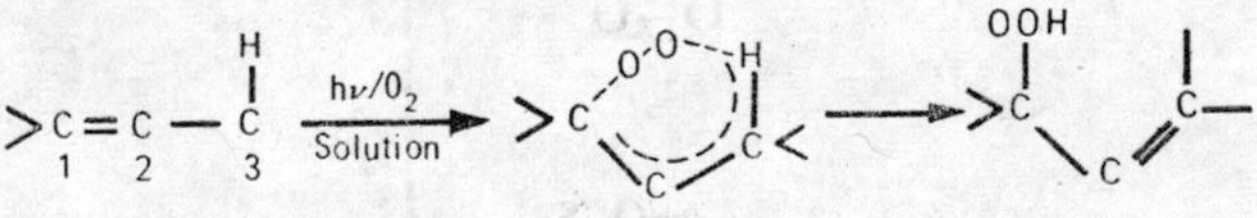

In general, tetraalkyl-substituted double bonds react at a faster rate than do trialkyl-substituted double bonds, which in turn react faster than dialkyl-substituted ones.

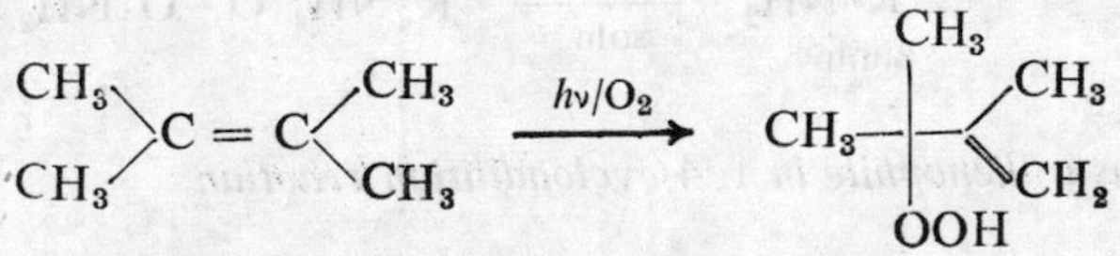

Tetramethyl ethylene (TME)

In many cases, stereoselective introduction of a hydroperoxy group occurs which makes this reaction very valuable in synthetic organic chemistry.

Few other reactions are:

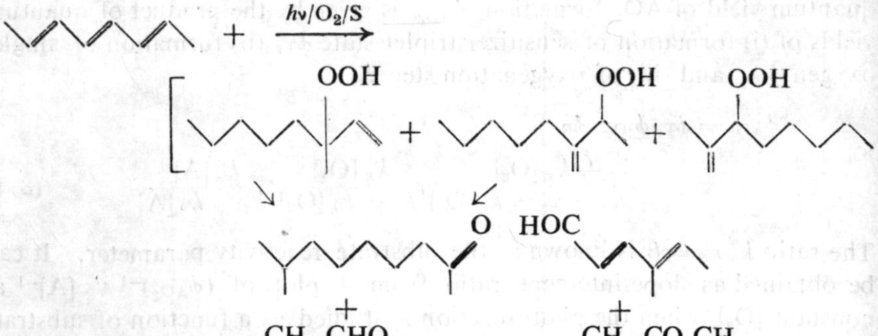

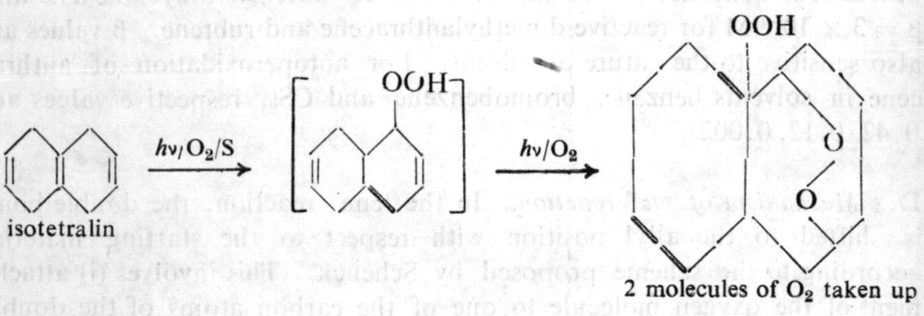

$$\underset{\text{sulphide}}{R{-}S{-}R'} \xrightarrow[\text{soln.}]{h\nu/O_2/S} R{-}\overset{\bullet}{\underset{\underset{O-O}{|}}{S}}{-}R' \xrightarrow{R-S-R'} \underset{\text{sulfoxide}}{2R{-}SO{-}R}$$

$$\underset{\text{dimethyl sulfoxide}}{2CH_3\cdot SO\cdot CH_3} \xrightarrow{h\nu/O_2/S} \underset{\text{sulfones}}{2\ CH_3\cdot SO_2\cdot CH_3}$$

$$\underset{\text{amine}}{2\ R\cdot NH_2} \xrightarrow[\text{soln.}]{h\nu/O_2/S} R\cdot NH_2\cdot O-O\cdot NH_2\cdot R$$

E. *Oxygen as a dienophile in* 1, 4-*cycloaddition reaction*

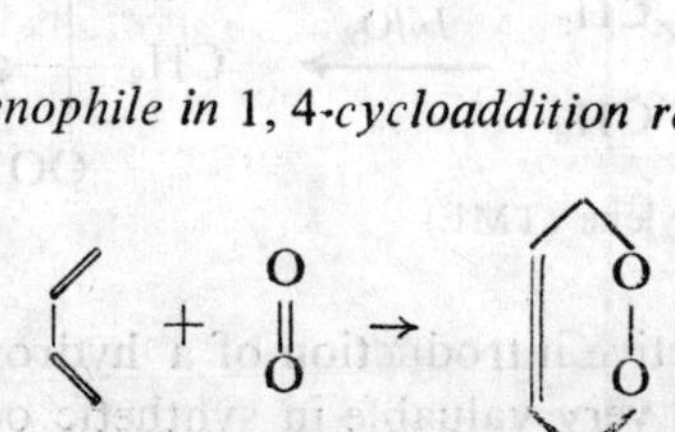

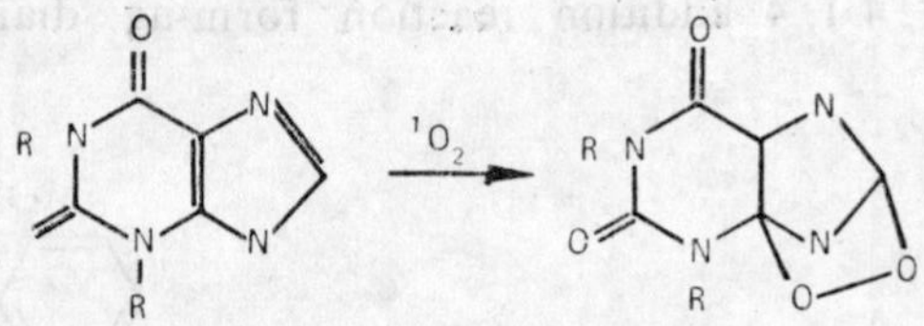

The endoperoxides may decompose in two ways: (1) evolution of molecular oxygen and regeneration of original substrate and (2) by the cleavage of —O—O— bond. The energy required for bond cleavage is about 146.3 kJ (35 kcal/mol^{-1}). The latter mode is favoured if the resonance energy for stabilization of the hydrocarbon upon regeneration is small. If the gain is large enough, singlet oxygen may be generated in the decomposition process. This explains the absence of photoperoxidation in phenanthrene and naphthalene where not much resonance energy is gained or lost.

With electron rich olefins 1, 2 cycloaddition forms relatively unstable dioxetanes which cleave to give carbonyl fragments:

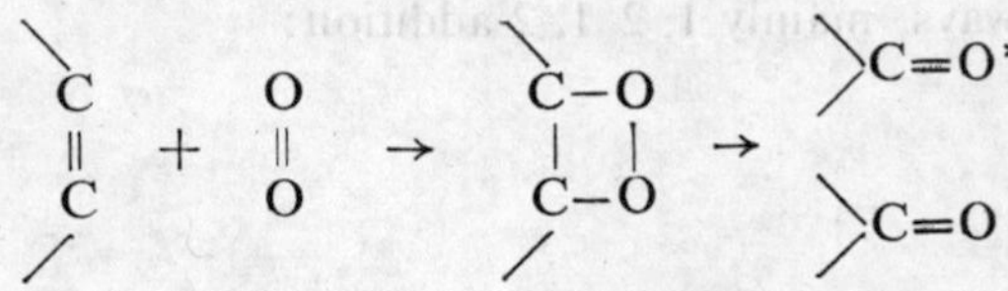

During the process of thermal decomposition, one of the carbonyls is produced in the excited state, often emitting to produce *chemiluminescence*.

8.3 CYCLOADDITION REACTIONS

8.3.1 Photodimerization

Photodimerization involves 1: 1 adduct formation between an excited and a ground state molecule. Olefinic compounds, aromatic hydrocarbons, conjugated dienes, α-unsaturated compounds are known to dimerize when exposed to suitable radiation. Photodimerization of olefinic compounds can occur by either (a) 1,2-1,2 addition, (b) 1,2-1,4 addition or (c) 1,4-1,4 addition.

Photodimerization of anthracene is one of the well documented reactions. It is thought to proceed by way of *excimer* formation (Section

6.5). It is a 1, 4-1, 4 addition reaction forming dianthracene as the single product

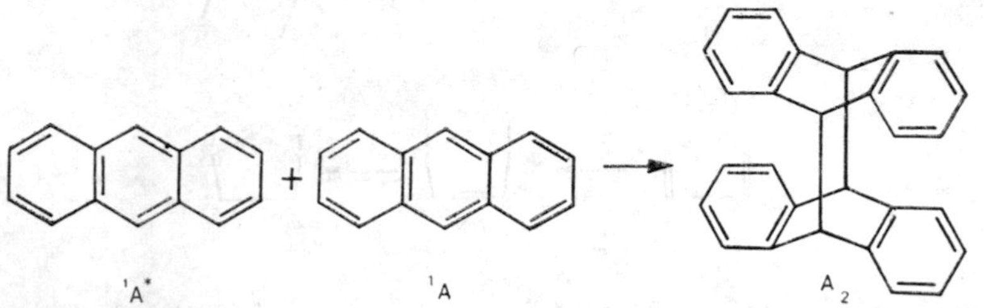

The reaction proceeds from the $^1S_1$ (π, π^*) state. The reaction is not quenched by oxygen. 9-substituted anthracenes can be dimerized but 9, 10 compounds do not dimerize because of steric hindrance (Table, 6.3).

In polyenes, dimerization occurs from the triplet states. The direct and photosensitized reactions are likely to give different products because of the inefficiency in $S_1 \xrightarrow{ISC} T_1$ transition. Irradiation of butadiene in the presence of triplet sensitizers with $E_T > 251$ kJ leads to dimerization in a number of ways, mainly 1, 2–1, 2 addition:

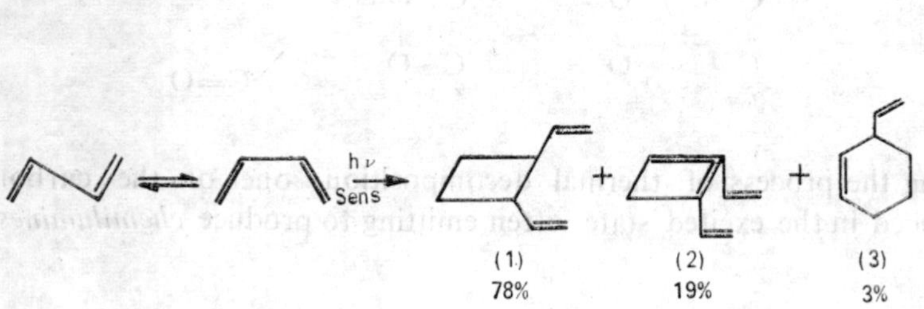

But for molecules whose triplet energy lies between 226 kJ and 251 kJ compound (3) is formed in large amounts. E_T of cis-butadiene is 226 kJ. This suggests that since rotation around the central bond in the excited state is a slow process it will not effectively compete with dimerization.

The photodimerization of α, β-unsaturated carbonyl compounds occurs from either an (n, π^*) or a (π, π^*) state which may be a singlet or a triplet. For example, coumarins photodimerize to give two main products as shown below:

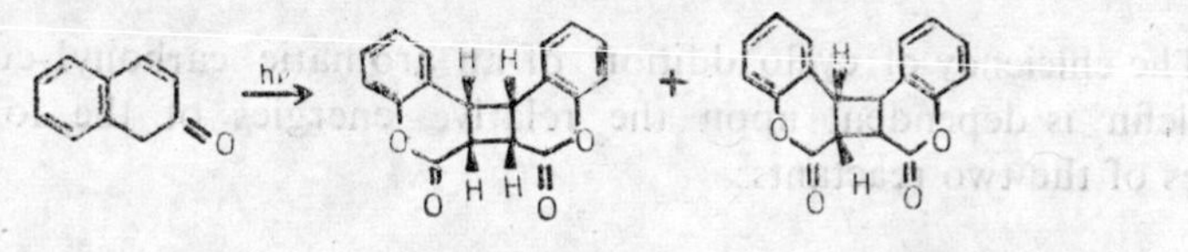

coumarin — (A) syn from S_1 (n, π^*) state in absences of sensitizer — (B) anti from T_1 (n, π^*) state in presence of benzophenone

Photodimerization of thymin to butane type dimers is one of the important causes of UV radiation damage to genetic material DNA in its native double helix form. The reaction occur from the triplet state as confirmed by acetone sensitized photodimerization of thymine. These dimers can be monomerized when irradiated by short wavelength radiations which are absorbed by the dimer only.

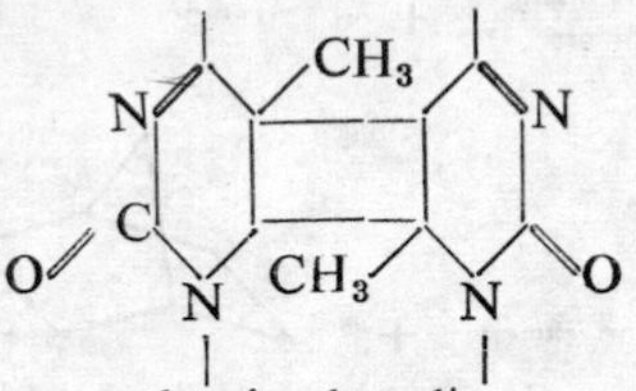

thymin photodimer

Four isomers are possible syn-syn, syn-anti, anti-syn and anti-anti

8.3.2 Oxetane Formation

The cycloaddition reactions to give oxetanes readily occurs when T_1 state is (n, π^*) in character and an electrophilic centre is created on carbonyl oxygen atom on photoexcitation (Section 8.1). *Paterno-Buchi reaction.*

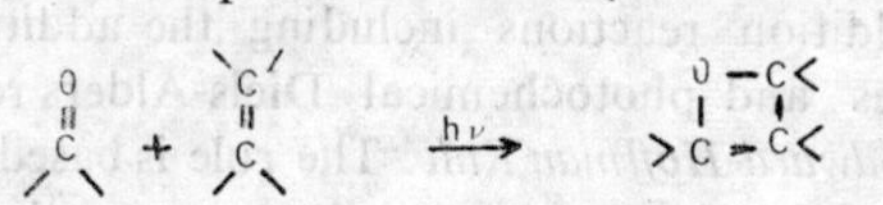

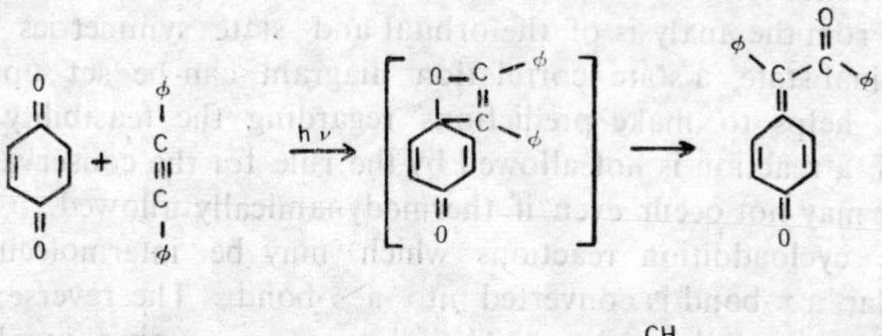

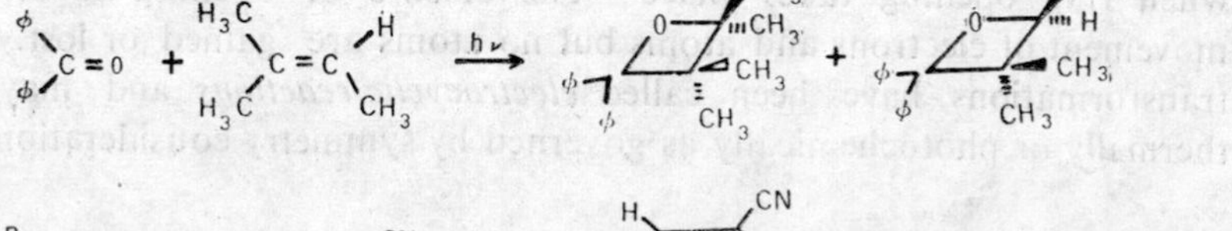

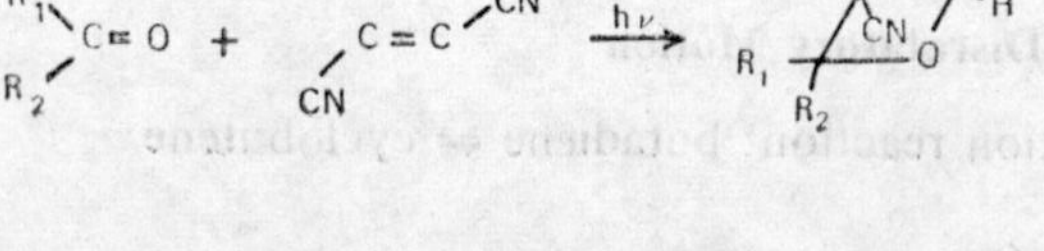

The efficiency of cycloaddition of an aromatic carbonyl compound to an olefin is dependent upon the relative energies of the lowest triplet states of the two reactants:

$$>C=O \qquad >C=C<$$
$$(i)\ T_1\,(n,\pi^*) \rightsquigarrow T_1\,(\pi,\pi^*) \longrightarrow \text{dimerization on energy transfer to olefin; no oxetane formation}$$

$$(ii)\ T(n,\pi^*) \not\rightsquigarrow T(\pi,\pi^*) \longrightarrow \text{No energy transfer: Oxetane formation}$$

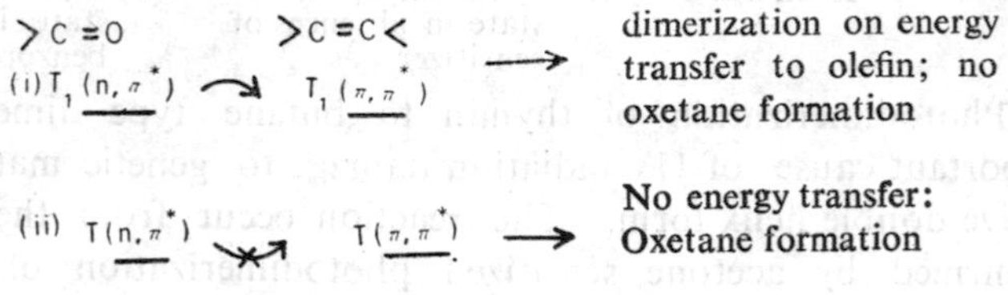

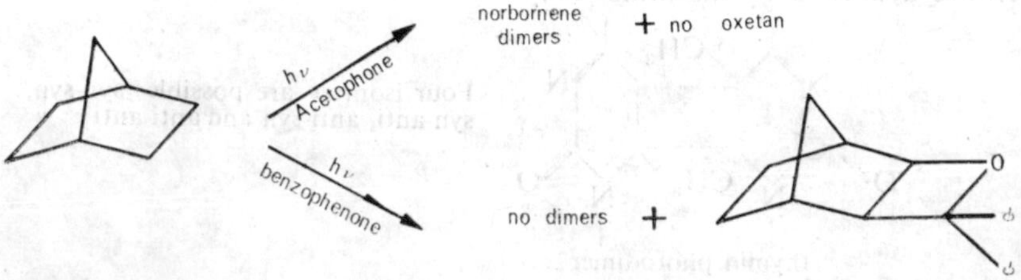

8.4 WOODWARD-HOFFMAN RULE OF ELECTROCYCLIC REACTIONS

The photochemical dimerization of unsaturated hydrocarbons such as olefins and aromatics, cycloaddition reactions including the addition of $O_2\ (^1\Delta_g)$ to form endoperoxides and photochemical Diels-Alders reaction can be rationalized by the *Woodward-Hoffman Rule*. The rule is based on the principle that the symmetry of the reactants must be conserved in the products. From the analysis of the orbital and state symmetries of the initial and final state, a state correlation diagram can be set up which immediately helps to make predictions regarding the feasibility of the reaction. If a reaction is not allowed by the rule for the conservation of symmetry, it may not occur even if thermodynamically allowed.

In these cycloaddition reactions which may be intermolecular, or intramolecular, a π-bond is converted into a σ-bond. The reverse occurs when ring opening takes place. The closure or opening involves the movement of electrons and atoms but no atoms are gained or lost. Such transformations have been called *electrocyclic reactions* and may occur thermally or photochemically as governed by symmetry considerations.

8.4.1 Conrotatory and Disrotatory Motion

Consider the cyclization reaction: butadiene $\rightleftharpoons$ cyclobutene

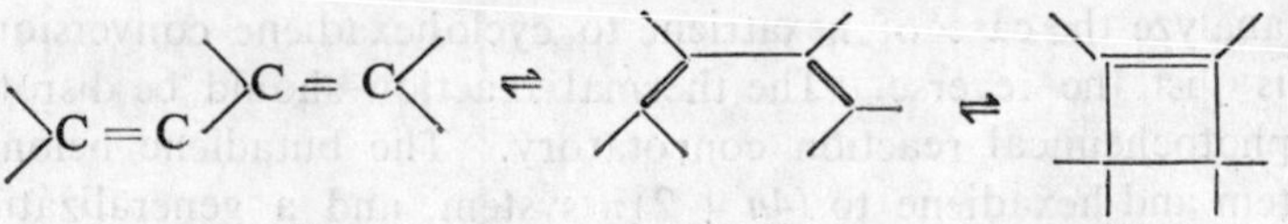

trans-butadiene cis-butadiene cyclobutene

The molecular orbitals of butadiene can be represented by the free electron (FE) model. The 4-C atoms give rise to 4 MOs (Sec. 2.10.1) as shown in Figure 8.5.

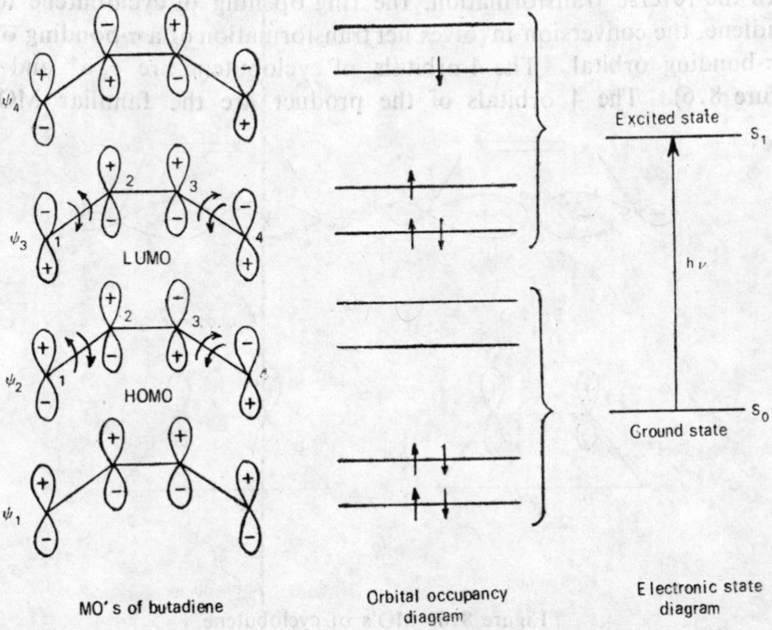

Figure 8.5 MO's of butadienee, orbital occupancy diagram and electronic state diagram.

The *highest occupied molecular orbitals* (HOMO) are of direct interest in describing the phenomenon of ring closure. In the ground state, the symmetry of HOMO is as described by ψ_2 (Figure 8.5). For ring closure between C-atoms 1 and 4, the orbital lobes must rotate either inwards or outwards. If both the lobes move in the same directions such that one appears to rotate inwards and the other outwards, the motion is called *conrotatory*. If the two lobes move in opposite directions such that both appear to move outwards or inwards, the motion is called *disrotatory*. From the diagram it is evident that conrotatory motion will cause a bonding situation by bringing the like phases of the lobes to overlap facilitating ring formation in the *thermal* reaction. On the other hand, in the excited state, the HOMO of interest is ψ_3. Now the *disrotatory motion* creates a bonding situation predicting ring closure in the *photochemical* reaction.

If we analyze the case of hexatriene to cyclohexadiene conversion, the situation is just the reverse. The thermal reaction should be disrotatory and the photochemical reaction conrotatory. The butadiene belongs to $(4n)\pi$ system and hexadiene to $(4n+2)\pi$ system, and a generalization of the systems may be attempted.

8.4.1 Reactant and Product Oribital Symmetry Correlation

In the reverse transformation, the ring opening of cyclobutene to give butadiene, the conversion involves net transformation of a σ-bonding orbital to π-bonding orbital. The 4 orbitals of cyclobutene are σ, σ* and π, π* (Figure 8.6). The 4 orbitals of the product are the familiar MO's of

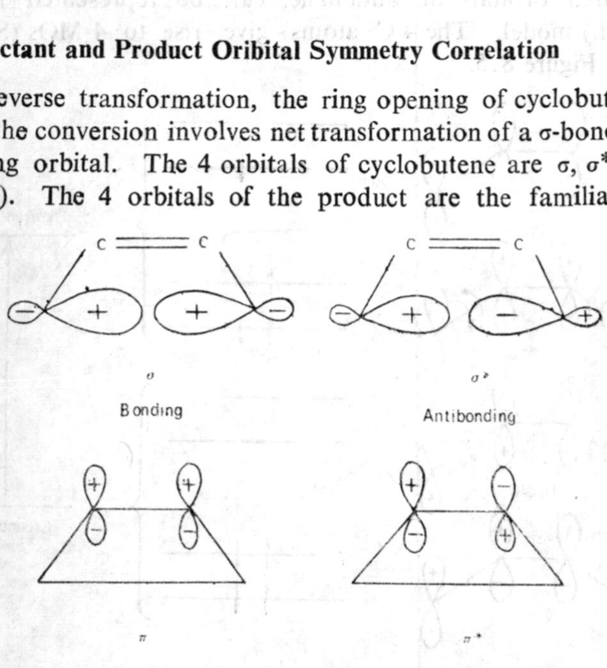

Figure 8.6 MO's of cyclobutene

butadiene illustrated in Figure 8.5. Each of the first set of orbitals passes over into the second set and it is instructive to examine how the sets of orbitals correlate during the transformation. The conrotatory and disrotatory modes follow different correlation paths because they preserve different symmetry elements as illustrated in Figure 8.7.

In the case of conrotatory mode, the symmetry is preserved with respect to C_2 axis of rotation. On 180° rotation along this axis, H_1 goes to H_3 and H_2 to H_4 and the new configuration is indistinguishable from the original. An orbital symmetric with respect to rotation is called *a* and antisymmetric as *b*. On the other hand, in the case of disrotatory mode, the elements of symmetry are described with respect to a mirror plane. The symmetry and antisymmetry of an orbital with respect to a mirror plane of reflection is denoted by *a'* and *a''* respectively (Section 2.9). The nature of each MO of cyclobutene with respect to these two operations is shown in the Table 8.4 for cyclobutene and butadiene.

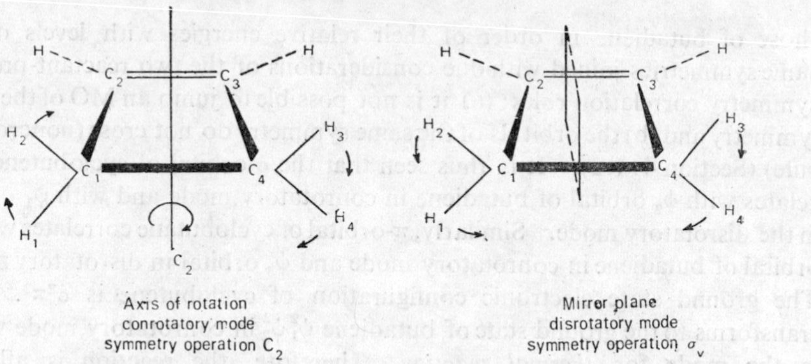

Figure 8.7 Symmetry operations for conrotatory and disrotatory modes of cyclobutene ⇌ butadiene transformation.

TABLE 8.3

	Cyclobutene				→	Butadiene			
	σ	σ^*	π	π^*		ψ_1	ψ_2	ψ_3	ψ_4
C_2-axis symmetry conrotatory mode	*a*	*b*	*b*	*a*	→	*b*	*a*	*b*	*a*
Mirror symmetry disrotatory mode	*a′*	*a″*	*a′*	*a″*		*a′*	*a″*	*a′*	*a″*

Thus for the conrotatory mode, σ and π^* of cyclobutene have the same symmetry as ψ_2 and ψ_4 of butadiene and σ^* and π as ψ_1 and ψ_3. For the disrotatory mode, σ and π correlate with ψ_1 and ψ_3 and σ^* and π^* with ψ_2 and ψ_4.

Below are arranged (Figure 8.8) the four orbitals of cyclobutene and

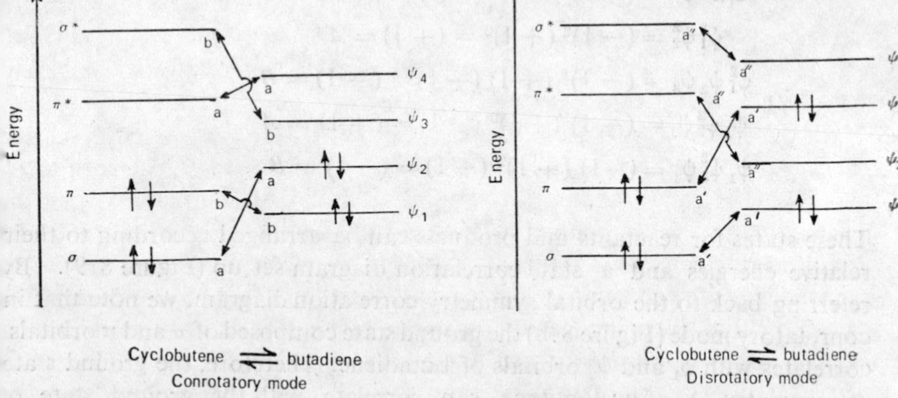

Figure 8.8 Reactant—product orbital correlation diagram for cyclobutene⇌butadiene transformation for (a) conrotatory and (b) disrotatory modes.

those of butadiene in order of their relative energies with levels of the same symmetries joined with due considerations of the two reactant-product symmetry correlation rules: (a) it is not possible to jump an MO of the same symmetry and (b) the orbitals of the same symmetry do not cross (noncrossing rule) (Section 4.4.2). It is thus seen that the σ-orbital of cyclobutene correlates with ψ_2 orbital of butadiene in conrotatory mode and with ψ_1 orbital in the disrotatory mode. Similarly, π-orbital of cyclobutane correlates with ψ_1 orbital of butadiene in conrotatory mode and ψ_3 orbital in disrotatory mode. The ground state electronic configuration of cyclobutene is $\sigma^2\pi^2$. This transforms to the ground state of butadiene $\psi_1^2\psi_2^2$ in conrotatory mode which is the mode for *thermal reaction.* Therefore, the reaction is allowed thermally, both in the forward and reverse directions. In the disrotatory mode, ground state cyclobutene transforms to doubly excited state $\psi_1^2\psi_3^2$ of butadiene. In the reverse reaction, the ground state of butadiene $\psi_1^2\psi_2^2$ correlates with the doubly excited state of cyclobutene $\sigma^2\pi^{*2}$. Both the transformations are *highly unfavourable for photochemical reaction.*

8.4.2 State Symmetry Correlation

The symmetry of the electronic energy states can be derived from group theoretical considerations (Section 2.9), putting the eigenvalue for the symmetric functions: $a = (+1)$ and for $b = (-1)$, and similarly for $a' = (+1)$ and $b' = (-1)$, the ground state of the cyclobutene is derived as: $a^2b^2 = (+1)^2(-1)^2 = 1$, which is totally symmetric and belongs to the symmetric species designated by the symbol A. For other energy states from the corresponding occupancy configurations for conrotatory mode we have:

$$\sigma^2\pi^2 = (+1)^2(-1)^2 = +1 = A$$

$$\sigma^2\pi\pi^* = (+1)^2(-1)(+1) = (-1) = B$$

$$\sigma\pi^2\sigma^* = (+1)(-1)^2(-1) = (-1) = B$$

$$\psi_1^2\psi_2^2 = (-1)^2(+1)^2 = (+1) = A$$

$$\psi_1^2\psi_2\psi_3 = (-1)^2(+1)(-1) = (-1) = B$$

$$\psi_1\psi_2^2\psi_3 = (-1)(+1)^2(-1) = (+1) = A$$

$$\psi_1\psi_2^2\psi_4 = (-1)(+1)^2(+1) = (-1) = B$$

These states for reactants and products can be arranged according to their relative energies and a state correlation diagram set up (Figure 8.9). By referring back to the orbital symmetry correlation diagram, we note that in conrotatory mode (Figure 8.8) the ground state composed of σ and π orbitals correlates with ψ_1 and ψ_2 orbitals of butadiene. Therefore, the ground state of symmetry A of cyclobutene can correlate with the ground state of butadiene of symmetry A (Figure 8.9 b). But the first excited state of

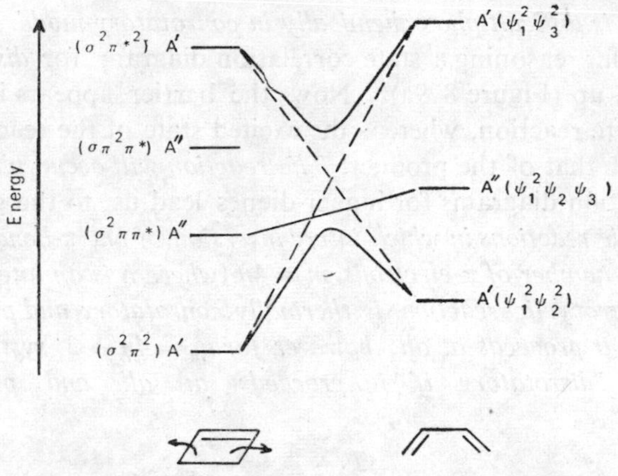

Disrotatory mode

(a)

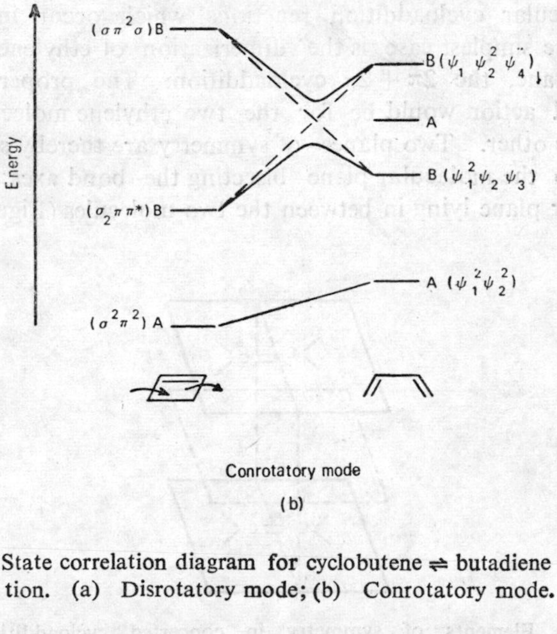

Conrotatory mode

(b)

Figure 8.9 State correlation diagram for cyclobutene ⇌ butadiene transformation. (a) Disrotatory mode; (b) Conrotatory mode.

cyclobutene $(\sigma^2\pi\pi^*)$ of symmetry B must correlate with $(\psi_1\psi_2^2\psi_4)$ state of symmetry B of butadiene. This is not the lowest state of B symmetry which is the $\psi_1^2\psi_2\psi_3$ configuration and correlates with a higher excited state in cyclobutene. On applying the noncrossing rule we find that the only way whereby the first excited state of cyclobutene can be correlated with the lowest state of B symmetry in butadiene is through an energy barrier. The conclusion that is drawn from such correlation diagram is that cyclobutene ⇌ butadiene transformation is energetically *allowed* *thermally but is difficult photochemically in conrotatory mode.*

By a similar reasoning a state correlation diagram for *disrotatory mode* can be drawn up (Figure 8.9a). Now the barrier appears in the thermal or ground state reaction, whereas the excited state of the reactant correlates smoothly with that of the product. *The reaction will occur photochemically.* Such correlation diagrams for higher dienes lead us to the generalization: *In electrocyclic reactions in which interconversion of one π-bond → one σ-bond occurs, if the number of π-electrons $m = 4n$ (where n is an integer, 0, 1, 2, 3, etc.), the electrocyclic reaction is thermally conrotatory and photochemically disrotatory if it proceeds at all; however for $m = 4n + 2$ system the process is thermally disrotatory if it proceeds at all and photochemically conrotatory.*

8.4.3 Cycloaddition Reactions

The concept of the conservation of orbital symmetry can be extended to intermolecular cycloaddition reactions which occur in a concerted manner. The simplest case is the dimerization of ethylene molecules to give cyclobutane, the $2\pi + 2\pi$ cycloaddition. The proper geometry for the concerted action would be for the two ethylene molecules to orient one over the other. Two planes of symmetry are thereby set up: σ_h-perpendicular to the molecular plane bisecting the bond axes: σ_v-parallel to the molecular plane lying in between the two molecules (Figure 8.10).

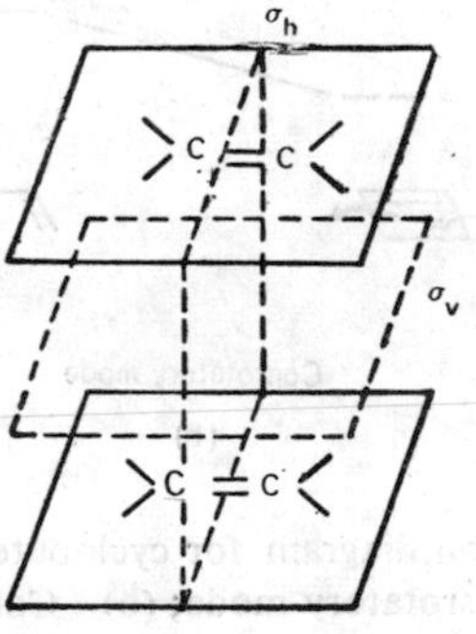

Figure 8.10 Elements of symmetry in concerted cycloaddition reaction: 2-ethylene ⇌ cyclobutane.

The symmetry properties of π-MO's of ethylene, formed by the in-phase and out-phase combination of two *p* orbitals in each molecule with respect to these planes of symmetry for the two reacting partners can be worked out. The same can be done for the product cyclobutane in which these π-bonds are converted into σ-bonds (Figure 8.11). The symmetry and antisymmetry are represented by *S* and *A* respectively and double symbols represent symmetry with respect to σ_h and σ_v planes respectively.

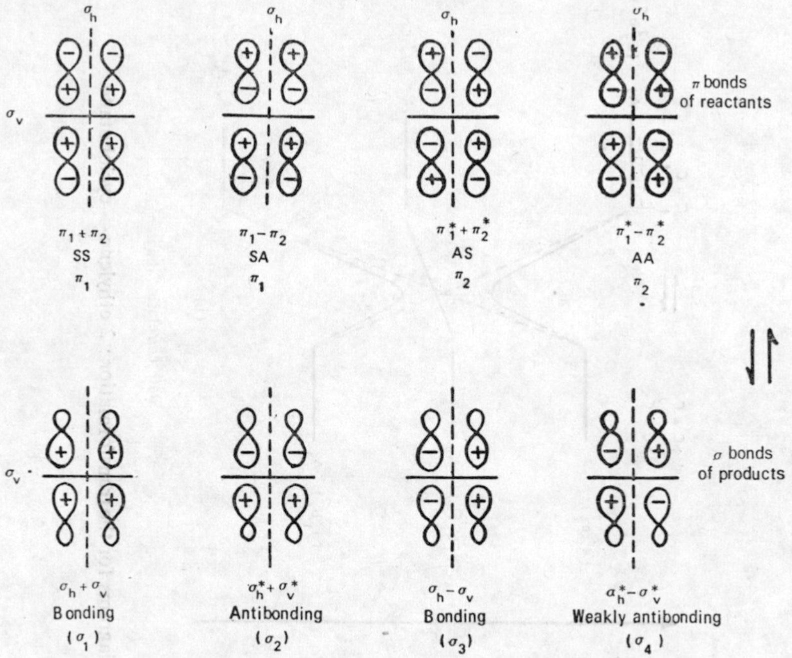

Figure 8.11 Symmetry properties of MO's for ethylene and cyclobutane: π-bonds of reactants and σ-bonds of products.

If SS and SA configurations of the reactants are represented as π_1^+ and π_1^- and AS and AA as π_2^+ and π_2^-, and the four configurations of the product as σ_1, σ_2, σ_3 and σ_4, then an orbital correlation diagram can be set up: *a* and *b* again stand for symmetric and antisymmetric functions with respect to the molecular plane and the levels are arranged according to their relative energies (Figure 8.12 a).

From the state correlation diagram (Figure 8.12 b) it is predicted that the ground state thermal reaction is not allowed because of large energy barrier imposed by symmetry restrictions. On the other hand, the excited states correlate smoothly and hence the reaction is *symmetry allowed photochemically*. The transition state postulated for this photochemical reaction is a dimer formed from one ground state molecule and another excited state molecule and has been termed as an *excimer* (Section 6.5). To generalize the selection rules for the above reactions in which two π-bond lead to two σ-bonds ($2\pi \rightleftharpoons 2\sigma$), the process will be allowed photochemically for $m_1 + m_2 = 4n$ and thermochemically for $m_1 + m_2 = 4n + 2$, where m_1 and m_2 are the numbers of π-electrons in the two reacting partners.

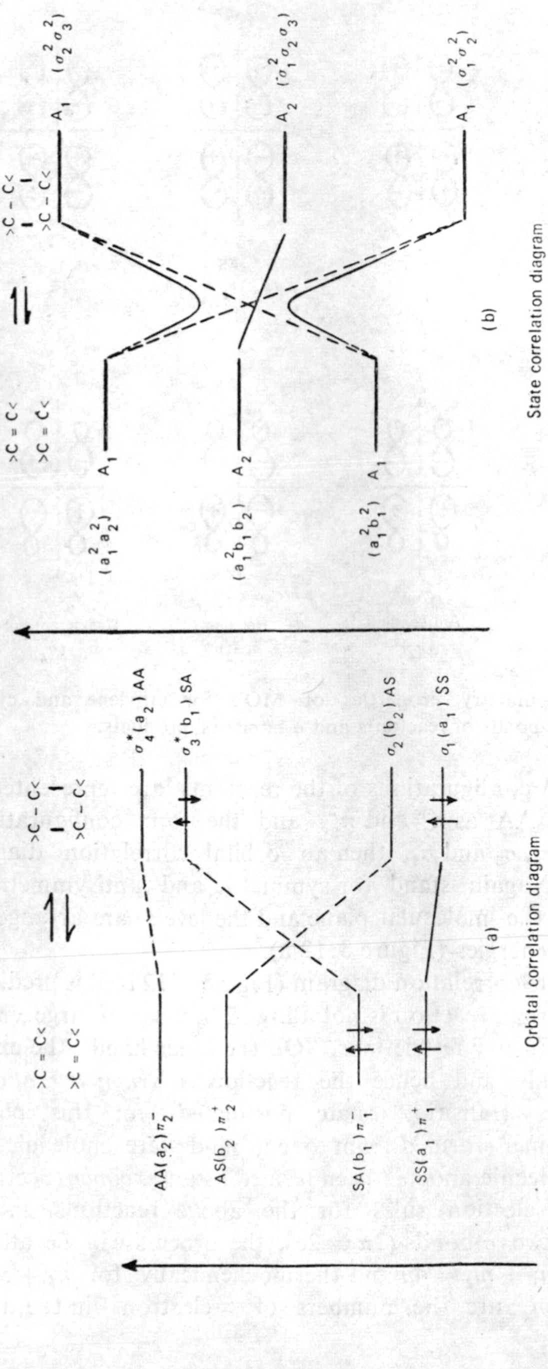

Figure 8.12 (a) Orbital and (b) state correlation diagrams for the transformation: 2 ethylene → cyclobutane.

These considerations help to understand why oxygenation of olefins and aromatic hydrocarbons occurs through excited singlet oxygens, i.e. photochemically and not by ground state triplet oxygen. In the photoperoxidation reactions both the spin conservation rule and symmetry conservation rule are important (Section 8.2.2). The lowest singlet state of O_2, the $^1\Delta_g$ state can react with the substrate to give the products in the ground state. But the transition state for $^1A_0 \ldots O_2(^3\Sigma_g) \rightarrow {}^3(AO_2)$ can only correlate with an excited state. Furthermore, the higher excited state $(^1A_0 \ldots O_2(^1\Sigma_g)$ must lead to the product in an excited singlet state.

8.5 CHEMILUMINESCENCE

Chemiluminescence is defined as the production of light by chemical reactions. To be chemiluminescent, a reaction must provide sufficient excitation energy and at least one species capable of being transferred to an electronically excited state. Thus a system of reaction coordinates favouring production of excited states rather than the ground states is essential. Hence only rather fast exergonic reactions, $-\Delta H$ ranging between 170 — 300 k J mol^{-1}, can be chemiluminescent in the visible range of the spectrum. These are in general, electron transfer or oxidation reactions. The quantum efficiency ϕ_{CL} of a chemiluminescent reaction

$$A + B \rightarrow C^* + D$$

is defined as

$$\phi_{CL} = \frac{\text{number of photons emitted}}{\text{molecules of A (or B) consumed}} = \phi_r^* \phi_f \qquad (8.2)$$

ϕ_{CL} depends on the chemical efficiency ϕ_r^* of the formation of excited product molecules and on the quantum yield of emission ϕ_f from this excited molecule.

Chemiluminescence is also generated by a radical-ion recombination mechanism as observed when polycyclic aromatic hydrocarbons in solution are electrolyzed (Figure 8.13). The anion contains an extra electron in the antibonding orbital whereas the cation is electron deficient in its highest bonding π-orbital. When the electron is transferred from the anion to the antibonding orbital of the cation with proper spin configuration, a singlet excited state of the compound is formed which can be deactivated by emission. Such transfer is possible in large hydrocarbon molecules because the geometries of the reactant ions and their excited states are not very different.

$$Ar^+ + Ar^- \rightarrow {}^1Ar^* + Ar$$
$$^1A^* \rightarrow Ar + h\nu$$

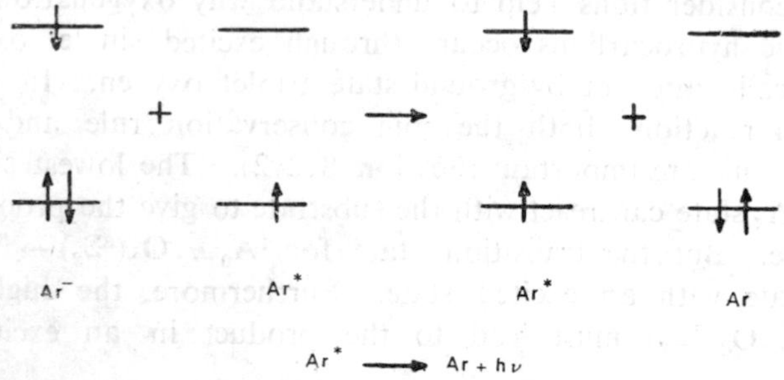

Figure 8.13 Chemiluminescence from excited state generated by radical-ion recombination.

The primary requirement of a chemiluminescent reaction is that it should be *energy-sufficient*. If the energy of the reaction is not enough to promote the product to first excited singlet state, the system is said to be *energy-deficient* and molecules in the triplet states may be formed. In such cases triplet-triplet annhilation of initially generated triplets may populate an excited singlet state.

$$Ar^{+} + Ar^{-} \rightarrow {}^{3}Ar^{*} + Ar$$
$${}^{3}Ar^{*} + {}^{3}Ar^{*} \rightarrow {}^{1}Ar^{*} + Ar$$

When generated electroehemically the emission is termed '*electro-chemiluminescence*' (Section 5.9c).

Chemiluminescence has been observed in oxidation reactions of hydrazides of which luminol (3–amino cyclicphthalhydrazide) is an important example. The general reaction sequence is

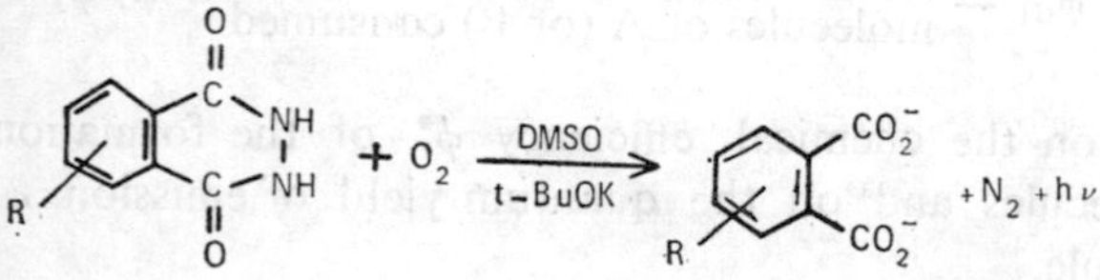

In luminol chemiluminescence and that of other cyclic hydrazides the corresponding dicarboxylate ion is the emitter. The luminol is oxidized by H_2O_2 and the oxidation is very sensitive to metal catalysts and pH of the solution.

Decomposition of oxetanes is still another chemiluminescent reaction. On the basis of Woodward-Hoffman rule of conservation of orbital symmetry, the concerted bond cleavage of dioxetane, a 4–membered ring peroxide, should yield one carboxyl moiety in the excited state

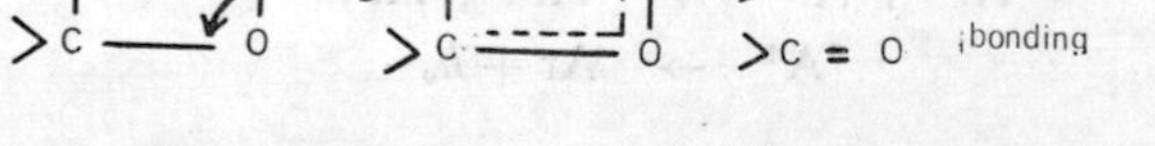

In presence of a suitable fluorescer, the emission from the fluorescer is observed.

$(CH_3)_2-C-C$ $\overset{H}{\underset{CH_3}{}}$ $\xrightarrow{\text{Eu-chelate}}$ emission of red Eu^{3+} line at 613 nm.

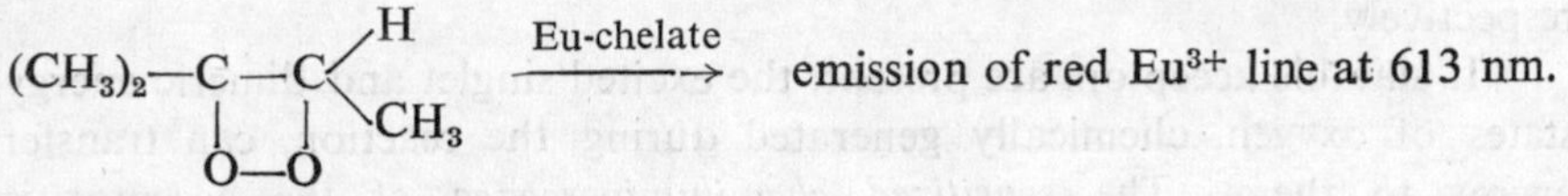

trimethyl 1, 2 dioxetane

oxalic ester

$+ H_2O_2 \rightarrow$ $\begin{array}{c} O=C-O \\ | \quad | \\ O=C-O \end{array}$ dioxetanedione

$\downarrow$ Fluorescer

Dioxitanedione—Fluorescer Complex

$\downarrow$

$2\ O=C=O+Fluorescer^*$

$\downarrow$

$Fluorescer + h\nu$

9,.10-diphenyl, 1, 4 dimethoxy anthracene endoperoxide

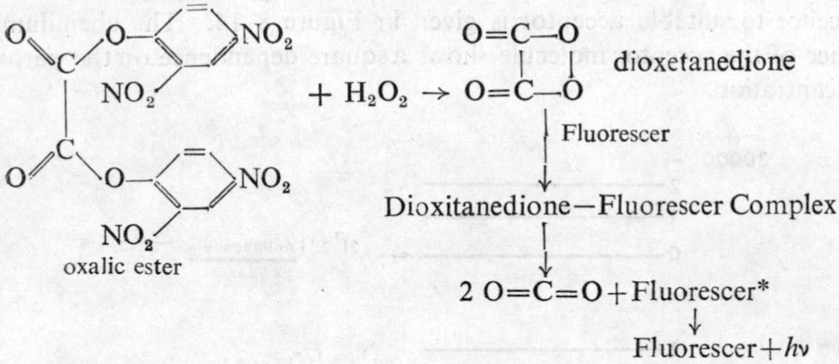

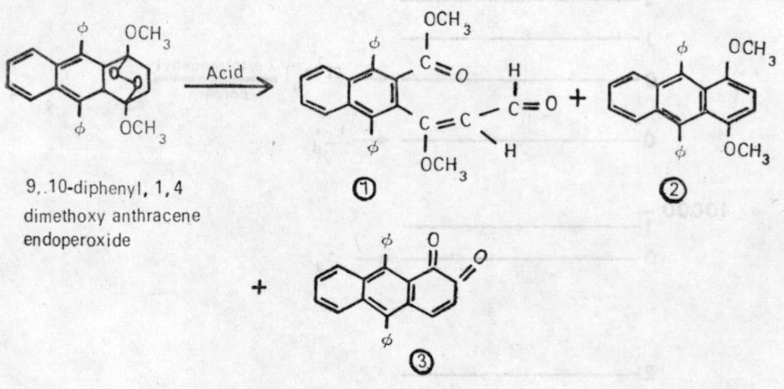

The emission spectrum is similar to fluorescence of 9,10–diphenyl–1,4–dimethoxy anthracene although it is not excited directly. The decomposition of the endoperoxide to (1) and (3) is the excitation producing step in acid-catalyzed cleavage and (2) is activated by energy transfer. The formation of $^1O_2$ by elimination reaction is a very minor reaction step.

An inorganic chemiluminescent reaction system involves the reaction between alkaline solution of H_2O_2 and either Cl_2 gas or OCl^- (hypochlorite ion). This reaction has been extensively studied by Kasha and his collaborators. The reaction produces a red glow with emission bands at 1270 nm, 762 nm and 633 nm which have been identified as emission from singlet oxygen $^1\Delta_g$ and $^1\Sigma_g$ and 'double-molecule' single quantum transition form $2[^1\Delta_g]$ (Section 8.2.2A). The emission from $[^1\Delta_g + {^1\Sigma_g}]$ and $2[^1\Sigma_g]$ double-molecule state is also observed at 478 nm and 366 nm, respectively.

If suitable acceptors are present, the excited singlet and dimeric energy states of oxygen, chemically generated during the reaction, can transfer energy to them. The *sensitized chemiluminescence* of the acceptor is observed. The electronic energy levels of molecular oxygen and excited singlet molecular oxygen pairs available for energy transfer in chemiluminescence to suitable acceptor is given in Figure 8.14. The chemiluminescence of the acceptor molecule shows a square dependence on the peroxide concentration.

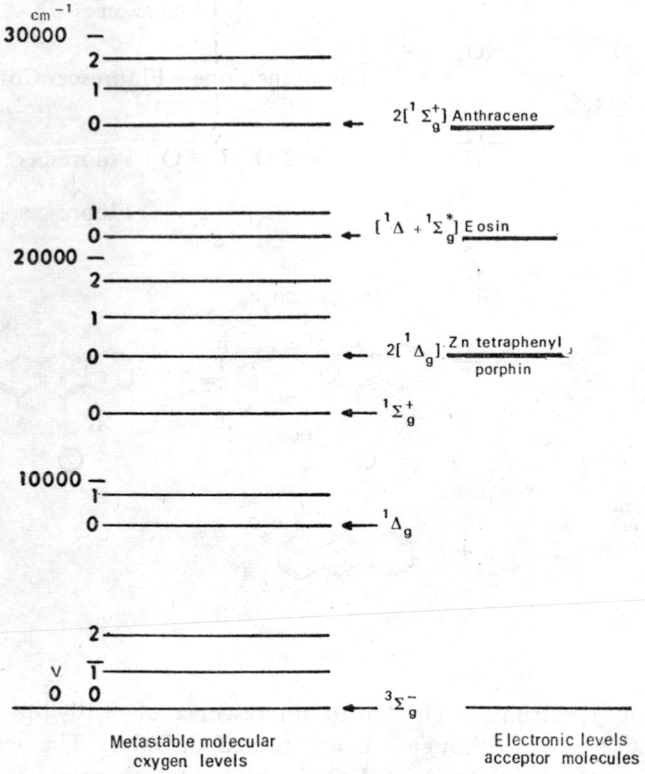

Figure 8.14 Sensitized emission from suitable acceptor in chemiluminescence reaction between $H_2O_2 + .OCl^-$ generating singlet oxygen and singlet oxygen pair states.

8.6 TRANSITION METAL COMPLEXES

The photochemistry of transition metal complexes is more varied than large organic molecules because of the possibility of different orbital pesty of excitation and of overall symmetry of the complex which may vary from O_h to C_s. In Fig. 2.2 are given energy level scheme for O_h (octahedral) symmetry for σ-bonding ligands. A set of metal orbitals and a set of ligand orbitals are available (Section 2.11). For π-bonding ligands a set of π, π* orbitals are also available. Basically four different types of low lying energy states can be described depending on whether the excitation is localized within the metal ion energy manifold or within the organic ligand energy manifold or delocalized. These are termed as follows:

(i) *d-d states or ligand-field states.* They arise from promotion of an electron frnm t_{2g} (nonbonding) to e_g (antibonding orbital) for octahedral symmetry, an exictation confined essentially to the metal ion. The energy difference is determined by Δ, the ligand field parameter. The Δ is a function of ligand field strengths, the position of the metal ion in the periodic table and the oxidation state of the ion. Further splitting of the T states may arise due to interelectronic repulsion terms. These states are labelled according to group symmetry nomenclature. Because of the Laporte forbidden nature, molar extinction values are low, $\epsilon \simeq 1–150$ $1\ mol^{-1}\ cm^{-1}$. The resulting increase in antibonding electron density decreases the net bonding in the complex and causes a lengthening of the metal ligand bonds.

(ii) *d-π* states.* These arise from excitation of a metal electron to a π* antibonding orbital located on the ligand system. This can be considered as transfer of an electron from the metal (M) to the ligand (L) and hence is termed CTML type. Since such transfers leave the metal ion temporarily in an oxidized state such states are related to the redox potentials of the complex. The (*d*, π*) state should lie at relatively low energy for easily oxidizable complex. Thus the change of central metal ion will considerably affect the position of the (*d*, π*) states. They have charge transfer character and high molar extinction, $\epsilon \simeq 10^4\ 1\ mol^{-1}\ cm^{-1}$. The reverse process, the charge transfer from the ligand to the metal may also occur, i.e. CTLM type transitions.

(iii) *π, π* states.* These states arise from localized transition within the ligand energy levels. They lie at relatively high energies. The metal ions perturb them only slightly but can drastically affect the photophysical processes, originating from them.

(iv) *π-d states.* Such states are expected to arise from a promotion of an electronic charge from ligand π-system to the higher orbitals of the metal (e-type for O_h symmetry). These are not very well established.

Each of these promotional types of energy states can further split by spin-orbital coupling interactions to give singlet and triplet states. For heavier elements the total angular momentum quantum number J becomes a 'good' quantum number (Section 2.5). Spin-orbital coupling interaction energy can vary from 500 cm^{-1} (Co^{3+}) to 4000 cm^{-1} (Ir^{3+}). The relative ordering of energy levels can be altered by replacing metal ions, exchanging ligands, modifying the ligand or by varying the geometry. These levels can also be modified by solvent effects.

8.6.1 Photophysical Processes

Very few inorganic complexes fluoresce. The classical examples are emission from rare-earths. For Sm^{3+}, Eu^{3+}, Gd^{3+}, Tb^{3+} when complexed with bidented ligands emissive transitions occur within the metal-energy levels (Section 6.6.10) which is formally Laporte forbidden. The levels are populated by intramolecular energy transfer in the complex when radiation is absorbed by the ligand. If the emitting state of the metal does not lie below the lowest triplet state of the ligand moiety, characteristic fluorescence and phosphorescence of the latter may be observed under suitable conditions. The ion UO_2^{++} is also highly luminescent.

Amongst the transition metal complexes, emission is confined to d^3 (Cr^{3+}, Co^{3+}) and d^6 systems (Ru^{2+}, Rh^{3+}, Os^{2+} and Ir^{3+}). These are *d–d* emitters and spin-forbidden phosphorescence generally predominates. The efficiency ϕ_f is low and is related to the spacing and coupling of lowest excited vibronic levels with the ground state vibronic levels. Fluorescence is generally broad, structured and exhibits large Stoke's shift which indicate geometrical distortion of the excited state. Phosphorescence is characterized by very small Stoke's shift and resolved vibrational structure at low temperature in glassy solutions and arises from $^2E_g \rightarrow {}^4A_{2g}$ transitions corresponding to $T_1 \rightarrow S_0$. In these complexes nonradiative internal conversion k_{IC} is relatively more efficient than nonradiative intersystem crossing. Thus spin-orbit coupling is not the only deciding factor for the loss of luminescence. The absence of emission for d^1, d^7, d^8, d^9 complexes and for most d^6 complexes is consistent with the existence in these complexes of low lying *d-d* states. Some photophysical processes in Cr (III) octahedral complexes are represented by idealized potential energy surfaces in Figure 8.15. To adequately represent the $3n-6$ vibrational modes a $3n-5$ dimensional hypersurfaces will be required which is not possible here.

Emission from transition metal complexes obey Kasha's rule and originate from the lowest excited state which are (i) $^3(\pi, \pi^*)$ state in $[Rh(phen)_3](ClO_4)_3$ in water-methanol glass (ii) $^3(d, \pi^*)$ state in $[Ru(bpy)_3]Cl_2$ in ethanol-methanol glass, and (iii) $^3(d\text{–}d)$ state, in solid $[RhCl_2(phen)_2]Cl$. Their characteristics differ in details and are given in Figure 8.16. Sometimes weak fluorescence is also observed, from Cr^{3+} complexes. $K_3[Co(CN)_6]$ is highly luminescent. The ϕ_p and τ_p are temperature

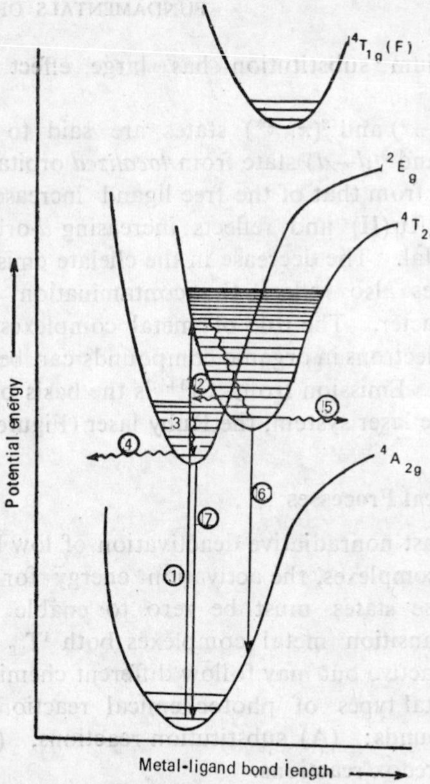

Figure 8.15 Idealized potential energy surfaces illustrating photophysical processes in octahedral complexes of Cr (III): (1) absorption, (2) intersystem crossing, (3) vibrational relaxation, (4) photoreaction from ^2E_g or nonradiative return to ground state, (5) photoreaction from $^4T_{2g}$ or nonradiative return to ground state, (6) emission from $^4T_{2g}$ and (7) emission from ^2E_g.

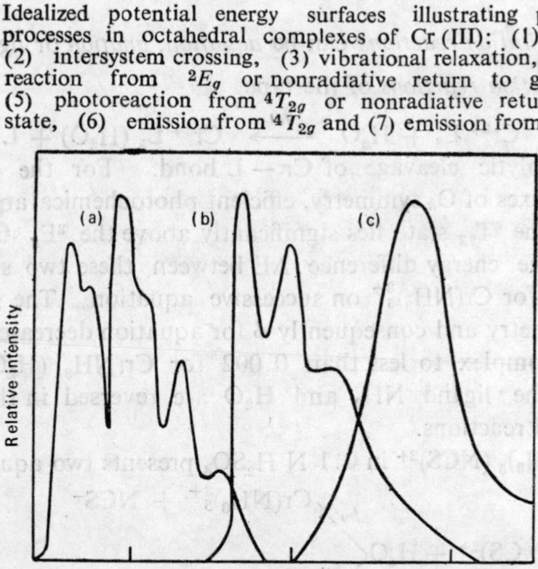

Figure 8.16 Characteristics of phosphorescence spectra from inorganic complexes. Emission from (a) $^3(\pi, \pi^*)$ state in $[Rh(phen)_3](ClO_4)$, (b) $^3(d, \pi^*)$ state in $[Ru(bpy)_3]Cl_2$ and (c) $^3(d{-}d)$ state in solid $[Rh(Cl)_2(phen)_2]Cl$.

dependent. Deuterium substitution has large effect on nonradiative processes.

Sometimes $^3(d-\pi^*)$ and $^3(\pi, \pi^*)$ states are said to be derived from *delocalized orbitals* and $^3(d-d)$ state from *localized* orbitals. The shift of the chelate emission from that of the free ligand increases in the sequence Rh(III) < Ir(III) < Ru(II) and reflects increasing *d*-orbital participation in the emission orbital. The decrease in the chelate emission lifetime from the free ligand values also reflect the contamination of the molecular orbitals with *d*-character. The role of metal complexes as quenchers of excited states of π-electrons in organic compounds can be rationalized from such considerations. Emission from Cr^{3+} is the basis of one of the most important solid state laser system, the Ruby laser (Figure 10.14).

8.6.2 Photochemical Processes

Because of the fast nonradiative deactivation of low lying energy states of transition metal complexes, the activation energy for the reactions that may occur from these states must be zero to enable them to compete effectively. For transition metal complexes both $^4T_{2g}$ and ^2E_g states can be photochemically active but may follow different chemical pathways.

Three fundamental types of photochemical reactions are known for coordination compounds: (A) substitution reactions, (B) rearrangement reactions, and (C) redox reactions.

A. *Photosubstitution reactions* can be *aquation*, *anation* or *ligand exchange*.

(i) *Photoaquation reactions* of the type

$$Cr^{3+} L_6 + H_2O \xrightarrow{h\nu} Cr^{3+} L_5 (H_2O) + L$$

involve heterolytic cleavage of Cr — L bond. For the specific case of Cr(III) complexes of O_h symmetry, efficient photochemical aquation reaction occurs when the $^4T_{2g}$ state lies significantly above the ^2E_g state. Efficiency decreases as the energy difference ΔE between these two states decreases. This happens for $Cr(NH_3)_6^{3+}$ on successive aquation. The system deviates from O_h symmetry and consequently ϕ for aquation decreases from 0.3 for hexammine complex to less than 0.002 for $Cr(NH_3)(H_2O)_5^{3+}$. Relative labilities of the ligand NH_3 and H_2O are reversed in the thermal and photochemical reactions.

The $Cr(NH_3)_5(NCS)^{2+}$ in 0.1 N H_2SO_4 presents two aquation modes :

$$Cr(NH_3)_5(NCS)^{2+} + H_2O \xrightarrow{h\nu} Cr(NH_3)_5^{3+} + NCS^- \quad \text{(i)}$$

$$Cr(NH_3)_5(NCS)^{2+} + H_2O \xrightarrow{h\nu} Cr(NH_3)_4(H_2O)(NCS)^{2+} + NH_3 \quad \text{(ii)}$$

Both $^4T_{2g}$ and ^2E_g states are photoactive. The ratio of ϕ_{NH_3}/ϕ_{NCS^-} is wavelength dependent, the ratios being 15 at 373 nm (quartet band); 22 at

492 nm (quartet band) and 8 at 652 nm (doublet band).

The *photosensitization* of this aquation can also be initiated by *energy transfer* from biacetyl and acridinium ion in the same solvent. The energy transfer reactions have helped to identify the details of this aquation reaction:

(a) *Biacetyl as sensitizer*: phosphorescence quenched; fluorescence not affected, release of NH_3; No apparent degradation of biacetyl; $\phi_{NH_3}/\phi_{NCS^-} > 100$.

(b) *Acridinium ion as sensitizer*: fluoresence quenched; both NH_3 and NCS^- released although NH_3 predominates; in presence of O_2, NCS^- release decreased, hence triplet state of the sensitizer is involved; ϕ_{NH_3}/ϕ_{NCS^-}, 33 for quartet state, 8 for doublet state.

Therefore it can be concluded that NH_3 release occurs from $^4T_{2g}$ state and NCS^- from ^2E_g state.

$Co(NH_3)_6^{3+}$ is stable in aqueous solution whereas $(Co(NH_3)_5\,X)^{2+}$ undergoes moderately rapid substitution of water for the acid group X^-. The rates of aquation reaction depends strongly on basicity of X^-, for example, nitrate: acetate is $10^3 : 1$. The difference in behaviour between Cr(III) and Co(III) ammines may be due to necessity for energy of activation in the transition state for reaction with the latter.

The photoaquation of $KCr(NH_3)_2\,(SCN)_4$ (Reinnecke's salt) and $Cr(urea)_6Cl_3$ can be used as efficient actinometers between the range 316 and 750 nm. For $Co(CN)_6^{3-}$ the photoreaction

$$Co(CN)_6^{3-} \xrightarrow{h\nu} Co(CN)_5\,(H_2O)^{2-} + CN^-$$

is without complications and occurs with high quantum yield $\phi = 0.31$.

(ii) *Photoanation reactions.* For example,

$$Cr(H_2O)_6^{3+} + NCS^- \xrightarrow{h\nu} Cr(H_2O)_5(NCS)^{2+} + H_2O$$

$$Cr(H_2O)_6^{3+} + Cl^- \xrightarrow{h\nu} Cr(H_2O)_5Cl^{2+} + H_2O$$

$$Fe(CN)_6^{4-} + dipyl \xrightarrow{h\nu} Fe(CN)_4\,(dipyl)^{2-} + 2CN^-$$

(iii) *Photosubstitution reactions.* The irradiation of monosubstituted carbonyl $M(CO)_{n-1}\,L$ in presence of excess of L, causes the formation of bisubstituted complexes: (M = metal, L = ligand). For example,

$$Mo(CO)_6 + L \xrightarrow{h\nu} Mo(CO)_5L + CO$$

where, $L = NH_3$, $(C_2H_5)NH_2$, $(CH_3)_2NH$, $(CH_3)_3N$, $(C_2H_5)_3N$ or other ammines. The pentacobaltic complex may undergo further substitution.

In the case of mixed ligands photoreactivity may be predicted empirically from considerations of the ligand field strength (Adamson's rule). For O_h complex, the axis having the weakest average field as determined by the position of its respective ligands in the spectrochemical series will be labilized. If the labilized axis contains two different ligands, the ligand of greater field strength will preferentially be replaced. For example, upon irradiation of $Cr(NH_3)_5Cl^{2+}$, the $NH_3—Cr—Cl$ axis is labilized rather than $NH_3—Cr—NH_3$. Furthermore, the stronger field ligand NH_3 is lost along $NH_3—Cr—Cl$ axis. Steric and solvation effects may cause deviations from these predictions. Lowering of symmetry of the complex will also affect the nature of reaction due to switching of states.

B. *Photorearrangement reactions* may involve: (i) geometrical isomerization, (ii) recemization, (iii) Linkage isomerization, and (iv) ligand rearrangement.

(i) *Geometrical isomerization*

$$\text{Cis-bis (glycinato) Pt (II)} \xrightarrow[\text{band}]{d\text{—}d} \text{trans-isomer}$$

$$\phi \simeq 0.13$$

$$\text{trans-} \not\longrightarrow \text{cis}$$

The photoprocess involves intramolecular twist mechanism without bond cleavage. The reactive intermediate is a triplet state of pseudo-tetrahedral geometry. The transformation is photochemically allowed but thermally disallowed from symmetry considerations.

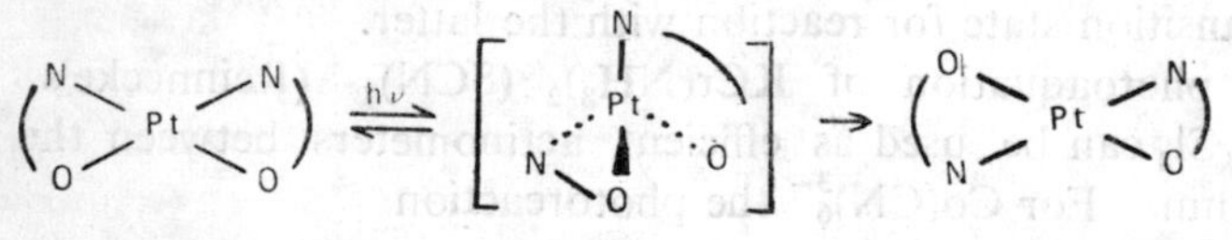

cis-singlet ground state | tetrahedral intermediate triplet state | trans-singlet ground state

(ii) *Racemization*

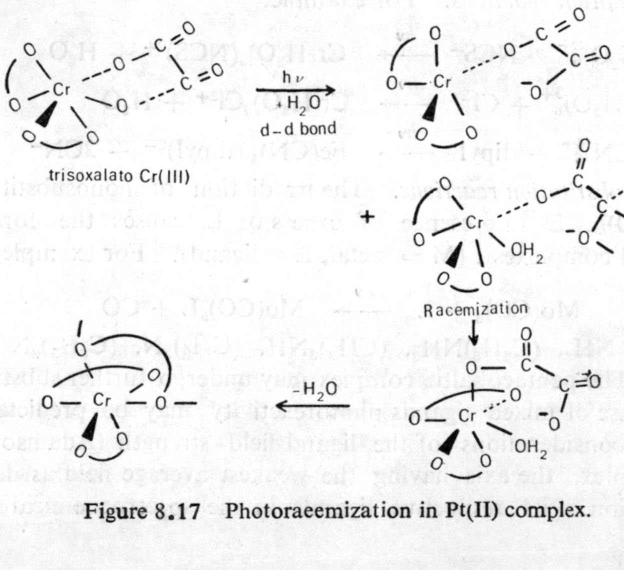

Figure 8.17 Photoraeemization in Pt(II) complex.

The reaction shows deuterium isotope effect.

(iii) *Linkage photoisomerization*

$$Co(NH_3)_5(NO_2)^{2+} \xrightarrow{h\nu \text{ (solid)}} (NH_3)_5Co(—ONO)^{2+}$$

The nitro-nitrito photoisomerization occurs in solution through an intramolecular mechanism involving the homolytic fission of Co—NO_2 bond and then cage recombination of the two fragments by means of a Co—ONO bond. This complex exhibits CT and CTLM character at 239 and 325 nm and a ligand field band at 458 nm.

$$Co(NH_3)_5(NO_2)^{2+} \xrightarrow[\text{in } 0.1N\ HClO_4]{\text{CT and d}-\text{d}} [Co(NH_3)_5^{2+} . NO_2]$$

$$\xrightarrow{\text{redox}} Co^{2+} + 5NH_3 + NO_2$$

$$\xrightarrow{\text{isomerization}} Co(NH_3)_5(ONO)^{2+}$$

(iv) *Photoexchange processes*

$$Co(CN)_5(NO)^{3-} + CN^{*-} \xrightarrow{h\nu} Co(CN^*)_5(NO)^{3-} + CN^-$$

(C) *Photo-oxidation-reduction* or *Redox-reactions.* A photo-oxidation-reduction process in solution may be *intramolecular* when the redox reaction occurs between the central metal atom and one of its ligands or *intermolecular* when the complex reacts with another species present in the solution.

When the central metal ion is strongly oxidizing, irradiation in the ligand field band (d-d) can lead to *intramolecular* reduction of the central metal ion

(i) $$2Fe^{III}(C_2O_4)_3^{3-} \xrightarrow[0.1N\ H^+]{h\nu} 2Fe(OX)_2^{2-} + OX^{2-} + 2CO_2$$

(ii) $$Co^{III}(NH_3)_5 I^{2+} \xrightarrow{h\nu} Co^{2+} + 5NH_3 + I$$

The photodecomposition of ferrioxalate in 0.1N H_2SO_4 (reaction (i)) is a very useful chemical actinometer for the near uv and the visible region upto 400 nm ($\phi \simeq 1.12$) — *ferrioxalate actinometer.* In some low-valence hydrated cations, such transition may bring about photo-oxidation:

$$Fe^{2+}.H_2O \xrightarrow{h\nu} Fe^{3+} + OH^- + H$$

Another example is Fe(thionine)$^{2+}$ $\underset{\text{dark}}{\overset{h\nu}{\rightleftarrows}}$ Fe^{3+} + semithionine which reverses in the dark. This reversible photoredox reaction has the possibility of conversion of light energy into electrical energy. The irradiation at the CTLM bands is more likely to result in photoreduction of the general type

$$(M^{z+}L_n)^{z+} \xrightarrow{h\nu} M^{(z-1)+} + (n-1)L + \text{oxidation products of } L$$

The excitation leads to homolytic M-L bond fission. Such a reactivity is not expected of d-d band and if redox reaction does occur under such excitation it is likely to hide a low lying reactive CT type band under its envelop.

It is not necessary that the reaction should always lead to decomposition of the ligand. If the central ion can give a stable complex with one lower (or higher) oxidation number, the product of a intramolecular redox reaction may be a complex with a central metal ion of different oxidation number.

$$IrCl_6^{2-} \xrightarrow[H_2O]{h\nu} IrCl_5(H_2O)^{2-} + Cl$$

Intermolecular photo-oxidation-reduction reactions involve a light initiated electron transfer between a complex and any other suitable molecule available in the medium. An oxidized or a reduced form of the complex may be obtained.

$$ML_n^{z+} + X \xrightarrow{h\nu} ML_n^{(z+1)+} + X^-$$
$$ML_n^{z+} + Y \longrightarrow ML_n^{(z-1)+} + Y^+$$

If the new complex species is unstable, a decomposition reaction may then follow. Examples:

$$Co(NH_3)_6^{3+} + I^- \xrightarrow{h\nu} Co(NH_3)_6^{2+} + I$$
$$Mo(CN)_8^{3-} + H_2O \xrightarrow{h\nu} Mo(CN)_8^{4-} + H^+ + OH^-$$

In the second example, no decomposition of the complex occurs. The excitation is in the ligand field band. When the new complex is stable, an electron may be ejected to the solvation shell leading to the formation of hydrated electron e_{aq}^-

$$Fe(CN)_6^{4-} + H_2O \xrightarrow{h\nu} Fe(CN)_6^{3-} + e_{aq}^-$$

This is a CTTS type of transition. The photoejection of electron may also occur if CTLM band increases the charge density towards the periphery of the complex.

(iii) *Chemiluminescence in ruthenium complexes.* Ru(III)-2, 2′ bipyridyl-5 methyl-1,10-phenanthroline and few other complexes when react with OH^- or hydrazide, generate enough thermal energy to promote it to an electronically excited state. The molecule relaxes by emission of radiation. In such reactions chemical energy is converted into electronic energy. The general reaction scheme is

$$\text{Metal (ligand)}^{(n+1)+} + e^{(-)} \text{ (from reductant)}$$
$$\rightarrow \text{Metal (ligand)}_x^{n+} + h\nu$$

8.6.3 Photochemistry of Metallocenes

Metallocenes act as good quenchers of triplets whose energies lie between 280 to 177 kJ/mol^{-1}. Quenching rates are nearly diffusion controlled. They are good sensitizers of cis-trans isomerization. The diamagnetic ferrcene is light stable to visible and uv but decomposes in presence of O_2 is an acidified solution. In halogenated solvents at room temperature, solutions of ferrocenes are stable in the dark but on exposure to light, decomposition occurs at a relatively rapid rate. In CCl_4, it forms a CT complex

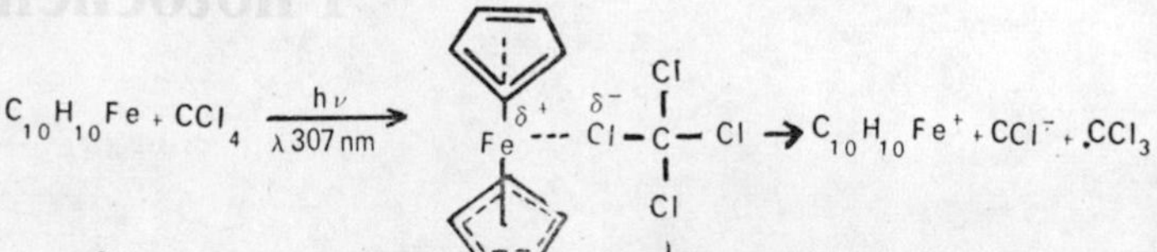

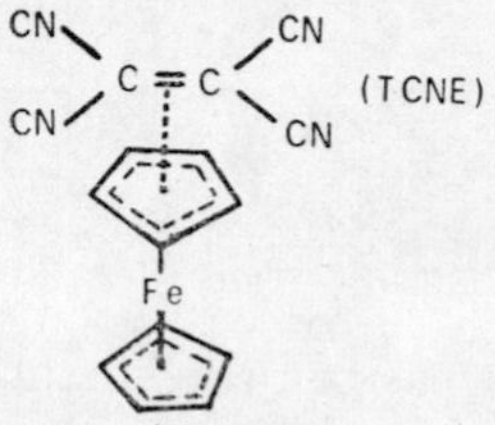

$$C_{10}H_{10}Fe + CCl_4 \xrightarrow[\lambda\,307\,nm]{h\nu} \overset{\delta^+}{Fe} \cdots \overset{\delta^-}{Cl} - \overset{\overset{Cl}{|}}{\underset{\underset{Cl}{|}}{C}} - Cl \rightarrow C_{10}H_{10}Fe^+ + CCl^- + CCl_3$$

(TCNE)

(I)

$$\overset{\delta^+}{Cr} \cdots \overset{\delta^-}{I}$$

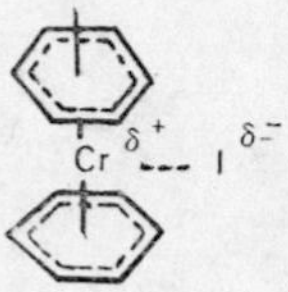

(II)

Tetracyanoathylene forms a π-type complex with ferrocene (I). Another metallocene is bis-benzenechromium iodide (II) a sandwitch type complex.

NINE

Some Current Topics in Photochemistry

9.1 ORIGIN OF LIFE

Photochemistry has played a vital role in earth's history. Long before life evolved, the atmosphere was of volcanic nature, consisting chiefly of methane, carbon dioxide, water vapour and nitrogen. Very short wavelength ultraviolet light from the sun dissociated these molecules into very reactive radicals. A little oxygen and ozone was formed, but kept at low concentrations by the radicals. Knowledge of the sequence of events is necessarily speculative, but seems to be as follows. First the radicals combined to give simple compounds. These dissolved in the sea and by further photoreaction formed aminoacids etc. Polymerization occurred, and colloidal matter formed. After an immense period of time, primitive cell structure evolved from the colloids and engaged in anaerobic metabolism (fermentation), reproducing themselves and consuming other molecules available. Pigments appeared, leading to chlorophyll formation. A critical period was now approached. Anaerobic photosynthesis produced oxygen, so that fermentation was replaced by oxidative respiration with a forty-fold increase in chemical energy gain. These events must have happened in shallow water (5-10 metre deep) where dangerous ultraviolet light was cut-off and yet visible light could penetrate. The optical asymmetry of biological compounds perhaps suggests that one single cell at this stage was the parent of all future developments.

A hazard still awaited primitive life at this point. Oxygen formation led to the photochemical production of ozone, lethal to organisms. Their existence in the depth of the sea must have been important. However, with oxidative respiration, cells fed and multiplied rapidly, until a high atmospheric concentration of oxygen was reached. Ozone formation was now confined to the upper atmosphere; its formation requires the very short wavelengths absorbed by oxygen but its photodecomposition is caused by longer wave ultraviolet capable of damaging proteins. The ozone layer by absorbing radiation of $\lambda < 300$ nm acts as a safety screen for existence of life on the earth. Therefore our concern about the possible effects of pollutants such as from the exhausts of supersonic aircraft which can disturb the ozone concentration in the upper atmosphere is very pertinent. Nearer the ground, photochemical reactions between combustion products of motor vehicles, oxides of nitrogen and sulpher, aerosol vapours etc. present hazards to the living system (Section 7.4.1).

9.2 MUTAGENIC EFFECT OF RADIATION

Nucleic acids and proteins are basic chemicals of life. The biopolymer deoxyribonucleic acid (DNA) is the information containing molecule of high molecular weight ($\simeq 10^{12}$ dalton) and is responsible for the propagation of life itself. It is composed of four bases, adenine (A), guanine (G), thymine (T) and cytosine (C) attached to sugar-phosphate backbone. Two such strands form a double-helix held together by H-bonds between the bases. The base pairing is very specific, adenine pairs with thymine (A—T) and guanine with cytosine (G—C), which assigns the double helix the character of a template.

All these bases absorb around 260 nm. Thymine and cytosine are most sensitive to irradiation. Two most important types of photochemical reactions that have been observed for these pyrimidine bases are photohydration and photodimerization. In *vivo* systems, interactions between proteins and nucleic acids can also be initiated by radiations of wavelength shorter than 300 nm.

Photohydration occurs at the 5, 6 positions of pyrimidine bases, giving 5-hydro-6-hydroxy compound. Cytosine and its various glycosides photohydrate readily in aqueous solution as detected from the disappearance of 260 nm peak and appearance of a new peak at 240 nm. On heating, loss of H_2O regenerates the original molecule. When this photoreaction occurs in DNA, base pairing is prevented leading to mutation.

Photodimerization occurs between pyrimidine bases when they are stacked in suitable geometry in the helical DNA molecule. Thymine is most susceptible to photodimerization which occurs at λ 280 nm. The reaction is photoreversible and monomerization occurs by irradiation with 240 nm radiation.

All these photoreactions caused by uv light prevent H-bonding between base pairs which is fundamental to the replication and propagation of life (Figure 9.1). Some of these damages can be repaired, others are permanent and lethal to life. Therefore the importance of ozone screen in cutting off these deleterious radiations from the solar spectrum.

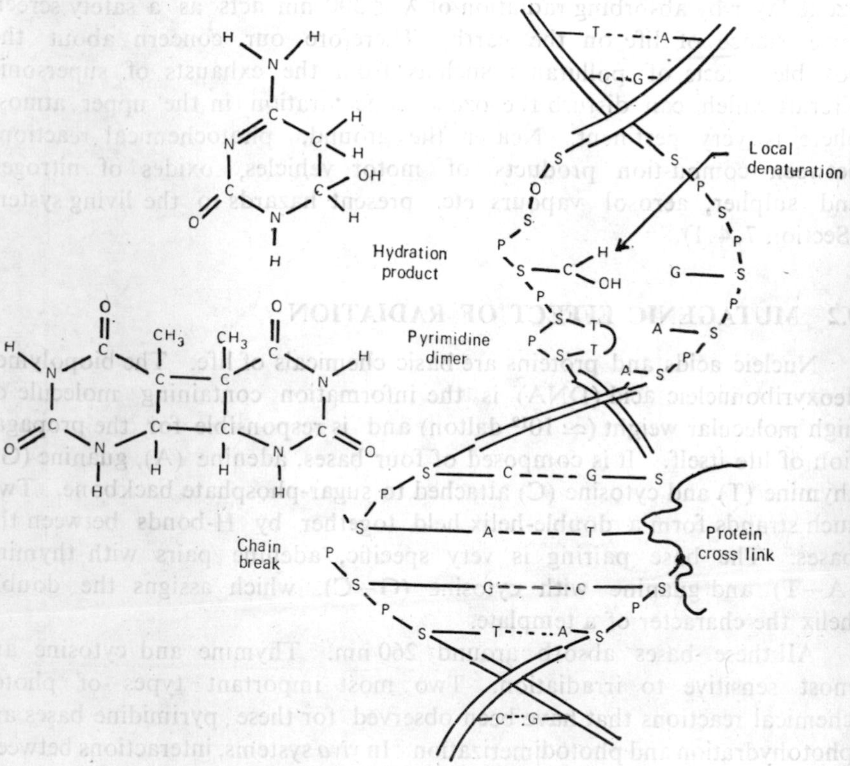

Figure 9.1 Mutagenic photochemical reactions in DNA helix, 5-hydro-6-hydroxy cytosine and thymin dimer.

9.3 PHOTOSYNTHESIS

In a broad sense photosynthesis in plants is a photoinduced electron transport reaction. Chlorophyll molecules in the green plants are the main light harvesting molecules. They are assisted by carotenoids and phycocyanins in this act. These molecules have absorption in the visible region covering the whole spectrum from blue to red. The energy absorbed by all these molecules is transferred to chlorophyll *a* (Chl *a*), which is the main light sensitive molecule, by mechanisms discussed in Section 6.6.4.

Chlorophyll *b* also transfers its energy to Chl *a*, which has two absorption peaks, one in the blue violet region ($\lambda = 448$ nm) and the other in the red region ($\lambda = 680$ nm) (Figure 9.2). *In vitro* it emits red fluorescence ($\lambda \simeq = 680$ nm). This characteristic fluorescence is the only fluorescence observed, whichever is the light absorbing molecule. This indicates that other molecules transfer excitational energy to Chl *a* and the photosynthetic process utilises a low energy quanta corresponding to red light only. *In vitro,* fluorescence efficiency of Chl *a* is $\phi_f \simeq 0.3$ but *in vivo,* it is much reduced. Any factor which tends to reduce photosynthesis enhances ϕ_f.

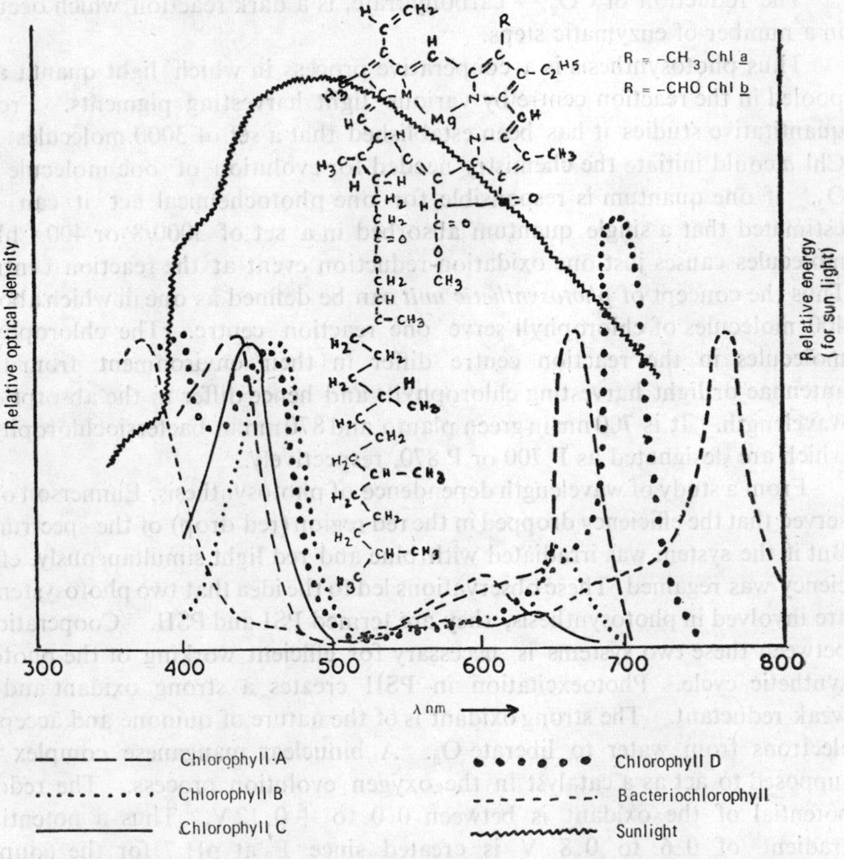

Figure 9.2 Chlorophyll molecules and absorption spectra of the pigments.

The net reaction of photosynthetic process in plants is given as

$$n\,CO_2 + n\,H_2O \xrightarrow[680]{h\nu} (CH_2O)_n + n\,O_2 \qquad (9.1)$$

chlorophyll

The energy equivalent of red light is about 190 kJ (45 kcal) mol^{-1}, but the actual energy requirement for the reduction of 1 mole of CO_2 to carbohydrate is 470 kJ ($\simeq$112 kcal) mol^{-1}. This shows that the mechanism of reduction of CO_2 is much more complex. In the initial act of electron transfer process from photoexcited Chl *a*, H_2O serves as an oxidisable substrate producing O_2 and also as an electron source for the reduction of CO_2. Two quanta are required to create one oxidized and one reduced species i.e. for transfer of one electron. Therefore for evolution of 1 molecule of O_2, four electron or 8 quanta of light are required

$$2H_2O \xrightarrow{8h\nu} 4H^+ + 4e^- + O_2 \quad (9.2)$$

The reduction of $CO_2 \rightarrow$ carbohydrate, is a dark reaction which occurs in a number of enzymatic steps.

Thus photosynthesis is a cooperative process in which light quanta are pooled in the reaction centre by various light harvesting pigments. From quantitative studies it has been established that a set of 3000 molecules of Chl *a* could initiate the chemistry needed for evolution of one molecule of O_2. If one quantum is responsible for one photochemical act it can be estimated that a single quantum absorbed in a set of 3000/8 or 400 Chl *a* molecules causes just one oxidation-reduction event at the reaction centre. Thus the concept of *photosynthetic unit* can be defined as one in which about 400 molecules of chlorophyll serve one reaction centre. The chlorophyll molecules in the reaction centre differ in their environment from the antennae or light harvesting chlorophylls and hence differ in the absorption wavelength. It is 700 nm in green planto and 870 nm in bactersiochlorophyll which are designated as P 700 or P 870, respectively.

From a study of wavelength dependence of photosynthesis, Emmerson observed that the efficiency dropped in the red region (red drop) of the spectrum. But if the system was irradiated with blue and red light simultaneously, efficiency was regained. These observations led to the idea that two photosystems are involved in photosynthesis, they are termed PSI and PSII. Cooperation between these two systems is necessary for efficient working of the photosynthetic cycle. Photoexcitation in PSII creates a strong oxidant and a weak reductant. The strong oxidant is of the nature of quinone and accepts electrons from water to liberate O_2. A binuclear manganese complex is supposed to act as a catalyst in the oxygen evolution process. The redox potential of the oxidant is between 0.0 to +0.12 V. Thus a potential gradient of 0.6 to 0.8 V is created since E_0' at pH 7 for the couple O_2/H_2O (reaction 9.2) is +0.8 V (Figure 9.3).

The trap P 700 in PSI is a special kind of Chl *a* and is present in the ratio of 1:300 in the reaction centre. Its redox potential is +0.4 V. In the primary photoact in PSI, a strong reductant and a weak oxidant are formed. The weak reductant of PSII and the weak oxidant of PSI are coupled through a number of different cytochromes of graded redox potentials ranging from $E_0' = 0.0$ to + 0.4 V. During the downhill electron transfer over the enzymatic chain some amount of energy is degraded, simultaneously

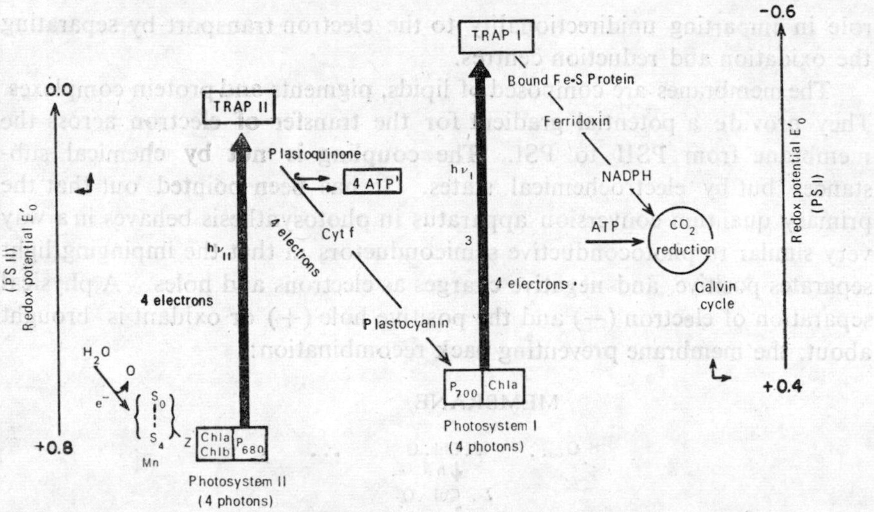

Figure 9.3 Photosynthetic cycle and reaction centres PS I and PS II.

storing a part of it as phosphate bond of ATP (adenosine triphosphate) through the reaction:

$$\underset{\text{(adenosine diphosphate)}}{\text{ADP}} + \underset{\text{(inorganic phosphate)}}{\text{P}_i} \rightarrow \underset{\text{(adenosine triphosphate)}}{\text{ATP}} \quad (9.3)$$

About 30 to 34 kJ (7 to 8 kcal) mol^{-1} is thus stored. The strong reductant transfers an electron through a number of enzymatic steps to NADP$^+$ (nicotinamide adenine diphosphate) to form NADPH. The intermediate enzymes are complexes of iron-sulpher proteins such as ferredoxin and hydrogenase. In absence of O_2 or low concentration of CO_2 hydrogenase can liberate molecular hydrogen. The reduced NADPH then utilizes the energy stored in high energy phosphate bond of ATP for the reduction of CO_2 to CH_2O in a number of dark steps. The reaction sequence in which CO_2 is first added to a five-carbon sugar was elucidated by M. Calvin and his associates and is described as *Calvin cycle*.

Photosynthesis in plants occurs in thylakoid membranes. These are lamellar structures, several microns in linear dimensions and are organized in stacks in green leaf chloroplast. There are regions of greater density which are called *grana* as different from *stroma*. By the action of detergent on isolated chloroplasts it has been demonstrated that PSI and PSII centres are located in different parts of the membrane. Chlorophyll molecules are flat molecules with a conjugated system of double bonds consisting of 4 pyrrol rings joined in a circle and attached to a long hydrocarbon chain (phytol chain) (Figure 9.2). Through this long chain the light absorbing molecule is attached to the lipoproteins of the membrane in regular ordered arrangement. The membrane plays a very important role in imparting unidirectionality to the electron transport by separating the oxidation and reduction centres.

The membranes are composed of lipids, pigments and protein complexes. They provide a potential gradient for the transfer of electron across the membrane from PSII to PSI. The coupling is not by chemical substances but by electrochemical states. It has been pointed out that the primary quantum conversion apparatus in photosynthesis behaves in a way very similar to photoconductive semiconductors in that the impinging light separates positive and negative charges as electrons and holes. A physical separation of electron (−) and the positive hole (+) or oxidant is brought about, the membrane preventing back recombination:

MEMBRANE

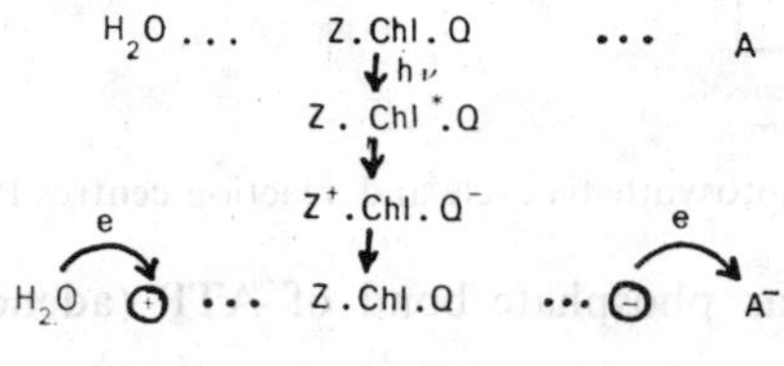

Thus in photosynthesis the initial events are a light stimulated charge separation. A cooperation of a suitable electron acceptor and a donor increases the efficiency of charge separation at excited chlorophyll molecules. Under these conditions, these molecules act as *photon-powered electron* pump for transfer of electrons 'uphill' from electronic levels of the donor to electronic levels of the acceptor. The oxidized chlorophyll molecule Chl^+ with positive hole (+) constitutes a strong oxidizing agent and should be capable of oxidizing compounds with fairly high redox potential (Figure 9.4).

The combination of excited electrons and the hole left behind is called an *exciton*. They are prevented from recombination by the hydrophobic interior of the membrane. The exciton migrates through the photosynthetic unit until it meets an oxidizing enzymatic centre which serves as a trap for the excited electron. The exciton is now reduced to a free hole which then migrates through the units until it reaches an electron donating centre. The primary photochemical oxidation-reduction process is thus completed, the oxidation product being formed at one enzymatic centre and the reduction product in another centre. Chlorophyll itself is not decomposed, but only sensitizes the reaction. The light energy is stored in the form of chemical energy in the reduced form of the oxidant.

Such ideas have initiated *in vitro* studies of chlorophyll-sensitized reactions in photosynthetic membranes from the laws of electrochemical kinetics. It has been possible to mimic the energy conversion and storage capacity of these membranes by experimental bilayer lipid membranes

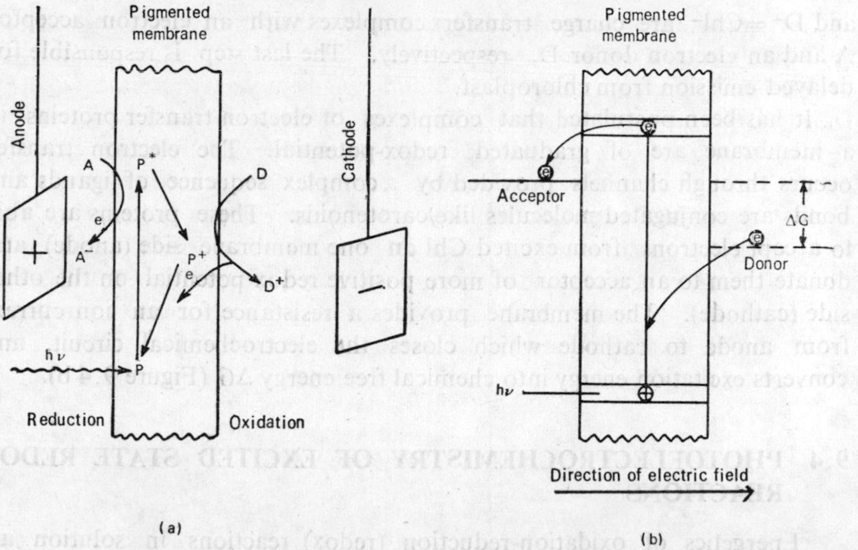

Figure 9.4 Membrane model for electron transfer reaction in photosynthetic cycle with acceptor A and donor D on either side of the membrane (a) P, P*, P+ are respectively normal state electronically excited state and oxidised form of pigment molecule. (b) Illustrating energy levels of ground and excited states of pigment molecule in the membrane and acceptor and donor molecules in solution. $\Delta G = nFE$ is theoretically available electrochemical free energy; e=electron, + =positive hole.

(BLM) formed from chloroplast lamella extracts. A photoelectrochemical cell can be constructed by placing two redox couples of graded potentials on either side of the membrane and dipping metal electrodes in each of them. On completing the circuit by suitable device, photocurrent is generated on excitation of the membrane by light energy.

The alternative sequence of reactions initiated by light can be summarized as follows:

$$\text{Chl} + h\nu \rightarrow \text{Chl}^{-+} \text{ (exciton formation)}$$

$$H_2O - \text{Chl}^{-+} \rightarrow e^-_{aq} + \text{Chl}^+$$

$$A - \text{Chl}^+ \rightarrow A^- - \text{Chl}^+ \rightarrow A^- + \text{Chl}^+$$

$$e_{aq} + A \rightarrow A^- + H_2O$$

$$H_2O - \text{Chl}^{-+} \rightarrow h^+_{aq} + \text{Chl}^-$$

$$D - \text{Chl}^{-+} \rightarrow D^+ + \text{Chl}^- \rightarrow D^+ + \text{Chl}^-$$

$$h^+_{aq} \text{ or } \text{Chl}^+ + \tfrac{1}{2}H_2O \rightarrow \text{Chl} + H^+ + \tfrac{1}{4}O_2$$

$$\text{Chl}^- + \text{Chl}^+ \rightarrow 2\text{Chl} + h\nu'$$

Here Chl denotes a photosynthetic pigment, (— +) is an electron-hole pair or exciton, e^-_{aq} is hydrated electron, h^+_{aq} is hydrated hole. A^-=Chl$^+$

and $D^{+}=Chl^{-}$ are charge transfer complexes with an electron acceptor A and an electron donor D, respectively. The last step is responsible for delayed emission from chloroplast.

It has been postulated that complexes of electron-transfer proteins in a membrane are of graduated redox-potential. The electron transfer occurs through channels provided by a complex sequence of ligands and bonds are conjugated molecules like carotenoids. These proteins are able to accept electrons from excited Chl on one membrane side (anode) and donate them to an acceptor of more positive redox potential on the other side (cathode). The membrane provides a resistance for an ion current from anode to cathode which closes the electrochemical circuit, and converts excitation energy into chemical free energy ΔG (Figure 9.4 b).

9.4 PHOTOELECTROCHEMISTRY OF EXCITED STATE REDOX REACTIONS

Energetics of oxidation-reduction (redox) reactions in solution are conveniently studied by arranging the system in an electrochemical cell. Charge transfer from the excited molecule to a solid is equivalent to an electrode reaction, namely a redox reaction of an excited molecule. Therefore, it should be possible to study them by electrochemical techniques. A redox reaction can proceed either by electron transfer from the excited molecule in solution to the solid, an anodic process, or by electron transfer from the solid to the excited molecule, a cathodic process. Such electrode reactions of the electronically excited system are difficult to observe with metal electrodes for two reasons: firstly, energy transfer to metal may act as a quenching mechanism, and secondly, electron transfer in one direction is immediately compensated by a reverse transfer. By using semiconductors or insulators as electrodes, both these processes can be avoided.

The energy levels and bands in molecules, metals and semiconductors are presented in Figure 9.5, which will help to understand these statements.

For isolated atoms quantum mechanics predicts a set of energy levels, of which lowest ones are occupied obeying Pauli's exclusion principle. On excitation with proper energy, electrons are promoted from the lower occupied energy levels to upper unoccupied levels. In solids, energy levels occur in bands. Because of the close proximity of atoms or ions fixed at the lattice points the energy levels of individual ions interact to give N closely spaced energy levels which form energy bands corresponding to each of the s, p, d, etc. levels. In metals these bands overlap such that the occupied and vacant levels lie contiguous to each other. In presence of a field, current can flow by thermal promotion of electrons to unoccupied levels. On the other hand, inorganic solids are insulators

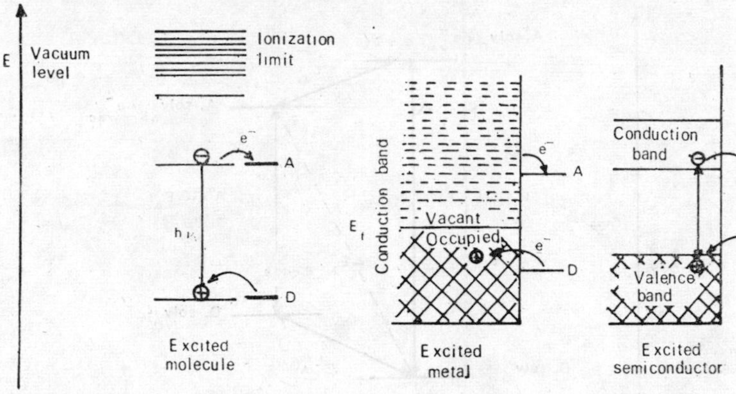

Figure 9.5 Energy levels and bands in molecules, metals and semiconductors.

because fully occupied (valence band) and empty (conduction band) energy bands are separated by forbidden zone. At ordinary temperature thermal energy is not enough to promote electrons to the conduction band from the occupied valence band. In semiconductors, imperfection or added impurity levels provide localized energy levels within this energy gap, either below the conduction band (*n*-type) or just above the valence band (*p*-type). In photoconductive semiconductors the energy gap can be bridged by photoexcitation, promoting an electron to the conduction band and leaving a positive hole behind. In *n*-type photoconductivity, charge carriers are negative electrons. In *p*-type photoconductivity, impurity levels accept an electron from the valence band injecting a positive hole in the band. The charge carriers are these positive holes which move towards the negative electrode by accepting an electron from the neighbour. A hole is created in the neighbour, which again accepts an electron from the next neighbour and so on.

Electron transfer is a fast reaction ($\simeq 10^{-12}$ s) and obeys the Franck-Condon Principle of energy conservation. To describe the transfer of electron between an electrolyte in solution and a semiconductor electrode, the energy levels of both the systems at electrode-electrolyte interface must be described in terms of a common energy scale. The absolute scale of redox potential is defined with reference to free electron in vacuum where E=0. The energy levels of an electron donor and an electron acceptor are directly related to the gas phase electronic work function of the donor and to the electron affinity of the acceptor respectively. In solution, the energetics of donor-acceptor property can be described as in Figure 9.6.

$$D\cdot solv_D \rightarrow A^{+}solv_D + e^{-}_{vac} \qquad \Delta G = +E_D = -I^{\circ}$$

$$A^{+}solv_D \rightarrow A^{+}solv_A{}^{*} \qquad \Delta G = -L_D$$

$$A^{+}solv_A + e^{-}_{vac} \rightarrow D\cdot solv_A{}^{*} \qquad \Delta G = -E^{0}_{A} = A^{\circ}$$

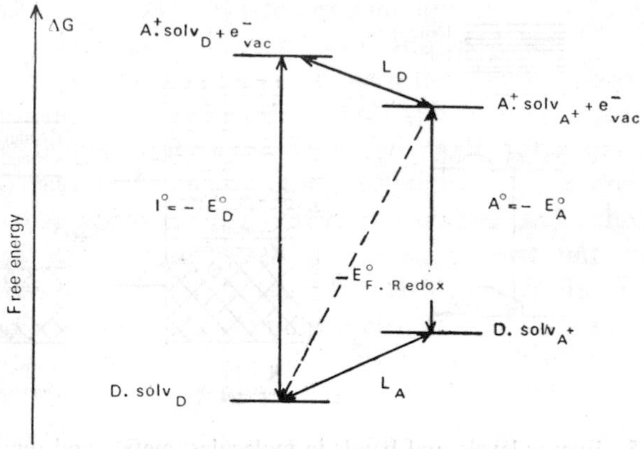

Figure 9.6 Free energy relations in defining electron donor-acceptor levels in solution of an electrolyte.

$$D.solv_{A^+} \rightarrow D.solv_D \qquad \Delta G = -L_A$$

$$I^\circ = A^\circ + L_D + L_A \qquad (9.4)$$

where I° is the ionization potential of donor, A° the electron affinity of acceptor, L_D and L_A are the respective reorganization energies of the solvation shell. Thus the redox reaction with electron transfer to vacuum level under thermal equilibrium condition is given as

$$D.solv_D \rightleftharpoons A^+solv_{A^+} + \bar{e}_{vac} \qquad \Delta G = \Delta G_{vac} \qquad (9.5)$$

If electron from vacuum is brought to an energy level $+\Delta G^\circ_{vac}$, an equilibrium situation will be obtained with $\Delta G^\circ = 0$. For a redox system in contact with a metal electrode, the equilibrium electrode potential is given by

$$Dsolv_D \rightleftharpoons A^+solvA^+ + E^\circ_{F_{redox}} \qquad \Delta G = 0$$

where E_{redox} is the Fermi level or filled level of the metal. The chemical potential of electrons in the metal, $E^\circ_{F_{redox}}$ we have from Figure 9.6.

$$E^\circ_{F_{redox}} = -\Delta G_{vac} = -I^\circ + L_D = -A^\circ - L_A \qquad (9.6)$$

The value $-\Delta G^\circ_{vac}$ represents the mean free energy of electrons in the redox electrolyte and is considered equal to the Fermi energy of the redox system. In the absolute energy scale the Fermi energy of electrons in a hydrogen electrode is at about −4.5 eV above the standard redox potential against standard hydrogen electrode (SHE) because more energy is required to remove an electron to vacuum than to SHE.

$$E_{F_{redox}} = -4.5 - E^\circ_{SHE_{redox}} \text{ (ev)} \qquad (9.7)$$

The density of occupied energy levels in solution can be identified with C_D, the concentration of the electron donor, and the density of unoccupied energy levels with C_A, the concentration of the electron acceptors.

Due to fluctuations of the interaction energy of the solute with its surroundings in a polar liquid, the donor and acceptor states cannot be represented by a single energy level, but must be described by a distribution function W (E) (Figure 9.7). The distribution functions represent time averages of electronic energies for single donor or acceptor molecules respectively, or the momentary energy distribution among a great number of molecules of the same kind. The electron transfer from or to the semiconductor will occur when the two levels are of nearly equal energy expressed in the same system of scale. If the difference between the energy levels of a donor in an electrolyte and the conduction band edge is not suitable for electron transfer in the ground state, the excitation of the donor may result in electron e^- injection. Similarly, hole h^+ injection may occur if the vacant energy level of the excited acceptor is in proper energy correlation with the

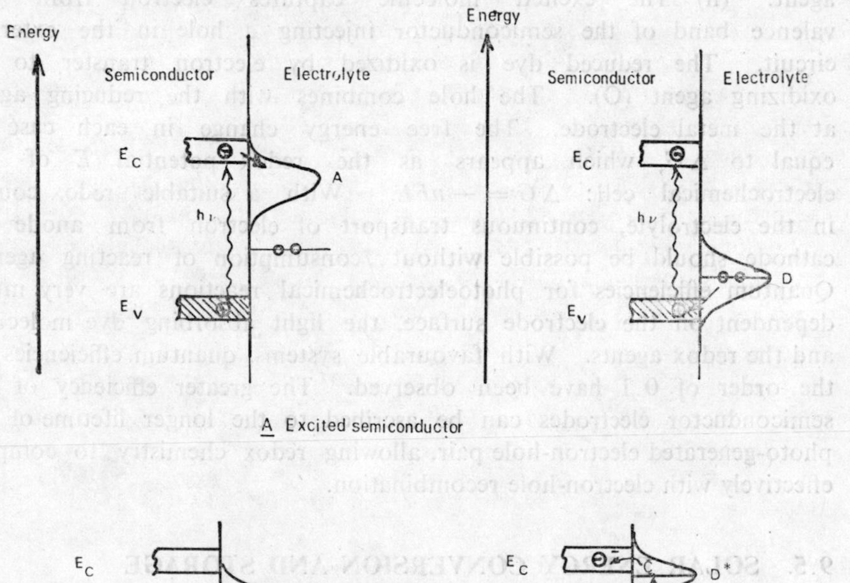

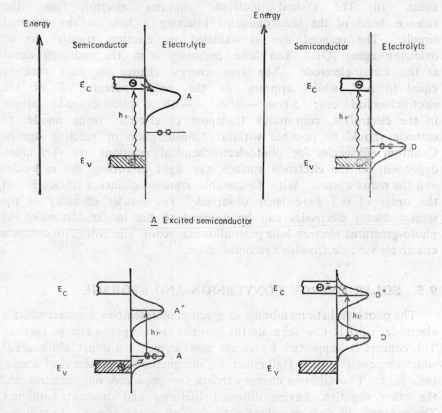

Figure 9.7 Energetics of donor-acceptor property in semiconductor-solution interface: (A) Photoexcited semiconductor: electron injection; hole injection; (B) Photoexcited donor or acceptor in solution: hole injection; electron injection.

valence band of the semiconductor (Figure 9.7). These changes can be studied in suitable galvanic cells since they are chemical changes due to oxidation or reduction of involved species.

In a way analogous to ordinary redox reactions between ground state molecules, it is possible to express photo-induced, redox reactions in terms of two redox couples which can proceed at two separate electrodes. The conversion of chemical energy into electrochemical energy can be compared for chemical and photochemical redox reactions which liberate energy in homogeneous electron-transfer process. Photosensitized redox reaction occur in two ways. (i) The excited molecule D* injects an electron in the semiconductor electrode which is transferred across the external circuit to reduce the oxidizing agent. The oxidized dye is returned to its original state by electron transfer from reducing agent. (ii) The excited molecule captures electron from the valence band of the semiconductor injecting a hole in the external circuit. The reduced dye is oxidized by electron transfer to an oxidizing agent (O). The hole combines with the reducing agent at the metal electrode. The free energy change in each case is equal to ΔG, which appears as the redox potential E of the electrochemical cell: $\Delta G = -nFE$. With a suitable redox couple in the electrolyte, continuous transport of electron from anode to cathode should be possible without consumption of reacting agents. Quantum efficiencies for photoelectrochemical reactions are very much dependent on the electrode surface, the light absorbing dye molecules and the redox agents. With favourable systems, quantum efficiencies of the order of 0.1 have been observed. The greater efficiency of the semiconductor electrodes can be ascribed to the longer lifetime of the photo-generated electron-hole pair, allowing redox chemistry to compete effectively with electron-hole recombination.

9.5. SOLAR ENERGY CONVERSION AND STORAGE

The photosynthetic membrane in green plant resembles a semiconductor electrode since it can separate the positive and negative charge carriers. This concept is supported by direct measurement on intact chloroplasts which demonstrate the Hall effect on illumination in presence of a magnetic field. Two different charge carriers are produced, one positive and the other negative, having different lifetimes and different mobilities. The capacity of such membranes to convert solar energy into chemical energy in an endoenergetic reaction ($\Delta G > O$) in photosynthetic cycle has stimulated interest in development of such model systems for storage and utilization of solar energy by direct conversion of light quanta. In photosynthesis solar energy is stored in two products, carbohydrate and oxygen. The energy is released when carbohydrates burn in air.

Solar energy quantum conversion devices are based on two objectives (1) conversion of light energy into thermal energy by suitable energy storing i.e. endergonic photochemical reactions, (2) conversion of light energy into electrical energy by suitable photoelectrochemical devices.

(1) The energy storing photoreactions occur with positive free energy change ($+\Delta G$) and are hence thermodynamically unstable (Figure 9.8).

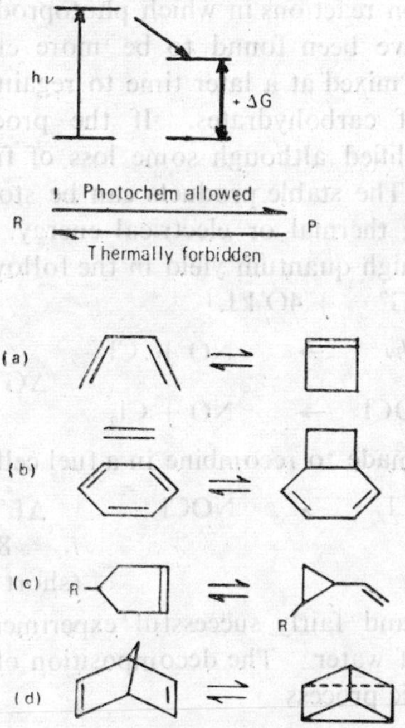

Figure 9.8 Endergonic and reversible electrocyclic reactions obeying Woodward-Hoffman rule. (a) Valence isomerization, (b) cycloaddition, (c) sigmatropic effect, and (d) norbornadiene to quadricyclene conversion.

If the reverse back reaction is prevented or is forbidden by other considerations, the energy remains stored in the photoproducts. Some simple photorearrangement reactions which are governed by Woodward-Hoffman rules have been found useful. These rules provide the stereochemical course of photochemical rearrangement based on symmetry properties of the highest occupied molecular orbital (HOMO) and the lowest unoccupied molecular orbital (LUMO) of the molecule (Section 8.6). A reaction which is photochemically allowed may be thermally forbidden. From the principle of microscopic reversibility, the same will be true for the reverse reaction also. Thermally forbidden back reaction will produce stable photoproducts. Such electrocyclic rearrangements are given in Figure

9.8. A reaction which can store about 260 cal per gram of material and has shown some promise is valence isomerization of norbornadiene (NBD) to quadricyclene (Figure 9.8 d). The disadvantage is that it does not absorb in the visible region and therefore the sunlight efficiency is poor. By attachment of chromophoric groups or by use of suitable sensitizers it might be possible to shift the absorption region towards the visible for better utilization of solar energy.

Photodecomposition reactions in which photoproducts can store energy separately as fuels have been found to be more efficient systems. The photoproducts can be mixed at a later time to regain the stored energy as in the combustion of carbohydrates. If the products are gases, the problem is much simplified although some loss of free energy occurs due to entropy increase. The stable products can be stored and then allowed to recombine releasing thermal or electrical energy. Photodecomposition of NOCl occurs with high quantum yield in the following two steps storing positive free energy $\Delta G^\circ = +40$ kJ.

$$NOCl + h\nu \rightarrow NO + .Cl$$

$$.Cl + NOCl \rightarrow NO + Cl_2 \qquad \Delta G^\circ = +40 \text{ kJ}$$

The products can be made to recombine in a fuel cell

$$NO + \tfrac{1}{2}Cl_2 \rightarrow NOCl \qquad \Delta E^\circ = 0.21 \text{ V} \qquad (9.8)$$

$$i_{sc} = 8.1 \times 10^3 \text{ A cm}^{-2}$$

(short circuit current)

Most interesting and fairly successful experiment is the photosensitized decomposition of water. The decomposition of water to H_2 and O_2 is a highly endothermic process

$$H_2O(l) \rightarrow H_2 + \tfrac{1}{2}O_2 \qquad \begin{matrix}\Delta H^\circ = +295 \text{ kJ} \\ \Delta G^\circ = +237 \text{ kJ}\end{matrix} \qquad (9.9)$$

Hydrogen is a nonpolluting fuel which burns in air to produce water again releasing large amount of heat. Alternatively, H_2 and O_2 may be made to recombine in a fuel cell generating electrical energy.

The thermodynamic breakdown energy for water is 1.23 eV. The electrochemical decomposition of water requires two electrons in consecutive steps.

At the cathode $\quad 2H^+ + 2e^- \rightarrow H_2$

At the anode $\quad 2OH^- \rightarrow O_2 + 2H^+ + 4e^-$

Therefore it should be possible to decompose water with two quantum photochemical process with the input of at least 237 kJ or 2.46 eV per molecule. For a one quantum process, the light of wavelength shorter than 500 nm can only be effective, which means poor utilization of solar spectrum. Semiconductor electrodes with suitable band gaps can act as quantum collectors and on irradiation with energy E greater than the band gap E_g can cause electrochemical oxidation or reduction of species in solution. On photolysis of water (0.1 N NaOH solution) using *n*-type TiO_2 semiconductor electrode as the anode and a platinised platinum electrode as the cathode, O_2 is evolved at the TiO_2 electrode on irradiation. At the dark electrode, H_2 is evolved (Figure 9.9). It is necessary to

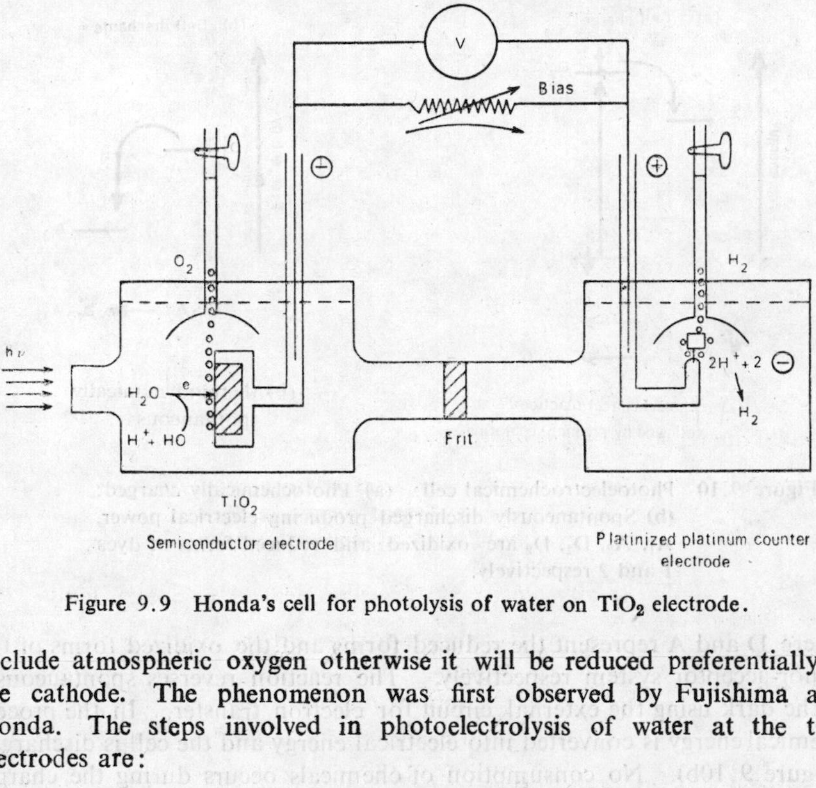

Figure 9.9 Honda's cell for photolysis of water on TiO_2 electrode.

exclude atmospheric oxygen otherwise it will be reduced preferentially at the cathode. The phenomenon was first observed by Fujishima and Honda. The steps involved in photoelectrolysis of water at the two electrodes are:

$$TiO_2 + 2h\nu \rightarrow 2e^- + 2h^+ \quad \text{(hole)}$$

$$2h^+ + H_2O \rightarrow \tfrac{1}{2}O_2 + 2H^+ \quad \text{(at } TiO_2 \text{ electrode)}$$

$$2e^- + 2H^+ \rightarrow H_2 \quad \text{(at } P_t \text{ electrode)}$$

Hydrogen and oxygen can be collected separately and combined again in a fuel cell to obtain electrical energy.

(2) Photoelectrochemical devices to convert light energy to electrical energy implies construction of a battery which undergoes cyclical charging and discharging processes. On illumination the cell is charged (Figure 9.10). Light energy is converted into chemical energy by driving a suitable redox reaction against the potential gradient

$$\underset{oxd_1}{A_1} + \underset{red_2}{D_2} \underset{\Delta}{\overset{h\nu}{\rightleftarrows}} \underset{red_1}{D_1} + \underset{oxd_2}{A_2}$$

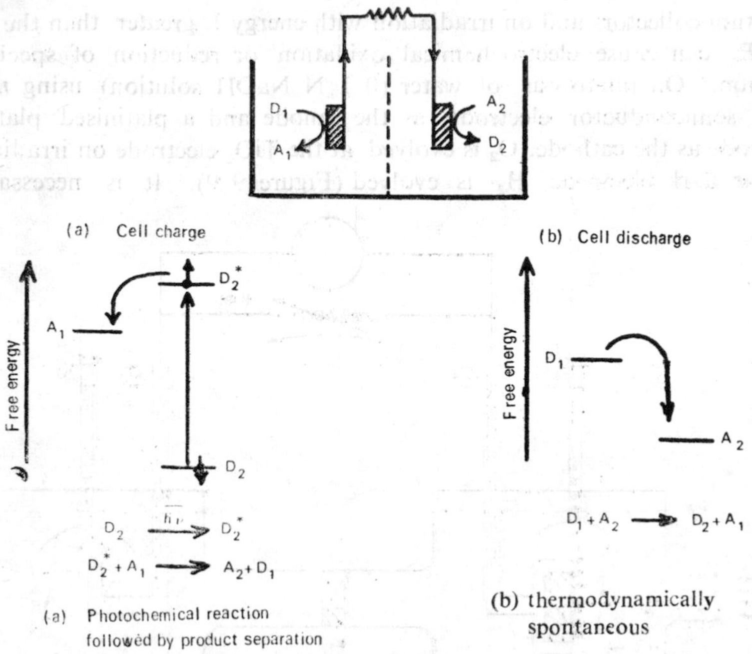

Figure 9.10 Photoelectrochemical cell: (a) Photochemically charged; (b) Spontaneously discharged producing electrical power. A_1, A_2, D_1, D_2 are oxidized and reduced forms of dyes 1 and 2 respectively.

where D and A represent the reduced forms and the oxidized forms of the donor-acceptor system respectively. The reaction reverses spontaneously in the dark using the external circuit for electron transfer. In the process chemical energy is converted into electrical energy and the cell is discharged (Figure 9.10b). No consumption of chemicals occurs during the charge-discharge cycle. Therefore in an ideal system, the system should function as a rechargeable battery without any loss of efficiency. These are known as *photogalvanic cells.* An example of such a reaction is the reversible photobleaching of thionine by Fe^{2+} ion (Figure 9.11). Thionine exists as a cation at low pH and can be represented as TH^+. Overall reaction is

$$TH^+ + 2Fe^{2+} + 2H^+ \underset{\text{dark}}{\overset{h\nu}{\rightleftharpoons}} TH_3^+ + 2Fe^{3+} \qquad (9.10)$$

The sequence of reaction at the two electrodes is:

$$TH^+(S_0) + h\nu(\lambda_{max} \simeq 600 \text{ nm}) \rightarrow TH^+(S_1)$$

$$H^+ + TH^+(S_1) \rightarrow TH_2^{2+}(T_1)$$

$$TH_2^{2+}(T_1) + Fe^{2+} \rightarrow TH_2^+ + Fe^{3+}$$

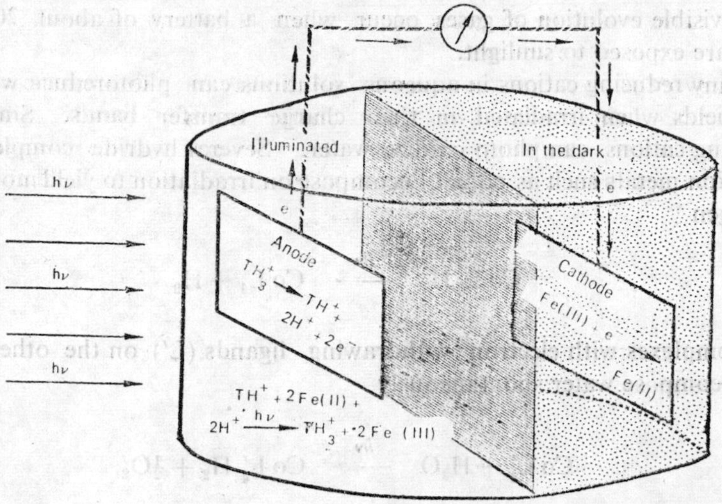

Figure 9.11 Primitive iron-thionine photogalvanic cell.

$$2TH_2^{+\cdot} \rightarrow TH^{+\cdot} + TH_3^+$$

$$TH_3^+ + Fe^{3+} \rightarrow TH_2^{+\cdot} + Fe^{2+} + H^+$$

$$TH_2^+ \rightarrow TH^+ + H^+ + e^- \text{ (illuminated electrode)}$$

$$Fe^{3+} + e^- \rightarrow Fe^{2+} \text{ (dark electrode)}$$

$$TH_2^{+\cdot} + Fe^{3+} \rightarrow TH^+ + H^+ + Fe^{2+}$$

$TH^+(S_1)$ = singlet state of dye cation; $TH_2^{2+}(T_1)$ = triplet state of protonated dye; $TH_2^{+\cdot}$ = semithionine half-reduced radical ion, TH_3^+ = leuco thionine.

The efficiency of photogalvanic devices for solar energy conversion is not very good mainly because of difficulties in preventing the spontaneous back recombination reactions. Furthermore they cannot utilize the total solar spectrum. Until now silicon solar cells have been found to be the most efficient quantum conversion devices. They are photovoltaic cells in which light energy is directly converted into electrical energy without the intermediate chemical reaction. The photocurrent is used to charge a storage battery. Since silicon of very high purity is required, they are fairly expensive.

From the point of view of light stability and range of absorptivity, inorganic redox systems might be more interesting. Photoinduced electron transfer in an aqueous solution of tris-(2, 2′–bipyridine) ruthenium (II) has been found to decompose water in to a mixture of H_2 and O_2. The complex can serve both as an electron donor and electron acceptor in the excited state. The efficiency is low because of barrier to electron transfer. When spread as a monolayer on glass slides after attaching to a surfactant chain, visible evolution of gases occur when a battery of about 20 to 30 slides are exposed to sunlight.

Many reducing cations in aqueous solutions can photoreduce water in low yields when irradiated in their charge transfer bands. Similarly, oxidizing cations can photo oxidize water. Several hydride complexes of transition metals such as cobalt decompose on irradiation to yield molecular hydrogen

$$Co\,L_4\,H_2 \xrightarrow{h\nu} Co\,L_4 + H_2$$

The complexes with electron withdrawing ligands (L′) on the other hand can decompose water spontaneously

$$Co\,L'_4 + H_2O \xrightarrow{h\nu} Co\,L'_4\,H_2 + \tfrac{1}{2}O_2$$

If suitable systems of ligands could be found such that light reaction is endergonic and dark reaction is exergonic, then the net result will be decomposition of water.

The abiological fixation of CO_2 and H_2O to reduced compounds which can be used as fuels will be a reaction of immense use. Until now this has not been very successful. The simplest of these compounds are formic acid, formaldehyde, methanol and methane:

$$CO_2 + H_2O \rightarrow HCOOH + \tfrac{1}{2}O_2 \qquad \Delta H^0 = 262 \text{ kJ mol}^{-1}$$

$$CO_2 + H_2O \rightarrow HCHO + O_2 \qquad \Delta H^0 = 560 \text{ kJ mol}^{-1}$$

$$CO_2 + 2H_2O \rightarrow CH_3OH + \tfrac{3}{2}O_2 \qquad \Delta H^0 = 713 \text{ kJ mol}^{-1}$$

$$CO_2 + 2H_2O \rightarrow CH_4 + 2O_2 \qquad \Delta H^0 = 690 \text{ kJ mol}^{-1}$$

All these reactions are highly endothermic. The products of these reactions are high calorific fuels. In nature fixation of CO_2 is a dark process although initiated by light induced chain of reactions. Except in some special plants like rubber, CO_2 is still not completely reduced. Carbohydrates are the common stored products. Rubber plants produce hydrocarbons.

A hypothetical photoelectrochemical cell based on membrane model of photosynthesis has been postulated (Figure 9.12). The membrane provides unidirectionality to electron flow and prevents spontaneous and direct recombination of photogenerated species. The recombination occurs through external circuit only converting light energy into electrical energy. These kinds of totally synthetic systems which emulate and simulate nature may very well in near future provide alternative sources of energy.

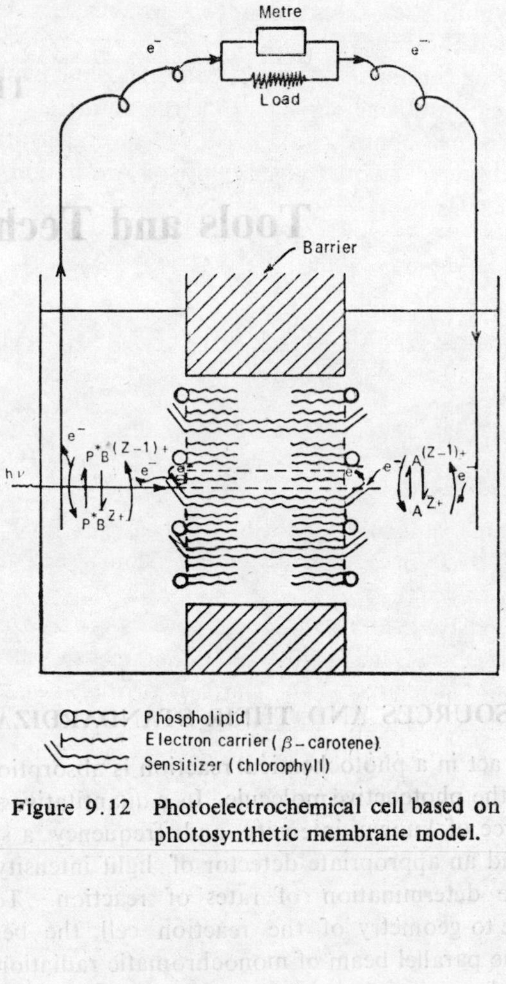

Figure 9.12 Photoelectrochemical cell based on photosynthetic membrane model.

TEN

Tools and Techniques

10.1 LIGHT SOURCES AND THEIR STANDARDIZATION

The primary act in a photochemical reaction is absorption of a quantum of radiation by the photoactive molecule. In a quantitative study, therefore, a radiation source of known intensity and frequency a suitable cell for the photolyte and an appropriate detector of light intensity are absolutely necessary for the determination of rates of reaction. To avoid experimental error due to geometry of the reaction cell, the best arrangement is to have a plane parallel beam of monochromatic radiation, incident upon a flat cuvette with proper stirring arrangement, as given in Figure 1.2.

The most convenient light source for the visible region is a high intensity tungsten lamp or xenon arc lamp. Introduction of halogen as in quartz iodine lamps improves the light output and increases the life of the lamp. Simple filters of proper band width or suitable monochromators can be used for obtaining a light beam of desired wavelength. High pressure mercury discharge lamps emit characteristic line frequencies in the visible which can be conveniently utilised with proper filter combinations. Medium pressure and low pressure mercury lamps are suitable for studies in the ultraviolet region. These lamps emit the strong resonance line at 253.7 nm and medium intensity triplets at 365 nm. A good account of these lamps is given in *Photochemistry* by Calvert and Pitts (1966). For studies in the ultraviolet regions, hydrogen or deuterium lamps provide continuous light sources. In this region quartz cuvettes and lenses are essential. High intensity laser sources are now available and have many specialized uses.

10.1.1 Actinometry

For the measurement of quantum yields, a knowledge of incident light flux I_0, i.e. number of quanta falling per unit time is required. A number of methods are available for the standardization of light sources. The procedure is known as *actinometry*. For absolute calibration, a standard lamp of known colour temperature, defined by Planck's Law, is used to standardize the detector which may be (i) a thermopile or a bolometer, (ii) a photocell or (iii) a photomultiplier. Thermopiles are thermocouples connected in series, and generate an emf on heating. Bolometers are thin blackened strips whose resistance changes on absorption of energy. Neither of these discriminate between the quality of radiation but integrate the total energy. A *photocell* consists of a photosensitive cathode and a collector anode enclosed in an evacuated bulb. Light quanta of energy greater than the threshold value of the metal composing the photocathode causes ejection of electrons which are collected by the anode, and current flows in the circuit. Under suitable conditions, the intensity of photocurrent thus generated will be linearly proportional to the incident light intensity. The light sensitivity of the photocell is wavelength dependent. Photocells of different cathode materials must be used for blue and red regions of the spectrum. It is nesessary therefore to calibrate the photocell against a thermopile or against a secondary standard to correct for the wavelength sensitivity. For measurement of small intensities, amplification becomes necessary.

There are another type of photocells known as *barrier layer* photocells which work on an entirely different principle. They are semiconductor devices in which impinging photons promote the electrons from the valence band to the conduction band across the energy gap. A photovoltage is generated which can be measured by a voltmeter. Such photovoltaic devices can have a large surface area and are easy to operate. They are commonly used in many simple colorimeters and fluorimeters and as light meters for cameras.

Photomultipliers are vacuum tube photocells with a sealed-in set of dynodes. Each successive dynode is kept at a potential difference of 100V so that photoelectrons emitted from the cathode surface are accelerated at each step. The secondary electrons ejected from the last dynode are collected by the anode and are multiplied so that a $10^6 - 10^7$ — fold amplification of electron flux is achieved. This allows simple devices such as microammeters to measure weak light intensities. Background thermal emission can be minimised by cooling the photomultiplier. The schematic diagrams of all these detector devices are given in Figure 10.1.

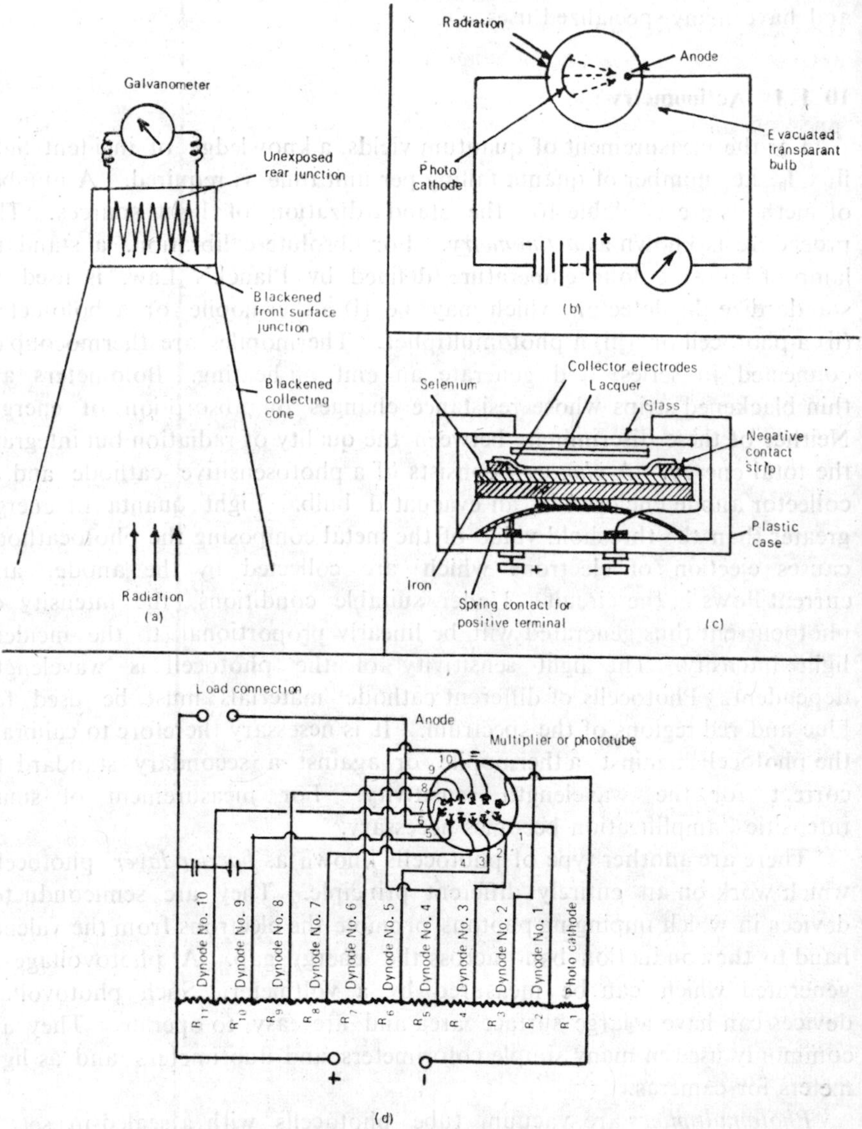

Figure 10.1 Devices for measurement of light intensity. (a) Thermopile; (b) vacuum tube photocell; (c) barrier-layer photocell; (d) photomultiplier.

10.1.2 Chemical Actinometry

A convenient method for the standardization of light sources in the laboratory is the use of suitable photochemical reactions whose quantum yields have been determined by a standard source. Conditions must be such that the light absorption remains constant during the exposure; hence the reaction must either be photosensitized or have so sensitive a product detection system that photochange is small.

The basic expression is the one which defines the quantum yield ϕ of a photochemical reaction:

$$\phi = \frac{\text{rate of reaction}}{\text{rate of absorption}}$$

$$= \frac{\text{no. of molecules decomposed or formed s}^{-1}\text{cm}^{-3}}{\text{no. of quanta absorbed s}^{-1}\text{cm}^{-3}}$$

$$= \frac{-dc/dt \text{ or } +dx/dt}{I_0 \times \text{fraction of light absorbed}}$$

The fractional light absorption can be measured in a separate experiment. Knowing ϕ and estimating the extent of decomposition, I_0, the incident intensity can be calculated in the units of einstein $cm^{-3}\,s^{-1}$ falling on the reaction cell. To avoid geometrical errors due to differences in absorptivity of the actinometer solution and the sample, the same cell is used for actinometry and for the reaction, under conditions of *equal optical densities.* There are a number of photochemical reactions which have been found suitable as actinometers. They are useful within their specific wavelength ranges.

1. *Ferrioxalate actinometer.* Photodecomposition of K-ferrioxalate was developed into an actinometer by Parker and Hatchard. It is one of the most accurate and widely used actinometers which covers a wavelength range between 250 nm to 577 nm. Irradiation of ferrioxalate solution results in the reduction of Fe^{3+} to Fe^{2+} which is estimated colorimetrically by using *o*-phenanthrolin as complexing agent. The OD at 510 nm of the deep red colour produced is compared with a standard. The quantum yield for Fe^{2+} formation is nearly constant within the wavelength range and shows negligible variation with temperature, solution composition and light intensity. The recommended actinometric solution, for wavelength upto 400 nm contains 0.006 M $K_3\,Fe(ox)_3$ in 0.1 N H_2SO_4. For longer wavelengths, a 0.15 M solution is more convenient. Quantum yields vary between 1.2 (λ 254–365 nm) to 1.1 at longer wavelengths.

2. *Uranyl oxalate actinometer.* This actinometer has a range of 208–435 nm with an average quantum yield of about 0.5. Since the UO_2^{++} ion acts as a photosensitizer for the oxalate decomposition the light absorption remains constant, but rather long exposures are required for final accurate oxalate titrations. It is now mainly of historical interest.

3. *MGL actinometer.* Malachite green leucocyanide (MGL) is

particularly useful in the range 220–300 nm where it absorbs strongly. On irradiation MGL is converted into ionized form MG^+ which has a very strong absorption at 662 nm. The quantum yield for production of MG^+ is 0.91 over the given range.

4. *Reinecke's salt actinometer.* The useful range for this actinometer extends from 316–735 nm and therefore is convenient in the visible region. Reinecke's salt is commercially available as ammonium salt $(NH_4)_3Cr(NH_3)_2(NCS)$. It should be converted into the K-salt. The ligand field band extends from 400 nm to 735 nm. On irradiation, aquation of the complex proceeds with release of thiocyanate. Quantum yields are calculated as moles of thiocyanate released per einstein of light absorbed. The concentration of the actinometer solution should be such as to absorb nearly 99% of the incident light. The pH is adjusted between 5.3–5.5. The quantum yield for the reaction over the visible range lies between 0.27 to 0.3.

10.2 MEASUREMENT OF EMISSION CHARACTERISTICS: FLUORESCENCE, PHOSPHORESCENCE AND CHEMILUMINESCENCE

Emission characteristics of a molecular system can be expressed by three types of measurements: (1) observation of emission and excitation spectra, (2) measurement of quantum efficiencies, and (3) determination of decay constants or radiative lifetimes.

(1) *Emission and excitation spectra.* Three types of geometrical arrangements with respect to the directions of excitation and observation are generally used for measurements, 90° arrangement, 45° arrangement and 180° arrangement (Figure 10.2). For dilute solutions, right angled observation is desirable. Errors due to geometry should be avoided and a suitable filter in the exciting and measuring beam should be used. For concentrated solutions which are optically dense 45° or 180° arrangements are necessary.

For recording of the emission spectrum, the emitted radiation is focussed on the slit of a monochromator and intensities measured at each wavelength. Since sensitivities of photocells or photomultipliers are wavelength dependent, a standardization of the detector-monochromator combination is necessary for obtaining true emission spectrum This can be done by using a standard lamp of known colour temperature whose emission characteristics is obtained from Planck's radiation law. The correction term is applied to the instrumental readings at each wavelength. Very often substances whose emission spectra have been accurately determined in the units of relative quanta per unit wavenumber intervals are

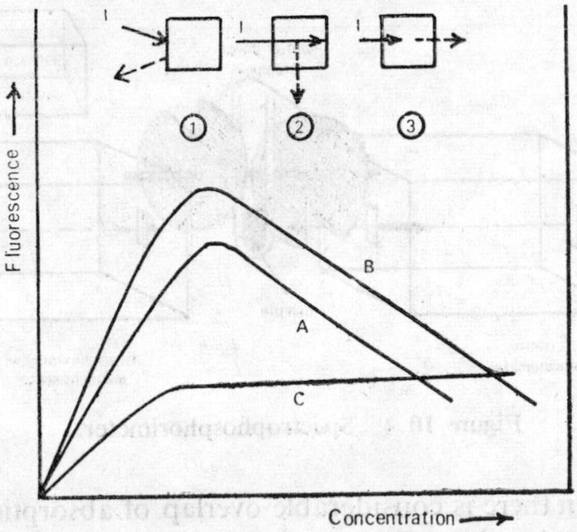

Figure 10.2 Different geometrical arrangements for the measurement of emission intensity: (1) front surface (45°) observation — A, (2) right angle (90°) observation — B (3) end-on (180°) observation — C.

used as secondary standards. The quinine sulphate in 0.1N H_2SO_4 and anthracene in benzene are commonly used standards. The instrument is called spectrophotofluorimeter (Figure 10.3). For measurement of phos-

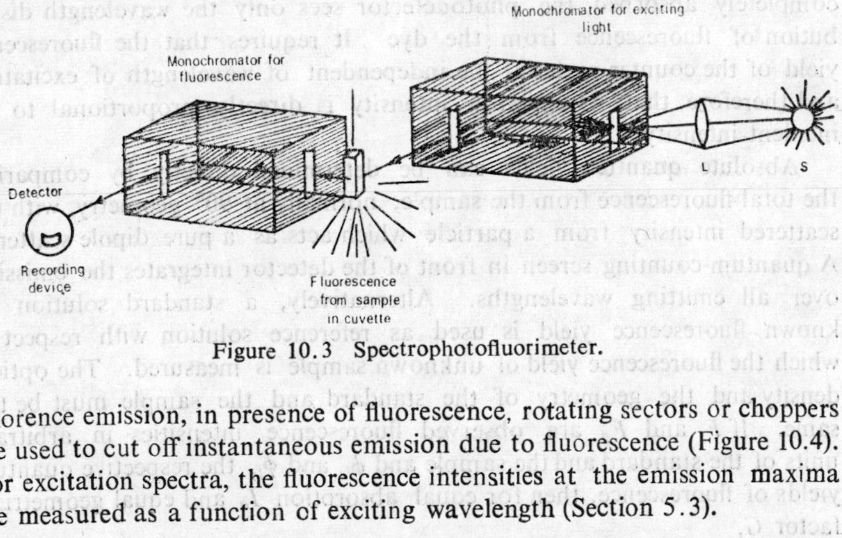

Figure 10.3 Spectrophotofluorimeter.

phorence emission, in presence of fluorescence, rotating sectors or choppers are used to cut off instantaneous emission due to fluorescence (Figure 10.4). For excitation spectra, the fluorescence intensities at the emission maxima are measured as a function of exciting wavelength (Section 5.3).

(2) *Quantum yields of emission.* For determination of quantum yields of fluorescence, areas under the emission spectra of very dilute solutions (OD $\simeq$ 0.03), expressed in units of relative quanta per unit wavenumber interval are obtained and compared with that of a standard solution of identical optical density under similar geometry of measurement. Dilute solutions are necessary to avoid distortion of the spectrum due to

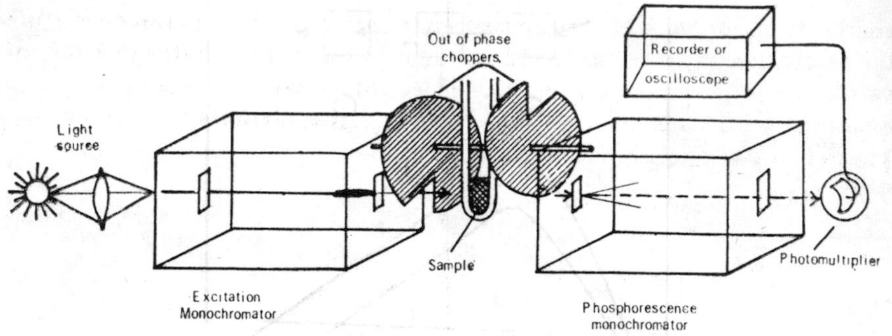

Figure 10.4 Spectrophosphorimeter.

reabsorption when there is considerable overlap of absorption and emission spectra. A correction for this effect may be required in such cases.

Wavelength dependence of detector response can also be compensated by using a fluorescent screen in front of the photocell or photomultiplier. This screen acts as a *quantum counter*. A concentrated solution of Rhodamin B in glycerol (3g per litre) or fluorescein in 0.01N Na_2CO_3 has been used for this purpose. Quantum counters work on the principle that whatever be the wavelength of radiation incident on the screen, if completely absorbed, the photodetector sees only the wavelength distribution of fluorescence from the dye. It requires that the fluorescence yield of the counter material be independent of wavelength of excitation and therefore that its emission intensity is directly proportional to the incident intensity.

Absolute quantum yields can be determined directly by comparing the total fluorescence from the sample, normally at 90° geometry, with the scattered intensity from a particle which acts as a pure dipole scatterer. A quantum-counting screen in front of the detector integrates the emission over all emitting wavelengths. Alternatively, a standard solution of known fluorescence yield is used as reference solution with respect to which the fluorescence yield of unknown sample is measured. The optical density and the geometry of the standard and the sample must be the same. If F_1 and F_2 are observed fluorescence intensities in arbitrary units of the standard and the sample and ϕ_{f_1} and ϕ_{f_2} the respective quantum yields of fluorescence, then for equal absorption I_a and equal geometrical factor G,

$$\frac{F_1}{F_2} = \frac{\phi_{f_1} I_a G}{\phi_{f_2} I_a G} \text{ and } \phi_{f_2} = \frac{F_2}{F_1}\phi_{f_1} \qquad (10.1)$$

Phosphorescence quantum yields can also be measured in the same way, but require choppers for eliminating fluorescence (Figure 10.4).

(3) *Determination of decay constants or lifetimes of excited states.* Decay

constants or lifetimes of the excited states are important parameters since the reactivities of these energy states depend on them. Rate constants of various photophysical and photochemical processes can be adduced from quantum yield data only if the mean radiative lifetimes (τ_0) are known. The defining relationships are (Section 5.3)

$$\frac{1}{\tau_0} = k_f = \phi_f/\tau_f \tag{10.2}$$

and

$$\frac{1}{\tau_f} = k_f + \Sigma k_i = \frac{1}{\tau_0} + \Sigma k_i \tag{10.3}$$

where k_f is the rate constant for the emission process and Σk_i is the summation of rate constant for all other photophysical or photochemical processes, unimolecular or bimolecular, originating from that state and competing with emission.

For atomic systems the integrated areas under the absorption and emission spectral curves provide values of τ_0 (Section 3.9). For molecules, various sources of error vitiate the result, specially when electronic bands overlap and one band can borrow intensity from the other. Existence of vibrational peaks quite often present problems in measuring the area under the absorption curve accurately. But if one can measure τ_f and ϕ_f, τ_0 can be calculated from the expression (10.2).

The various methods for the measurement of τ_f, the actual lifetime, can be classified as *steady state* and *nonsteady state* methods. In the steady state methods are included lifetime determinations from fluorescence quenching data and depolarization studies, the theoretical aspect of which have already been discussed in Sections 6.4 and 4.10, respectively.

10.2.1 Nonsteady State Methods for Determination of Fluorescent Lifetimes

The nonsteady state methods may be conveniently divided into two categories (A) pulse methods and (B) phase-shift methods.

(A) *Pulse methods.* In the pulse method, the sample is excited intermittently by *pulsed* or *chopped* light source and the decay of emitted luminescence is observed by various techniques, depending on the life-time ranges, after a preset delay times.

Becquerel phosphoroscope. The earliest apparatus constructed by Becquerel for the measurement of phosphorescence consists of a pair of rotating sectors mounted on a common shaft Figure 10.4. The openings in the sectors are so arranged that the sample is alternatively illuminated and then viewed during cut-off periods. For solutions and for 90° illumination a modification is made, the sample is placed within a rotating cylinder with cut-out slots (Figure 10.5). During the course of rotation of the cylinder,

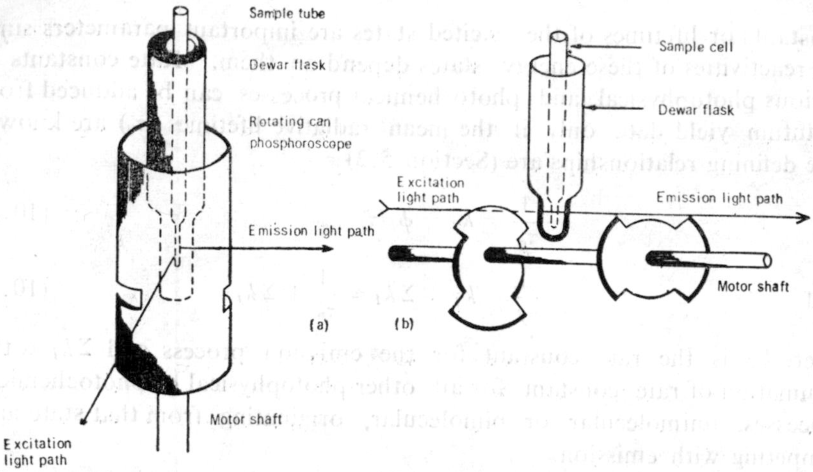

Figure 10.5 Becquerel phosphoscope with (a) coaxial cylinders; (b) a pair of rotating sectors mounted on a common shaft.

when the window faces the light source, the sample is excited and when it faces the detector, the emission from the sample is recorded. The detector output may either be fed to an ammeter or to an oscilloscope. The time between excitation and detection can be varied by changing the speed of rotation and thus complete decay curve can be plotted. In case an oscilloscope is used, the signal is applied to the y-axis of the oscilloscope and the phosphoroscope is synchronized with the oscilloscope sweep. The oscilloscope then displays the exponential decay curve of phosphorescence which can be photographed. The shortest lifetime that can be measured by this type of phosphoroscope is 10^{-4} s only.

Kerr cell technique. For the measurement of very small decay periods in the nanoseconds range, mechanical shutters are replaced by electrical devices called *electro-optic shutters* such as a Kerr cell. The Kerr cell originally consisted of a glass vessel containing highly purified nitrobenzene and two electrodes placed on either side of the cell to which high voltages (10–40 kV) could be applied. Later devices also work on the same principle. The electric field causes a small degree of orientation of the molecules so that the cell becomes doubly refracting like an uniaxial crystal. When placed between crossed polaroids, the cell becomes transparent to the incident light when the field is on and opaque when the field is off. The shutter frequency is governed by the frequency of the oscillating voltage. Two such cells are used, one for the exciting radiation and the other for the emitted radiation. By varying the distance between the two shutters and the fluorescent cell and by proper synchronization, complete decay curve can be obtained. The distances are converted into time scale using the time of flight of light in the apparatus.

Other nanosecond instrumental techniques. In recent times very sensitive

tive and sophisticated instruments have been developed for measurements of fluorescent decay times in the nanosecond regions. Pulsed systems utilize short-duration flash lamps that are either *gated* or *free running*. These flash lamps are useful in all aspects of photochemistry and photophysics where short-lived transients are under investigation. The *free running* lamp is essentially a relaxation oscillator in which the discharge takes place at approximately the breakdown voltage of the lamp under the given conditions of gas pressure and electrode geometry. The charging resistors and capacitors set up the *relaxation oscillator*. These lamps are classified as high pressure or low pressure lamps. On the other hand a *gated* lamp is fired by a switch such as a thyratron tube and discharge takes place at a voltage two to three times higher than the breakdown voltage. The gated lamps have greater flexibility since the lamp pressure, voltage and frequency, all are independent variables and the changes in flashing gas do not require adjustment of RC (resistance-capacitance) time constant. The discharge voltage is determined by thyratron breakdown voltage and is initiated by a positive pulse at the grid.

Stroboscopic method. In this method the photomultiplier is also gated or pulsed having an on-time of nanosecond or subnanosecond duration during which it operates at very high gain. The flash lamp and the detector systems are synchronized such that a suitable delay can be introduced between the two. The gated photomultiplier samples the photocurrent each time the lamp fires and the photocurrent is thus proportional to the

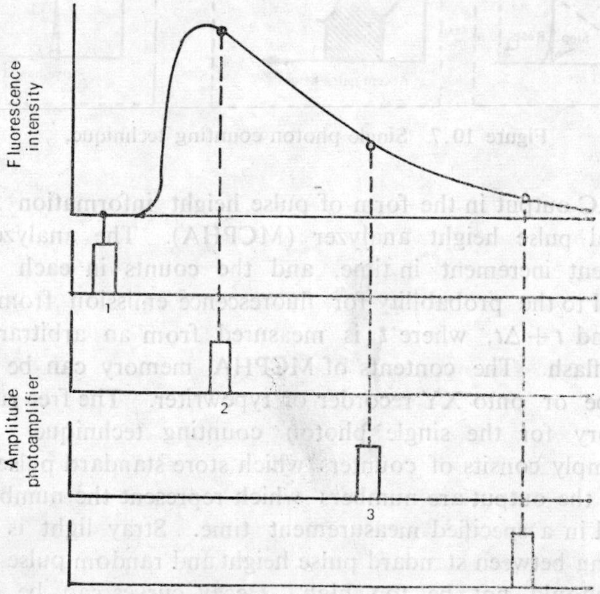

Figure 10.6 Decay curve by stroboscopic method.

fluorescence intensity between t and $t + \Delta t$, the width of the photomultiplier pulse. A complete decay curve can thus be obtained (Figure 10.6).

Single photon counting technique. This technique is becoming popular for measuring low light intensities as is necessary in luminescence studies. The great advantage is that the method eliminates disturbance due to noise and stray light. The technique measures the time of emission of individual fluorescence photons, the reference zero time being the initial rise of the flash lamp light or an electrical pulse related in time to the flash lamp discharge. The time coordinate of arrival of each recorded photon with reference to a fixed time zero is converted into an amplitude of the resultant pulse in an time-to-amplitude converter (TAC) circuit. Each time the lamp flashes, a synchronization pulse is sent to TAC and starts the time sweep. If a stop pulse is received from the photomultiplier during the time sweep, a TAC output pulse is generated with amplitude proportional to the time ($t_{stop} - t_{start}$) (Figure 10.7).

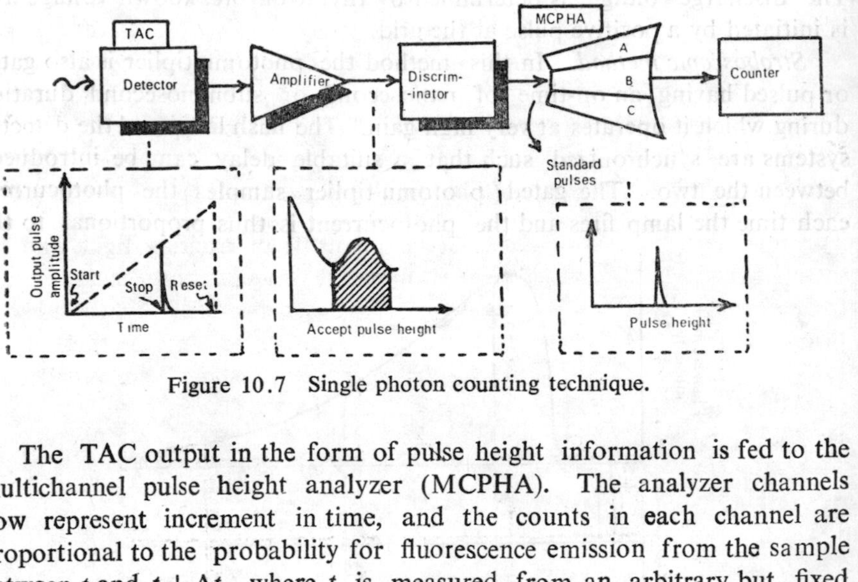

Figure 10.7 Single photon counting technique.

The TAC output in the form of pulse height information is fed to the multichannel pulse height analyzer (MCPHA). The analyzer channels now represent increment in time, and the counts in each channel are proportional to the probability for fluorescence emission from the sample between t and $t + \Delta t$, where t is measured from an arbitrary but fixed point each flash. The contents of MCPHA memory can be read onto a punched tape or onto XY recorder or typewriter. The free running lamp is satisfactory for the single photon counting technique. The signal analyzers simply consits of counters which store standard pulses of specific heights and the output are numbers which represent the number of pulses accumulated in a specified measurement time. Stray light is avoided by discriminating between standard pulse height and random pulse height. The count rate should not be too high. Decay curves can be obtained by plotting counts/channel against channel number. Counting times may be lengthy for very weak intensities.

(B) *Phase-shift methods.* The phase shift method for determining fluorescence lifetimes is based on the principle that if fluorescence is excited by suitably modulated light source, emitted radiation will also be similarly modulated. With reference to a scattering substance, emission from a fluorescent substance will introduce a time lag due to finite time between absorption and emission. This, by definition is the lifetime of the excited state. The time lag will cause a phase-shift relative to the exciting light. Phase fluorimetry requires a modulated light source and a phase sensitive detector.

The differential equation for decay in intensity of fluorescence I, is

$$\frac{dI}{dt} = -k_1 I \tag{10.4}$$

where k_1 includes all first order rate constants which deactivate the excited molecule. Photoactivation occurs at a rate proportional to the intensity of exciting light. When excitation is by light varying in periodic manner, the equation becomes,

$$\frac{dI}{dt} = -k_1 I + k_2 J(t) \tag{10.5}$$

where intensity of exciting light $J(t)$ may be represented by a Fourier series,

$$J(t) = \frac{a_0}{2} + (a_n \cos n\omega t + b_n \sin n\omega t) \tag{10.6}$$

The first term $a_0/2$ indicates average intensity of exciting light and the second term gives the periodic variation about this average, ω is the angular frequency of modulation. The value of k_2 is determined by the absorption coefficient and concentration of the fluorescer and by the thickness of the solution used. In any given experiment it is constant. On substituting equation (10.6) in (10.5), we have

$$\frac{dI}{dt} + k_1 I = k_2\left[\frac{a_0}{2} + \sum (a_n \cos n\omega t + b_n \sin n\omega t)\right] \tag{10.7}$$

which is linear first order differential equation and can be solved by usual methods giving

$$I = \frac{k_2}{k_1}\frac{a_0}{2} + k_2 \frac{\sum a_n (k_1 \cos n\omega t + n\omega \sin n\omega t)}{k_1^2 + n^2\omega^2} + k_2 \frac{\sum b_n (k_1 \sin n\omega t + n\omega \cos n\omega t)}{k_1^2 + n^2\omega^2} + Ce^{-k_1 t} \tag{10.8}$$

By making further substitutions

$$\sin \theta_n = \frac{n\omega}{(k_1^2 + n^2\omega^2)^{1/2}}, \quad \cos \theta_n = \frac{k_1}{(k_1^2 + n^2\omega^2)^{1/2}}$$

$$I = \frac{k_2}{k_1}\frac{a_0}{2} + k_2 \frac{\sum a_n \cos (n\omega t - \theta_n)}{(k_1^2 + n^2\omega^2)^{1/2}} + k_2 \frac{\sum b_n \sin (n\omega t - \theta_n)}{(k_1^2 + n^2\omega^2)^{1/2}} + Ce^{-k_1 t}$$

Thus fluorescence intensity is represented by another Fourier series plus an additional term in t. This final term, however, becomes negligibly small as t becomes large compared to $k_1(=1/\tau)$ and can be neglected. For a given component of exciting light, a component of the same frequency is obtained in the fluorescence light but retarded in phase by an angle θ_n which is related to the rate constant k_1 and the average lifetime τ by the relation

$$\tan \theta_n = \frac{n\omega}{k_1} = n\omega\tau \tag{10.9}$$

For sinusoidally varying light, all of the coefficients a_n for $n > 1$ are zero so that only the first component of Fourier series remains. By proper choice of time base, either a_1 or b_1 can be made equal to zero so that the exciting light is

$$J(t) = \frac{a_0}{2} + a_1 \cos \omega t \tag{10.10}$$

and fluorescence intensity is

$$I(t) = \frac{k_2}{k_1}\frac{a_0}{2} + \frac{k_2}{k_1^2 (1+\omega^2\tau^2)^{1/2}} a_1 \cos(\omega t - \theta) \tag{10.11}$$

The phase relationship is given by

$$\tan \theta = \omega\tau = 2\pi\nu\tau$$

where θ is the phase angle between $I(t)$ and $J(t)$.

A variety of methods exist for modulating light sources. From the relationship $\tan \theta = 2\pi\nu\tau$ we can see that for measurable phase shift, frequencies from 2 to 20 Mc should be employed for lifetimes in the range of 10^{-8} to 10^{-10} s. The devices or methods used for light modulation are: (a) Kerr cells, (b) Pockels cell, (c) ultrasonic modulators, and (d) rf discharges through gases at the desired frequency. Methods (a), (b) and (c) can be used to modulate any light source. The first phase fluorimeter, built by Gaviola as early as 1926, employed two Kerr cells, one for excitation and the other for observation. Pockels cells are a solid state analog of Kerr cells. An oscillating electric field is impressed upon KH_2PO_4 (KHP) or KD_2PO_4 (KDP) crystals placed between crossed polarizers, so that a rapid shutter effect is obtained.

Another method is the 'water tank method' in which ultrasonic standing waves are generated by a suitable crystal oscillator in a small tank of water or a water-alcohol mixture. During the course of propagation of the standing wave (consisting of regions of compression and expansion), twice during a cycle the tank is completely transparent or isotropic. But when the standing wave forms a diffraction grating the incident light is scattered into a diffraction pattern. Normally the zero order band is selected for excitation of the sample. When a mixture of 19% absolute alcohol and 81% distilled water is used, the temperature coefficient for the ultrasonic velocity is minimized. The modulation frequency thus obtained is double the crystal frequency and hence the modulator acts as a frequency doubler. In the method of Bailey and Rollefson, modulated light source is allowed to fall on two sample cells each placed in front of a photomultiplier for 90° observation. The zero phase shift is first adjusted by using scattering solution in both the cells. The outputs of two photomultipliers are fed into the X and Y axis of an oscilloscope with proper circuitry. When the second scattering solution is replaced by the fluorescer solution, a phase shift is introduced between the two due to finite lifetime of the excited state. The detection is made more sensitive by generating a beat frequency in the Kc range by feeding 5.2 Mc frequency from a piezoelectric crystal oscillator to the output signals from the photomultipliers. A block diagram of such an instrument is given in Figure 10.8.

10.2.2 Pico-second Studies

Techniques have now been developed to study decay rates in pico-second ranges such as vibrational relaxation and radiationless transitions ($\tau = 10^{-12} - 10^{-14}$ s) by using high intensity laser pulses (see Section 10.4).

10.3 TECHNIQUES FOR STUDY OF TRANSIENT SPECIES IN PHOTOCHEMICAL REACTIONS

For a mechanistic study of photochemical reactions it is necessary to know the identity of the excited states and the unstable intermediates formed on photoexcitation. Free radicals and unstable species are produced as primary photoproducts in many organic and inorganic reactions. Most radicals are extremely reactive species. This extreme reactivity may be utilized to detect the species by adding substances which are known to have specific reactivity towards them and thus act as *scavengers*. By rapid thermal reaction with the specific free radicals they suppress the secondary reactions. For example, photolysis of a number of organic compounds such as alkyl esters, iodides, some aliphatic ketones, etc. are postulated to generate alkyl free radicals as an important primary process. Molecular iodine is an effective scavenger of alkyl radicals: $R^{\cdot} + I_2 \rightarrow RI + I^{\cdot}$. The alkyl iodide can easily be detected, and the iodine atoms recombine to form molecules. Azide ion and N_2O are good scavengers of singlet oxygen; *p*-nitroso-N, N-dimethylaniline is a selective scavenger for $OH^{\cdot}$ radical in alkaline aqueous solution. One of the most useful forms of radical scavenger are species which can undergo free radical initiated polymerization. The polymer can be isolated and analyzed to determine the identity of the initiator. The intermediate species may also be identified by optical, EPR or mass spectroscopy and other techniques if suitably designed to study short lived transients, under steady state conditions.

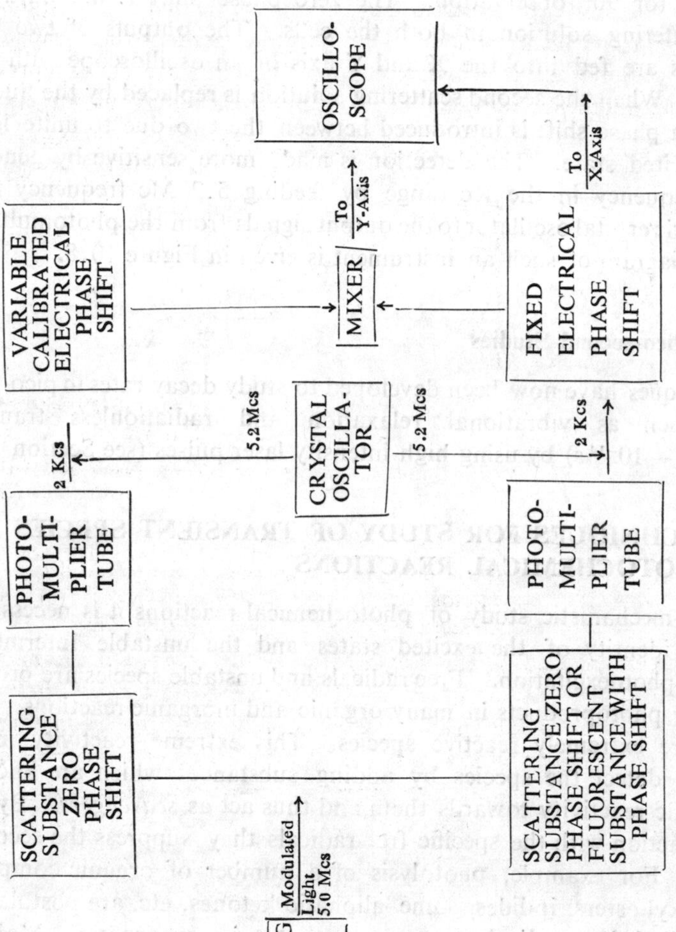

Figure 10.8 A block diagram of phase-shift method for measurement of life-time.

10.3.1 Nonsteady State Methods

The decay constant of the transient species can be measured directly by using nonstationary state methods. In these methods the system is irradiated intermittently so that transients are created in the light period and decay in the dark period. The simplest form of the arrangement is the use of sector wheel in which a number of sectors are cut (Figure 10.9). The cut and uncut portions can be of equal size or may vary in a definite

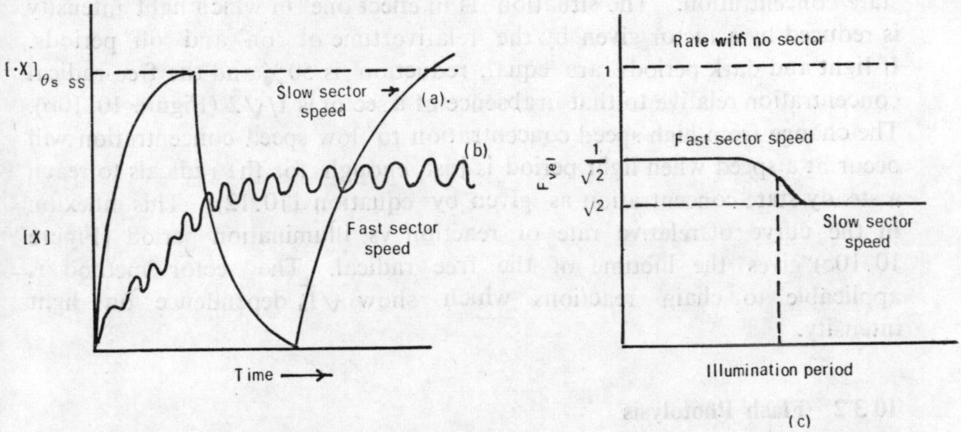

Figure 10.9 Sector technique for the determination of free radical decay times: (a) Slow sector speed; (b) fast sector speed; (c) rate of product formation as a function of illumination time.

ratio. The chopping frequency is determined by the speed of the rotating wheel and number of open sectors cut in it. Variation of the speed of rotation has interesting consequences on the rate of free radical reaction. Consider a generel reaction scheme,

$$X_2 + h\nu \xrightarrow{k_1} 2X\cdot$$

$$X\cdot + A \xrightarrow{k_2} B + Y\cdot$$

$$Y\cdot + X_2 \xrightarrow{k_3} XY + X\cdot$$

$$2X\cdot \xrightarrow{k_4} X_2$$

From stationary-state treatment, the concentration of the radical intermediate $[X\cdot]_{ss}$ is expressed as (I_a = intensity of absorption; ϕ = quantum yield)

$$\text{rate of formation of } X\cdot = \text{rate of disappearance of } X\cdot$$

$$2I_a\phi + k_3[Y\cdot][X_2] = k_2[X\cdot][A] + k_4[X\cdot]^2$$

A similar equation can be set for $Y\cdot$,

$$k_2[X\cdot][A] = k_3[Y\cdot][X_2]$$

On using these two relations we have for stationary state concentration of the radical $X\cdot$,

$$[X\cdot]_{ss} = \sqrt{\frac{2I_a\phi}{k_4}} \tag{10.12}$$

At a slow rotation speed, the effect of the disc is to cause the reaction to proceed in a sequence of on-off periods. The free radicals generated in the light period, decay in the dark period to the final product. The steady state concentration of the radical remains the same as under continuous incident intensity I_a (Figure 10.10a), only the rate of reaction is reduced in proportion to the ratio of light and dark periods. At sufficiently high speeds, the 'on' and 'off' periods may not be long enough for the free radicals to reach a steady state concentration. The situation is in effect one in which light intensity is reduced by a factor given by the relative time of 'on' and 'off' periods. If light and dark periods are equal, reduction is 50% and the free radical concentration relative to that in absence of a sector is $1/\sqrt{2}$ (Figure 10.10b). The change from high speed concentration to low speed concentration will occur at a speed when light period is just enough for the radicals to reach a steady state concentration as given by equation (10.12). This inflexion in the curve of relative rate of reaction vs illumination period (Figure 10.10c) gives the lifetime of the free radical. The sector method is applicable to chain reactions which show $\sqrt{I_a}$ dependence on light intensity.

10.3.2 Flash Photolysis

The stationary state concentrations of intermediates in a static system are very small to be detected by common spectroscopic techniques. But if a strong flash of light is used, a large concentration of the intermediates may be generated which can easily be subjected to spectroscopic analysis. Essentially it is a relaxation method, the flash duration must suitably match the decay constant of the intermediates. The technique was developed by Norrish and Porter in early fifties.

The technique consists of subjecting the system to a high intensity flash (photolysis flash) and monitor it, as soon as possible, by single beam absorption spectroscopy. The monitoring flash is fired after a preset time delay and is directed perpendicular to the photolysis flash. Light from the monitoring flash passes through reaction cell and into a spectrograph (Figure 10.10). The absorption spectra of the transient species formed due to photolysis flash are photographed. The decay of a transient is observed by repeating the experiment using a set of different delay times between the photolyzing and the spectroscopic flash. This technique is known as *flash spectroscopy*, and gives time resolved spectra.

After having thus obtained the regions of absorption changes in the overall spectrum, decay or formation rates of the transients can be obtained by *flash kinetic spectrophotometry*. A continuous beam of monitoring light passes through the reaction vessel and light at the wavelength of maximum absorption due to the transient is selected by a dispersing element for transmission to the photodetector. The output is fed to a DC oscilloscope which displays the decay of absorption by the transient species and may be photographed if so desired (Figure 10.10b).

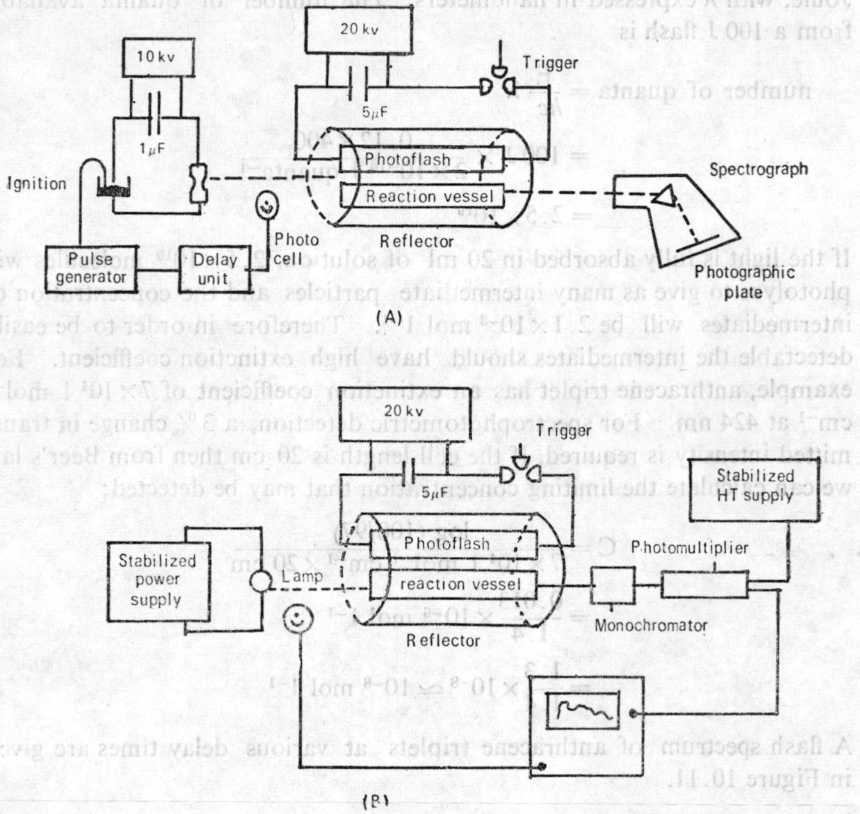

Figure 10.10 Instrumentation for flash photolysis. (A) Flash spectroscopy; (B) flash kinetic spectrophotometry.

The duration of the photolyzing flash sets the upper limit to lifetime of transients which may be detected by flash photolysis. An upper limit of 1 and 10 μs is set by use of a flash lamp. High intensities are obtained by charging a bank of capacitors at high voltages (0.5 kV – 20 kV). When the capacitors are discharged a flash of energy between 20-2000 J is released in a time period ranging from 1 μs to 100 μs or more. The discharge time depends on the capacitance of the condenser, the type of gas inside the flash tube, its pressure and distance between the electrodes. Breakdown in the flash tube is initiated by suitable devices such as closure of a spark gap or use of a thyratron.

Theoretically one may calculate the number of quanta available from a flash of known energy and colour temperature. If the colour temperature is 6000 K, from Planck's radiation law it can be estimated that 12%

of the total radiation density will fall within 100 nm centered at 400 nm. Each quantum of light at this wavelength has energy $hc/\lambda = 2\times10^{-16}/400$ Joule, with λ expressed in nanometers. The number of quanta available from a 100 J flash is

$$\text{number of quanta} = \frac{E}{hc}\lambda$$

$$= 100\ \text{J} \times \frac{0.12\times400}{2\times10^{-16}\ \text{J quanta}^{-1}}$$

$$= 2.5\times10^{19}$$

If the light is fully absorbed in 20 ml of solution, 2.5×10^{19} molecules will photolyze to give as many intermediate particles and the concentration of intermediates will be 2.1×10^{-3} mol 1^{-1}. Therefore in order to be easily detectable the intermediates should have high extinction coefficient. For example, anthracene triplet has an extinction coefficient of 7×10^4 1 mol^{-1} cm^{-1} at 424 nm. For spectrophotometric detection, a 3% change in transmitted intensity is required. If the cell length is 20 cm then from Beer's law we can calculate the limiting concentration that may be detected;

$$C = \frac{\log(100/97)}{7\times10^4\ 1\ \text{mol}^{-1}\ \text{cm}^{-1}\times20\ \text{cm}}$$

$$= \frac{0.013}{1.4}\times10^{-6}\ \text{mol}\ 1^{-1}$$

$$= \frac{1.3}{1.4}\times10^{-8} \simeq 10^{-8}\ \text{mol}\ 1^{-1}$$

A flash spectrum of anthracene triplets at various delay times are given in Figure 10.11.

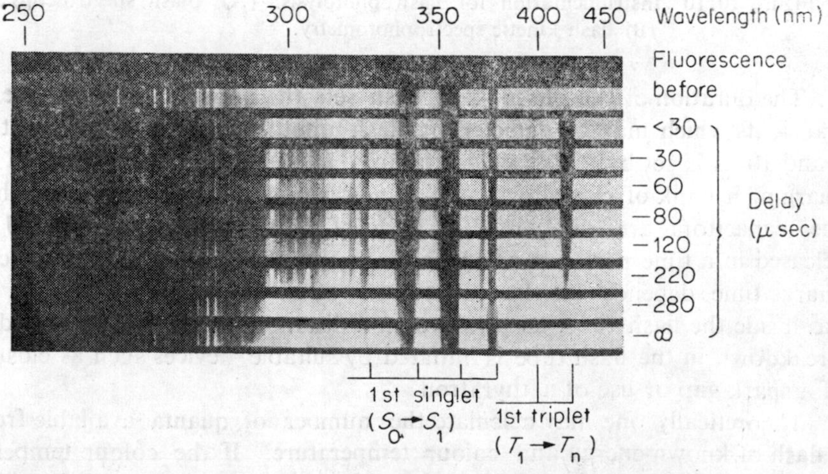

Figure 10.11 Flash spectrum of anthracene triplet at different delay times.

10.4 LASERS IN PHOTOCHEMICAL KINETICS

Recently the limits of photolyzing flash times have been improved by many orders of magnitude by using *Q*-switched laser pulses. A laser is a very special source of light. Laser light is of extremely narrow line width because lasing occurs usually on single atomic or molecular transitions. It is highly collimated and coherent (Section 3 2 1). Power densities as high as 10^{19} W/cm² can be produced. Pulsed laser as short as 10^{-12} s has been achieved. A single-frequency laser designed for spectroscopy can often be 0.001 Å wide, and can excite single electronic-vibration rotation state. Laser with short pulses can allow very short lifetimes to be determined accurately. The flash photolysis may be applied to species lasting a few nanoseconds and emission studies may give information of events in the picosecond (10^{-12} s) range.

Conditions for laser action. The principle of laser action has been discussed in Section 3.2, the primary condition being to create population inversion between the energy levels emitting the laser radiation. The population inversion cannot be achieved if only two energy levels are available since under ideal conditions at the most only 50% molecules can be promoted to the upper energy state. To achieve population inversion it is necessary to have a system with three energy levels E_1, E_2 and E_3 such that $E_1 < E_2 < E_3$. If a strong photon source of frequency ν_{13} is incident on the system, a large number of absorbing units (atoms, molecules or ions) will be promoted from E_1 to E_3. If the state E_3 has a short lifetime and is coupled with the state E_2 which is a long lived metastable state, the return from E_2 to E_1 will be forbidden. Thus atoms are pumped from E_1 to E_3 and transferred from E_3 to E_2 but return from E_2 to E_1 is small. Under this condition the population of E_2 may attain a higher value than that of E_1 and a *lasing*

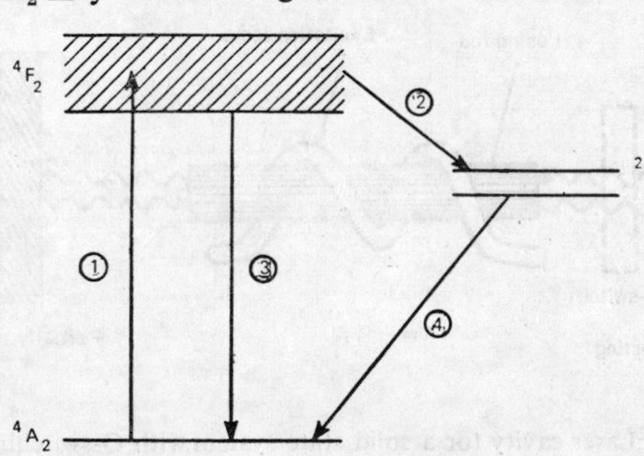

Figure 10.12 Energy level scheme for ruby laser: Cr(III) ion in Al_2O_3 matrix. (1) Pump radiation. (2) Intersystem crossing $^4F_2 \rightsquigarrow {}^2E$. (3) Fluorescence radiation $^4F_2 \rightarrow {}^4A_2$, $k_f = 3\times10^5$ s^{-1}. (4) Laser radiation $^2E \rightarrow {}^4A_2$, $\lambda = 694.3$ nm, $k_p = 2\times10^2$ s^{-1}.

condition may be established between the two energy states. Now if a radiation of frequency ν_{21} is incident on the system, it will stimulate the emission of radiation from the state E_2 to E_1. The radiation of energy $h\nu_{13}$ is known as *pump radiation* and that of $h\nu_{21}$ as *laser radiation*. The energy level scheme for a three level system such as Ruby (Cr^{3+} in Al_2O_3 matrix) is shown in Figure 10.12. It is important to note that for laser action, pumping must be very intense in order to counteract strong tendency towards spontaneous emission.

It is easier to establish lasing condition, if the final state is not the ground state. Thus when a four level scheme is utilized the system is pumped from $E_1 \rightarrow E_4$, relaxes from $E_4 \rightsquigarrow E_3$ and lases from $E_3 \rightarrow E_2$, finally returning to E_1 by nonradiative process. The pumping frequency will be ν_{14} and lasing frequence ν_{32}. A much less pump power will be required to stimulate the laser action. The pumping and lasing energy levels may belong to two different atoms or molecules, the system being coupled by energy transfer process (eg. He—Ne gas laser).

Once this emissive condition is established, the coherent emitted radiation is made to stimulate further emission from the excited atoms in an optical cavity in order to amplify the intensity of the phase coherent radiation.

A continuous laser operates by continually 'pumping' atoms or molecules into the excited state from which induced decay produces a continuous beam of coherent radiation. The He—Ne laser is an example of continuous system. Another mode of operation is to apply an energy pulse to the system, exciting a considerable fraction into the excited state. When all these molecules or atoms are induced to decay simultaneously, intense but exteremely short pulse of coherent radiation is emitted. The ruby laser falls in this category.

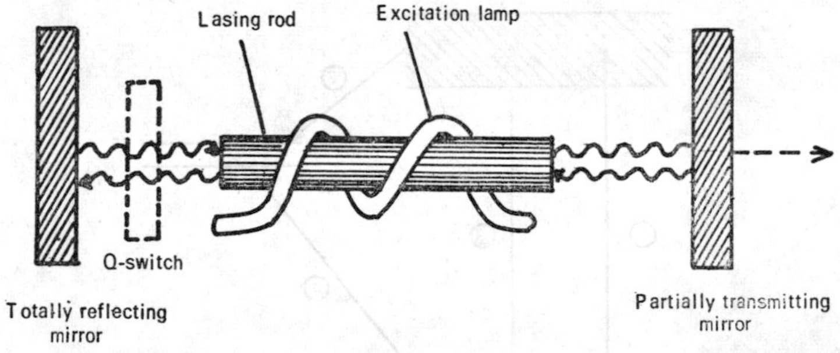

Figure 10.13 Laser cavity for a solid state system with Q-switching technique.

The laser cavity for solid state ruby laser (Figure 10.13) consists of a ruby rod with excitation lamp fixed parallel to it. At the end of the cavity there is a totally reflecting mirror and at the opposite end, a partially transmitting mirror. The excitation lamp 'pumps' the lasing material and causes *population inversion* at the emitting state. Much of the energy from this upper state may be lost through ordinary fluorescence, depending on the lifetime of the state. However, some of the photons move along a path between the mirrors and stimulate emission of more photons from excited material during each trip. An ocillation is thus set up and intensity of the beam builds up. On each trip, some of the energy passes out of the cavity through the partially transparent mirror. To effect a Q-switching technique, the path between the two mirrors in the cavity is optically blocked with a cell of bleachable dye, Pockel cell or some other device. This prevents oscillations within the cavity and allows relevant energy states to be populated much in excess of the lasing threshold. The switch is then opened temporarily allowing the stimulated emission to build up rapidly with high amplification due to the excess excitation. This results in a giant pulse of nanosecond duration and energy of several joules. By passing the giant pulse through frequency doublers such as KDP crystals, the wavelength of the emitted pulse is transformed from 694 nm to 347 nm with an efficiency of about 20%. The neodymimium glass laser can similarly be made to emit at 265 nm. By such frequency doubling processes, wavelengths suitable for photochemical reactions can be generated.

In the spectroscopic monitoring method developed by Porter and Topp, the UV pulse is passed through a beam splitter which allows a small portion of the light to pass through and a fraction to be directed to the reaction vessel. The light which passes through is reflected back from the mirror set at a suitable distance which controls the delay time based on speed of light. The reflected beam again passes through the beam splitter and a portion of the light is reflected to a scintillator solution whose fluorescence provides a spectral continuum in the wavelength region longer than the exciting light. The short fluorescence lifetime of the scintillator solution assures the time shape of the monitoring pulse to be nearly the same as the laser pulse. This induced flash now travels through the reaction solution and into the spectrograph recording the spectrum of the transient at the pre-chosen delay (Figure 10.14).

By using the biphotonic pulse coincidense technique (Section 3.11), the vibrational relaxation processes and emission from the S_1 state of azulene has been studied by Rentzepis. Pulsed nitrogen laser (337 nm) and Cd-He continuous lasers (325, 442 nm) are now commercially available and have considerable power in the UV. Tunable dye lasers can be used, from 400 nm to 700 nm region using diffraction grating and Fabry-Perrot interferometer inside the laser cavity. These lasers are useful light sources in high resolution spectroscopy. The laser produces monochromatic radiation which excite definite states whereas flash lamps emit a wavelength continuum which often allow different reaction to occur

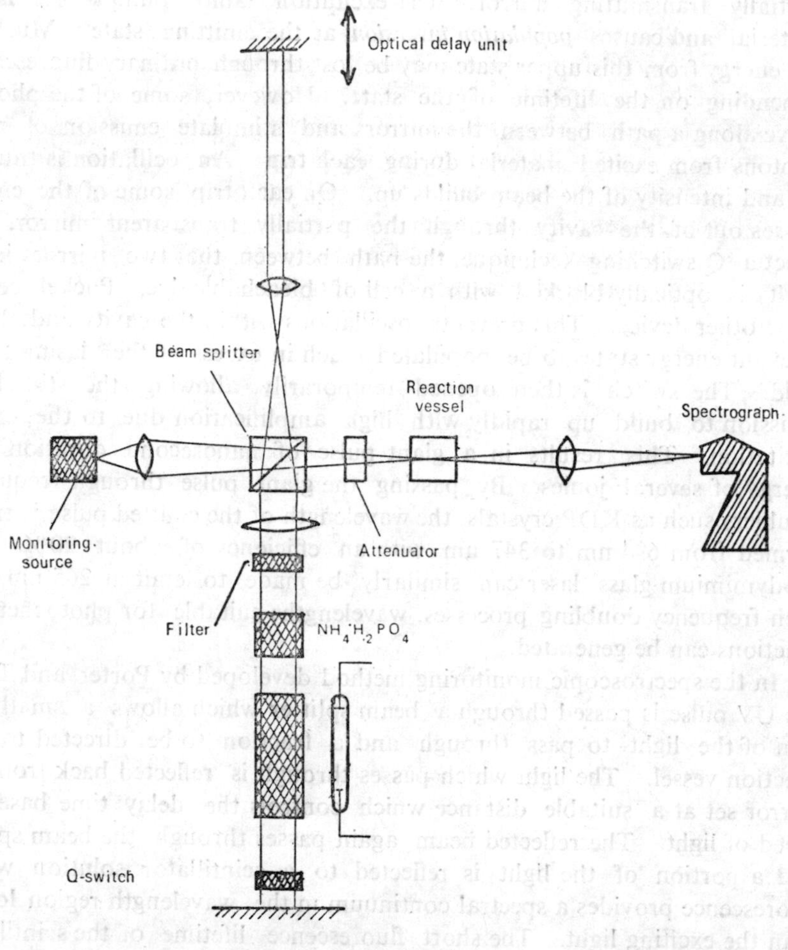

Figure 10.14. The nanosecond flash spectrographic apparatus of Porter and Topp.

simultaneously complicating the interpretation of the results.

With special mode-locking technique, it has been possible to obtain high peak power light pulses (10^6-10^{13} watt) with a half-width of several picoseconds. Using this technique, Novak and Windsor have obtained $S_2 \leftarrow S_1$ absorption and extinction values, and singlet-triplet intersystem crossing rate by measuring time dependent decrease of $S_2 \leftarrow S_1$ absorption and simultaneous increase of corresponding $T_2 \leftarrow T_1$ transition. Recently higher energy states of charge-transfer complexes have been studied by this technique.

The monochromaticity of laser beams from a tunable source permits selective photochemistry, even isotopically selective ones. For example, it is possible to excite by UV laser only one isotope $H_2{}^{13}CO$ of formaldehyde in a mixture with ^{12}C and ^{13}C. The $H_2{}^{13}CO$ can dissociate to H_2 and ^{13}CO. In a mixture of NH_3 and ND_3, the deuterated ammonia can be perferentially decomposed. The most interesting application that has been reported is in the use of enrichment of fissionable uranium ^{235}U. A beam of uranium vapour is allowed to emerge from a vacuum chamber and is crossed with a narrow laser beam from a tunable dye laser and also a broad band UV source. The laser beam selectively excites ^{235}U and not ^{238}U. The UV radiation then ionises the photoexcited ^{235}U only. The $^{235}U^+$ ion thus generated is deflected by charged plates and collected.

APPENDIX I

Mathematical Equation for the Combination of Two Plane Polarized Radiation

Since waves are composed of simple harmonic motion (S.H.M.), they can be analytically represented as

$$x = a \cos \theta \tag{1}$$

where x is the displacement of the wave from the mean position, θ is known as the phase angle and a is the amplitude. The phase angle $\theta = \omega\tau$ where ω = angular velocity defined as number of radians covered per unit time. If ν is the frequency of rotation and if per rotation 2π radian is covered, then

$$\omega = 2\pi\nu, \theta = 2\pi\nu t$$

and

$$x = a \cos 2\pi\nu t \tag{2}$$

(A) Compounding of two waves moving in the same direction but with different amplitudes and phase angle:

$$x_1 = a \cos \theta \tag{3}$$

$$x_2 = b \cos (\theta + \delta) \tag{4}$$

since both are moving in the same direction

$$x = x_1 + x_2$$
$$= a \cos \theta + b \cos (\theta + \delta)$$
$$= a \cos \theta + b \cos \theta \cos \delta - b \sin \theta \sin \delta$$
$$= (a + b \cos \delta) \cos \theta - b \sin \theta \sin \delta \tag{5}$$

Let $(a + b \cos \delta) = c \cos \gamma$; $b \sin \delta = c \sin \gamma$

Substitution in (5) gives

$$x = c \cos \gamma \cos \theta - c \sin \gamma \sin \delta$$
$$= c \cos (\theta + \gamma) \qquad (6)$$

The resultant motion is again a S.H.M. with c as amplitude and $(\theta + \gamma)$ as the phase angle where γ is the phase difference between the resultant wave and the first wave.

(B) The compounding of two waves of equal periods moving at right angles to each other, differing in amplitude and phase.

Let

$$x = a \cos \theta \qquad (7)$$
$$y = b \cos (\theta + \delta) \qquad (8)$$

$y \uparrow$ $\longrightarrow x$

The displacements x and y are at right angles to each other. Now

$$\frac{y}{b} = \cos \theta \cos \delta - \sin \theta \sin \delta$$

$$= \frac{x}{a} \cos \delta - \sqrt{1 - \frac{x^2}{a^2}} \sin \delta$$

$$\frac{y}{b} - \frac{x}{a} \cos \delta = - \sqrt{1 - \frac{x^2}{a^2}} \sin \delta$$

Squaring $\quad \frac{y^2}{b^2} + \frac{x^2}{a^2} \cos^2 \delta - \frac{2xy}{ab} \cos \delta = \sin^2 \delta - \frac{x^2}{a^2} \sin^2 \delta$

Transposing $\quad \frac{y^2}{b^2} - \frac{2xy}{ab} \cos \delta + \frac{x^2}{a^2} = \sin^2 \delta \qquad (9)$

This is a general equation for an ellipse.

Cases of Interest: Polarized Radiation

CASE 1. The waves are at right angles but the phases are the same:

$$\delta = 0, \sin \delta = 0, \cos \delta = 1$$

The general equation becomes

$$\frac{y^2}{b^2} - \frac{2xy}{ab} + \frac{x^2}{a^2} = 0$$

$$\left(\frac{y}{b} - \frac{x}{a}\right)^2 = 0; \qquad \boxed{y = \frac{b}{a} x} \qquad (10)$$

which is an equation for a straight line with a slope b/a.

CASE 2. The waves are at right angles but with a phase difference, $\delta = \pi$, i.e. the phase difference is 180°,

$$\sin\delta = 0,\ \cos\delta = -1$$

$$\boxed{y = -\frac{b}{a}x} \tag{11}$$

These two cases imply that two linearly polarized waves at right angles to each other, differing in amplitude and differing in phase by 0, π or a multiple of π, *compound* to give a resultant wave which is also *linearly polarized* but the plane of polarization lies at an angle $\tan^{-1}(b/a)$ or $\tan^{-1}(-b/a)$ to one of them, depending whether the phase difference is an even or an odd multiple of π.

CASE 3. $\delta = \frac{\pi}{2}$, $\sin\delta = 1$, $\cos\delta = 0$ and $a \neq b$

The general equation becomes

$$\frac{x^2}{a^2} + \frac{y^2}{b^2} = 1 \tag{12}$$

which is an equation for an *ellipse* and expresses that when amplitudes are different and phase angles differ by $\pi/2$ or a multiple of $\pi/2$, the resultant wave is *elliptically polarized.* When phases are equal, $a = b$, the equation reduces to that for a *circle.*

$$x^2 + y^2 = a^2 \tag{13}$$

That is, the resultant wave is *circularly polarized.*

Decomposition of Linearly Polarized Wave

The decomposition of linearly polarized wave is the reverse of compounding of two plane polarized waves of the same phase angle ($\delta = 0$). Depending on the slope $\tan^{-1}(b/a)$, the amplitudes a and b of the two waves, will differ and can be computed. For fluorescence depolarization studies, these amplitudes will correspond to $I_{\perp}$ and $I_{\parallel}$ components of the emitted radiation.

APPENDIX II

Low Temperature Glasses

Composition parts by volume	Classification	Useful temperature range[a]
3–Methylpentane	Hydrocarbon	
Isopentane, 3–methylpentane (1:6 to 6:1)	Hydrocarbon	
2, 3–Dimethyl pentane	Hydrocarbon	
Isopentane, methylcyclohexane (1:6 to 6:1)	Hydrocarbon	
Polyisobutylene	Hydrocarbon polymer	− 80 to − 100°C
Propane, propene (1:1)	Hydrocarbon (olefin)	Liquid at 77°K
3–methylpentane, piperylene (2:1)	Hydrocarbon (diene)	
Methylcyclohexane, toluene (1:1)	Hydrocarbon (aromatic)	
2–Methyl tetrahydrofuran	Ether	
Di–n–propyl ether, isopentane (3:1)	Ether	
Glycerol triacetate	Ester	
Ethanol, methanol (4:1 to 1:1)	Alcohol	
Ethanol, diethyl ether, isopentane (EPA) (2:5:5) (1:4:4)	Alcohol	

Composition parts of volume	Classification	Useful temperature range
Ethanol, diethylether, toluene (1:2:1)	Alcohol	
Glycerol	Alcohol	
Glycerol (with 0.5% water)	Alcohol	
Glucose	Alcohol	
Sulfuric acid	Acid	– 60 to – 80°C
Phosphoric acid	Acid	– 60 to – 80°C
Triethanolamine	Amine	– 60 to – 80°C
Trimethylamine, diethylether, isopentane (2:5:5)	Amine	
Water, ethylene glycol (1:2)	Aqueous alcohol	– 80 to – 150°C
Water, propylene glycol (1:1)	Aqueous alcohol	– 78 to – 90°C
Water, ethanol, methanol (18:19:191) (18:91:91)	Aqueous alcohol	
Ethyliodide, diethyl ether isopentane (1:2:2)	Halide	
Ethyliodide, methanol, ethanol (1:4:16)	Halide	
Ethyliodide, ethanol, diethyl ether, toluene (EEET) (1:1:1:1)		

[a]If no temperature is listed, the glass is either good to liquid nitrogen temperatures, or no lower temperature limit has been reported.

APPENDIX III

Photokinetic Scheme for Determination of Quantum Yields

The photokinetic scheme for concentration quenching assuming excimer formation and emission is

			Rate
$A + h\nu_a$	$\rightarrow$	A^*	I_a
A^*	$\rightarrow$	$A + h\nu_f$	$k_f\,[A^*]$
A^*	$\rightarrow$	A	$k_{IC}\,[A^*]$
A^*	$\rightarrow$	^3A	$k_{ISC}\,[A^*]$
$A^* + A$	$\rightarrow$	A_2^*	$k_{ex}\,[A^*][A]$
A_2^*	$\rightarrow$	$A^* + A$	$k_{ed}\,[A_2^*]$
A_2^*	$\rightarrow$	$2A + h\nu_a$	$k_{ef}\,[A_2^*]$
A_2^*	$\rightarrow$	$2A$	$k_{eq}\,[A_2^*]$

From photostationary state approximation:

$$\frac{d\,[A^*]}{dt} = 0 \text{ and } \frac{d\,[A_2^*]}{dt} = 0$$

The quantum yields for excimer emission ϕ_E, for monomer emission in concentrated solution ϕ_M and in dilute solution $\mathring{\phi}_M$ can be derived as follows from steady state concentrations of $[A^*]$ and $[A_2^*]$:

Rate of formation of A^* = rate of decay of A^*

$$I_a + k_{ed}[A_2^*] = (k_f + k_{IC} + k_{ISC} + k_{ex}[A])\ A^*$$

$$[A^*] = \frac{I_a + k_{ed}[A_2^*]}{k_f + k_{IC} + k_{ISC} + k_{ex}[A]}$$

Similarly for $[A_2^*]$:

$$k_{ex}[A][A^*] = (k_{ed} + k_{ef} + k_{eq})\ [A_2^*]$$

$$[A_2^*] = \frac{k_{ex}[A][A^*]}{k_{ed} + k_{ef} + k_{eq}}$$

Let $\quad a = k_{ed} + k_{ef} + k_{eq}$

$$b = k_f + k_{IC} + k_{ISC} + k_{ex}[A]$$

Then

$$[A_2^*] = \frac{k_{ex}[A]}{a} \cdot \left(\frac{I_a + k_{ed}[A_2^*]}{b}\right)$$

$$= \frac{I_a\, k_{ex}[A]}{ab} + \frac{k_{ed}\, k_{ex}[A][A_2^*]}{ab}$$

$$[A_2^*]\left(1 - \frac{k_{ed}\, k_{ex}[A]}{ab}\right) = \frac{I_a\, k_{ex}[A]}{ab}$$

$$[A_2^*]\,(ab - k_{ed}.k_{ex}[A]) = I_a\, k_{ex}[A]$$

$$[A_2^*] = \frac{I_a\, k_{ex}[A]}{ab - k_{ed}.k_{ex}[A]}$$

and

$$\phi_E = \frac{k_{ef}[A_2^*]}{I_a} = \frac{k_{ef}.k_{ex}[A]}{ab - k_{ed}.k_{ex}[A]}$$

Again

$$[A^*] = \frac{I_a + k_{ed}[A_2^*]}{b}$$

$$= \frac{I_a}{b}\left(1 + \frac{k_{ed}\, k_{ex}[A]}{ab - k_{ed}\, k_{ex}[A]}\right)$$

$$\phi_M = \frac{k_f[A^*]}{I_a}$$

$$= \frac{k_f}{b}\left\{\frac{ab - k_{ed}\, k_{ex}[A] + k_{ed}\, k_{ex}[A]}{ab - k_{ed}\, k_{ex}[A]}\right\}$$

$$= \frac{k_f\, a}{ab - k_{ed}\, k_{ex}[A]}$$

$$\phi_M^\circ = \frac{k_f}{k_f + k_{IC} + k_{ISC}}$$

Bibliography

Some of the important books on photochemistry are listed below. The first five books are undergraduate level texts. All these books are referred to in the section-wise bibliography by the alphabets used to list them.

A. E.J. Bowen, *Chemical Aspect of Light*. Oxford: Clarendon Press, 1946.

B. C.H.J. Wells, *Introduction to Molecular Photochemistry*. London: Chapman & Hall, 1972.

C. R.P. Wayne, *Photochemistry*, London: Butterworths, 1970.

D. R.P. Cundall and A. Gilbert, *Photochemistry*. London: Thomas Nelson, 1970.

E. A. Cox and T.J. Kemp, *Introductory Photochemistry*. London: McGraw-Hill, 1971.

F. N.J. Turro, *Molecular Photochemistry*. New York: W.A. Benjamin, 1966.

G. J.P. Simons, *Photochemistry and Spectroscopy*. New York: Wiley, 1971.

H. J.G. Calvert and J.N. Pitts, Jr., *Photochemistry*. New York: Wiley, 1966.

I. C.A. Parker, *Fluorescence in Solution*. Amsterdam: Elsevier, 1968.

J. D.M. Hercules (ed.), *Fluorescence and Phosphorescence Analysis: Principle and Applications*. New York: Wiley, 1967.

K. R.S. Becker, *Theory and Interpretation of Fluorescence and Phosphorescence*. New York: Wiley, 1969.

L. E.J. Bowen (ed.), *Luminescence in Chemistry*. London: Van-Nostrand, 1968.

M. J.B. Birks, *Photophysics of Aromatic Molecules*. New York: Wiley, 1970.

N. C. Reid, *Excited States in Chemistry and Biology*. London: Butterworth, 1959.

O. P. Pringsheims, *Fluorescence and Phosphorescence*. New York: Wiley, 1949.

P. G.G. Guilbault, *Fluorescence: Theory, Instrumentation and Practice*. New York: Marcel Dekker, 1967.

Q. E.C. Lim, *Molecular Luminescence*. New York: W.A. Benjamin, 1969.

R. S.P. McGlynn, T. Azumi and M. Kinoshita, *Molecular Spectroscopy of the Triplet States.* Englewood Cliff, N.J.: Prentice-Hall, 1969.

S. J.M. Fitzgerald (ed.), *Analytical Photochemistry and Photochemical Analysis.* New York: Marcel Dekker, 1971.

T. A.A. Lamela (ed.), *Creation and Detection of Excited States,* Vol. 1, Pt. A. New York, Marcel Dekker, 1971.

U. F. Daniels. *Photochemistry in the Liquid and Solid States.* Subcommittee on the Photochemical Storage of Solar Energy—NAS—NRC/USA. New York: Wiley, 1957.

V. V. Balzani and V. Carasatti, *Photochemistry of Coordination Compounds.* New York: Academic, 1970.

W. W.A. Noyes, G.S. Hammond & J.N. Pitts, Jr. (eds.), *Advances in Photochemistry,* Vol. 1 (1963), Vol. 2 (1964), Vol. 3 (1964), Vol. 4 (1966), Vol. 5 (1968), Vol. 6 (1968), Vol. 7 (1969), Vol. 8 (1971). New York: Wiley.

X. E.F.H. Brittain, W.D. George and C.H.J. Wells, *Introduction to Molecular Spectroscopy: Theory and Experiment.* New York: Academic Press, 1972.

Y. E.C. Lim (ed.), *Excited States,* Vol. 1 (1974), Vol. 2 (1975). New York: Academic Press.

Z. *Specialist' Report on Photochemistry,* Vols. 1–6. London: Chemical Society.

Section 1

General references A, H

1. J.M. Mellor, D. Phillip and K. Salisbury, "Photochemistry: new technological application", *Chemistry in Britain* **10(5)**, (1974), 160.
2. R.K. Clayton, *Light and Living Matter,* Vol. 1. New York: McGraw-Hill, 1971.
3. Reference I.
4. Reference H.
5. *Scientific American,* **219**, Sep. 1968—an issue devoted to light and its interaction with matter.

Section 2

General references A, F, G, H, J, K, M

1. G. Herzberg, *Atomic Spectra and Atomic Structure.* New York: Dover, 1944.
2. G. Herzberg, *Molecular Spectra and Molecular Structure,* Vol. 1. New York: Van-Nostrand, 1950.
3. G. Herzberg, *Electronic Spectra and Electronic Structure of Polyatomic Molecules.* New York: Van-Nostrand, 1966.
4. J. Walter, G.E. Kimball and W. Eyring, *Quantum Chemistry.* New York: Wiley, 1944.
5. Reference G, Section 1.5a.
6. Reference X.
7. J.N. Pitts, F. Wilkinson and G.S. Hammond, "Vocabulary of photochemistry", Reference W. 1, Ch. 1.
8. M. Kasha, "The Nature and Significance of n—η* transitions", in W.D. McElroy and B. Glass (eds.), *Light and Life.* Johns Hopkins, 1961.
9. F.A. Cotton, *Chemical Applications of Group Theory,* 2nd ed. New York: Wiley, 1971.
10. M. Orchin and H.H. Jaffe, "Symmetry, point groups and character tables, parts I, II and III", *J. Chem. Edu.* **47,** 1970, 246.

11. Reference **K**, Ch. 3,
12. Reference **G**, Section 1.3.
13. H.B. Gray, *Electrons and Chemical Bonding*. New York: W.A. Benjamin, 1964.

Section 3

General references A, H
1. G.M. Barrow, *Introduction to Molecular Spectroscopy*. New York: McGraw-Hill, 1962.
2. L. Pauling and E.B. Wilson, *Introduction to Quantum Mechanics with Application to Chemistry*. New York: McGraw-Hill, 1935, Ch. XI.
3. J.P. Simons, Reference G, Section 1.5.
4. M. Kasha, M.A. El Bayoumi and N. Rhodes, *J. Chem. Physique* **1961**, 916.

Lasers

5. R.A. Smith, Masers and Lasers, *Endeavour*, **21**, April, 1962, 108.
6. A.L. Schawlow, "Optical masers", *Scien. Amer*., June 1961; July 1963.

Luminiscence Lifetimes

7. P. Pringsheim, "Duration of the luminescence process", Reference **0**, p. 3.
8. J.B. Birks, Reference **M**, Ch. 4.
9. R.S. Becker, Reference **K**, Ch. 7.
10. N.J. Turro, Reference **F**, Ch. 3.
11. D.M. Hercules, Reference **J**, Ch. 1.
12. M. Kasha, "*Theory of molecular luminescence*", in Reference **P**, Ch. 4.
13. M. Kasha, "Ultraviolet radiation effects" in M. Burton, J.S. Kirby-Smith and J.L. Magee, *Comparative Effects of Radiation*. New York: Wiley, 1960.
14. Reference **W**, **1**, 1963, 1.
15. E.F.J. Brittain, W.D. George and C.H.J. Wells, Reference **X**.

Multiphoton Absorption

16. W.M. McClain, *Acc. Chem. Res.* **7(5)**, 1974, 129.
17. G. Lakshminarayanan, *Science Reporter*, **12**, 1975, 299.
18. W.L. Peticolas, "Multiphoton spectroscopy", *Ann. Rev. Phys. Chem.*, **18**, 1967, 233.

Section 4

1. Reference A.
2. Reference H.
3. G. Herzberg, *Molecular Spectra and Molecular Structure*, Vol. 1, 2nd ed. New York: Van-Nostrand, 1950.
4. Reference F, Ch. 3.
5. Reference X.

Crossing of Potential Energy Surfaces

6. E. Teller, *J. Phys. Chem.*, **41**, 1937, 109.
7. K. Razi Naqvi, "Emission spectra environmental effect and excited dipole moments", *Chem. Phys. Letts*, **15**, 1972, 634.

Emission Spectra, Environmental Effect and Excited State Dipole Moments

8. Reference **M.**
9. Reference **H.**
10. Reference **I.**
11. Reference **K.**
12. N. Mataga, Y. Kaifu and M. Koizumi, "Solvent effect upon fluorescence spectra and the dipole moment of excited molecules", *Bull. Chem. Soc.*, Japan, **29(4)**, 1956, 465.
13. E. Lippert, "Dipole moment and electronic structure of excited molecules", *Z. Natruforsch.*, **10a**, 541, 1955.
14. J. Czekella, *Z. Electrochem. Ber. Bunsenges. Physik. Chem.*, **64**, 1960, 1221.
15. S. Basu, *Adv. Quantum Chem.*, **1**, 1964, 145.
16. W. Liptay, "Dipole moments and polarizabilities of molecules in the excited electronic states", Reference **Y** 1, p. 129.

Excited State Acidities

17. Th. Förster, "Elementary processes in solution", Reference **U**, p. 10; also *J. Phys. Chem.*, 1959.
18. C.A. Parker, Reference **I.**
19. E.L. Wehry and L.B. Rogers, Reference **J**, Ch. 3.
20. Reference **N.**
21. G. Jackson and G. Porter, *Proc. Roy. Soc.* (London), **A260**, 1961, 13.

Excited State Redox Potentials

22. L.I. Grossweiner and A.G. Kepka, *Photochem. and Photobio.*, **16**, 1972, 307.
23. D. Mariele and A. Maurer, *Chem. Phys. Lett.* **2**, 1968, 602.
24. H. Berg, *Z. Chem.*, **2**, 1962, 237.

Polarized Emission

25. G. Weber, "Polarization of the fluorescence of solution", Reference **J**, Ch. 8.
26. Reference **O**, see F 118, p. 366.
27. Reference **I.**
28. F. Dorr, "Polarized light in spectroscopy and photochemistry", Reference **T.**

Excited State Geometrics

29. K.K. Innes, "Geometrics of molecules in excited electronic states", Reference **Y, 2,** Ch. 1.
30. Reference **J**, Ch. 1.

Flash and Laser Spectroscopy

31. G. Porter, "Flash spectroscopy", *Angew Chem. Int. Edit*, **80**, 1968, 852.
32. G. Porter, "Lasers in chemistry", *Chemistry in Britain*, **6**, 1970, 248.
33. K.W. Chambers and I.M. Smith, *J. Chem. Edu.*, **51**, 1974.
34. Stephen R. Leon, *J. Chem. Edu.*, 63, 1976, 13.

Section 5

General references **F, H, I, J, K, L, M, N, O**

1. E.J. Bowen, "Photochemistry of aromatic hydrocarbons in solutions", Reference **W, 1,** 1963, 21.

2. E.J. Bowen, "Light emission from organic molecules", *Chemistry in Britain*, **2**, 1966, 249.

3. E. Lippert, "Photophysical primary steps in solutions of aromatic compounds", *Acc. Chem. Res.*, **3**, 1970, 74.

4. A.T. Gradyashko, A.N. Sevchenke, K.N. Solovyov and M.P. Tsvirko, "Energetics of photophysical processes", *Photochem. Photobio.*, **11**, 1970, 387.

Radiationless and Radiative Transitions

5. C.M.O' Donnell and T.S. Spencer, "Some considerations of photochemical reactivity", *J. Chem. Edu.*, **49**, 1972, 822.

6. J. Jortner, S.A. Rice and R.M. Hochstrasser, Reference, **W**, **7**, 1969, 149.

7. D. Phillip, J. Lemaine and C.S. Burton, Reference **W**, **5**, 1968, 329.

8. M.W. Windsor and J.P. Novak, Reference **Q**, p. 365.

9. B.R. Henry and W. Siebrand, Reference **Q**, p. 423.

0. G.W. Wilson, "Molecular electronic radiationless transitions", Reference **Y**, **1**, 1974, 1.

1. J.M. Bridges, "Fluorescence of organic compounds", Reference L, Ch. 6.

2. C.J. Seliskar, O.S. Khalil and S.P. McGlynn, "Luminescence characteristics of polar aromatic molecules", Reference **Y**, **1**, 1974, 231.

Triplet State

3. N.J. Turro, "The triplet state", J. *Chem. Edu.*, **46**, 1969, 2.

4. S.P. McGlynn, F.J. Smith and G. Cilento, "Some aspects of the triplet state", *Photochem. Photobio.*, **3**, 1964, 269.

5. Reference **R**.

6. P.J. Wagner and G.S. Hammond, Reference **W**, **5**, 1968, 21.

7. Reference G, Section 2.5d.

8. A.A. Lamola and N.J. Turro, "Energy transfer and organic photochemistry", in A. Weissbergerled, *Techniques of Organic Chemistry*, Vol. 14. New York: Wiley, 1969.

9. W.A. Noyes and I. Unger, "Singlet and triplet states: benzene and simple aromatic compounds", Reference W, **4**, 1966, 49.

Delayed Fluorescence

0. C.A. Parker, "Phospherescence and delayed fluorescence in solution", Reference W, 2, 1964, 305.

1. F. Wilkinson and A.R. Horrocks, "Phosphorescence and delayed fluorescence of organic substances", Reference L, Ch. 7.

Temperature Effect

2. E.J. Bowen and J. Sahu, "The effect of temperature on fluorescence of solutions", Reference U, p. 55.

Section 6

General references **A, F, G, H, I, J, K, L, M, N, O**

1. P. Pringsheim, Reference O, Part 1, G, p. 89.

2. Reference G, Section 2.3b.

3. Reference M, Sections 7-7, 9.11.

4. B. Stevens, "Molecular association in aromatic systems", Reference **W**, **8**. 1971, 161.

5. S. Nagakura, "Electron donor-acceptor complexes in their excited states", Reference **Y**, **1**, 1975, 321,

6. M. Ottolenghi, "Charge-transfer complexes in the excited states—laser photolysis studies", *Acc. Chem. Res* , **6**, 1973, 153.
7. M.R.J. Daek, "Charge transfer complexes and photochemistry", *J. Chem. Edu.*, **50**, 1973, 169.
8. Th. Forster, "Excitation transfer", in M. Burton, J.S. Kirby-Smith and J.L. Magee (eds.), *Comparative Effects of Radiation*. New York: Wiley, 1960.
9. F. Wilkinson, "Electronic energy transfer between organic molecules in solution", Reference W, 3, 1964, 241.
10. Reference L, Ch. 8.
11. A.A. Lamola and N.J. Turro, "Energy transfer and organic photochemistry", Chs. I and II in A. Weissberger (ed.) *Techniques of Organic Chemistry*, Vol. 14. New York: Wiley, 1969.
12. P.J. Wagner, "Energy transfer kinetics in solution", Reference T, Ch. 4.
13. P.S. Engel and C. Steel, "Photochemistry of azo-compounds in solution", *Acc. Chem. Res.*, **6**, 1973, 275.
14. D. Rehm and A. Weller, "Redox potentials and electron transfer processes", *Ber. Bunsegeses Physik. Chem.*, **73**, 1969, 8343; *Isael J. Chem.* **8**, 1970, 259.
15. K.H. Grellmann, A.R. Watkins and A. Weller, "The electron transfer mechanism of fluorescence quenching in polar solvents", *J. Phys. Chem.*, **76**, 1972, 469, 3132.
16. R.A. Marcus, *J. Chem. Phys.*, **43**, 1965, 679.
17. N. Filipescu, "Intramolecular energy transfer between non-conjugated chromospheres", Reference Q, p. 697.
18. J.A. Hudson and R.M. Hedges, "Luminescence and intramolecular energy transfer in rigid model compounds", Reference Q, p. 667.
19. M. Kleinerman, "Energy transfer and electron transfer in some lanthanide complexes", Reference Q, p. 281.
20. A.B.P. Sinha, "Fluorescence and laser action in rare earth chelates" in C.N.R. Rao and J.R. Ferrao (ed.), *Spectroscopy in Inorganic Compounds*, Vol, II. New York: Academic Press, 1971.
21. M. Kasha, H.R. Rowles and M.A. El Bayoumi, *Pure and Appl. Chem.*, **11**, 1965, 371.
22. E.G. McRae and M. Kasha in *Physical Processes in Radiation Biology*, L. Augenstein, B. Rosenberg and R. Mason (eds.). New York: Academic Press, 1964, p. 22.
23. H. Kuhn, "Interaction of chromospheres in monolayer assemblies", *Pure & Appl. Chem.*, **27**, 1971, 421.

Section 7

General references A, F, H, N
1. G. Porter, "Reactivity radiationless conversion and electron distribution in the excited state" in *Reactivity of Photoexcited Organic Molecules, Proc Thirteenth Conf. on Chem.* at the Univ. of Brussels, October 1965. New York: Wiley, 1967, p. 80.
2. Th. Förster, "Photochemical reactions", *Pure & Appl. Chem.*, **24**, 1970, 443.
3. G.S. Hammond, "Reflection on photochemical reactivity", Reference W, 7, 1969, 373.
4. H.E. Zimmerman, "A new approach to mechanistic organic photochemistry", Reference W, 1, p. 183, 1963.
5. G. Quinkert, "An route to multisurface chemistry", *Angew Chem. Int. Edit.*, **14**, 1975, 12.
6. W.G. Dauben, L. Salem and N.J. Turro, "A classification of photochemical

reactions", *Acc. Chem. Res.*, **8**, 1975, 1.
7. J. Michl, "Physical basis of qualitative MO arguments in organic photochemistry", in *Topics in Current Chemistry*, Vol. 46, Photochemistry. Berlin: Springer-Verlag, 1971.
8. J. Michl, "The role of biradicoloid geometries in organic photochemistry", *Photochem. Photobio.*, **25**, 1977, 141.
9. D.E. Hoare and G.S. Pearson, "Gaseous Photooxidation reactions", Reference **W**, **3**, 1964, 83.
0. J.R. McNesby and H. Okabe, "Vacuum ultraviolet photochemistry", Reference **W**, **3**, 1964, 157.
1. G.C. Pimental, "Chemical lasers", *Scientific American*, April 1966.
2. R.P. Wayne, "Singlet molecular oxygen", Reference **W**, **7**, 1969, 11.
13. J. Pitts, "Photochemistry of polluted toposhere", *Science*, **192**, 1976, 111.
14. H.E. Gunning and O.P. Strauss, "Isotope effects and the mechanism of energy transfer in mercury photosensitization", Reference **W**, **1**, 1963, 209.
15. G Quinkert, "Thermally reversible photoisomerization", *Angew Chem. Int. Edit*, **11**, 1972, 1072.
16. R. Dessauer and J.P. Paris, "Photochromism". Reference **W**, **1**, 1963, 275.
17. H. Meir, "Photochemistry of dyes", in **K**. Venkataraman (ed.) *Chemistry of Synthetic Dyes*, Vol. 4. New York: Academic Press, 1971, Ch. 7.

Section 8

General references **F**, **H**
1. N.J. Turro, "Photochemical reactions of organic molecules", in A. Weissburger (ed.) *Techniques of Organic Chemistry*, Vol. 14. New York: Wiley, 1969.
2. R. Srinivasan, "Photochemistry of dienes and trienes", Reference **W**, **4**, 1966.
3. P.J. Wagner and G.S. Hammond, "Properties and reactions of organic molecules in their triplet states", Reference **W**, **5**, 1968, 21.
4. J.S. Seventon, "Photochemistry of organic compounds—selected aspects of olefin photochemistry", *J. Chem. Edu.*, **41**, 1969, 7.
5. K. Golnick, "Type II photooxygenation reactions in solution", Reference **W**, **6**, 1968, 1.
6. B. Stevens, "Kinetics of photoperoxidation in solution", *Acc. Chem. Res.*, **6**, 1973, 90.
7. D.R. Kearn, "Physical and chemical properties of singlet molecular oxygen", *Chem. Rev.*, **71**, 1971, 395.
8. G.S. Foote, "Mechanisms of photosensitized oxidations", *Science*, **162**, 1968, 963.
9. D.L. Kearn and A.U. Khan, "Sensitized photooxygenation reactions and the role of singlet oxygen", *Photochem. Photobio.*, **10**, 1969, 193.
10. K.D. Gundermann, "Recent advances in research on the chemiluminescence of organic compounds", in *Topics in Current Chemistry*, Vol. 46: Photochemistry. Berlin: Springer-Verlag, 1974.
11. A.J. Bard, K.S.B. Santhanam, S.A. Cruser and L.R. Faulkner, "Electrogenerated chemiluminescence", Reference **P**, Ch. 14.
12. R.B. Woodward and R. Hoffman, "The conservation of orbital symmetry", *Angew Chem. Int. Edit*, **8**, 1969, 781.
13. M. Orchin and H.H. Jaffe, *Importance of Antibonding Orbitals*. Boston: Houghton Miffilin, 1967.
14. E.H. White, J.D. Miano, C.I. Watkins and E.J. Breaux, "Chemically produced excited states", *Angew Chem. Int. Edit*. **13**, 1974, 229.
15. V. Balzani and V. Carassiti, *Photochemistry of Coordination Compounds*.

New York: Academic Press, 1970.
16. M. Keith de Armond, "Relaxation of excited states in transition metal complexes", *Acc. Chem. Res.*, **7**, 1974, 309.
17. G.A. Crosby, "Spectroscopic investigation of excited state of transition metal complexes", *Acc. Chem. Res.*, **8**, 1975, 231.
18. W.A. Runciman, "Luminescence of inorganic substances", Reference **L**, Ch. 5.
19. A.W. Adamson, "Some aspects of photochemistry of coordination compounds", *Coordination Chemistry Review*, **3**, 1968, 169.
20. W.L. Waltzer and R.G. Sutherland, "The photochemistry of transition metal coordination compounds—a survey", *Chem. Soc. Rev.*, **1**, 1972, 241.
21. C. Kutal, "Mechanistic inorganic photochemistry Part I—reactivity of the excited states of Cr (III) complexes", *J. Chem. Edu.*, **52**, 1975, 502.
22. P. Grutch and C. Kutal, "Mechanistic inorganic photochemistry Part II—application of intermolecular energy transfer", *J. Chem. Edu.*, **53**, 1976, 437.
23. D. Valentin, Jr., "Photochemistry of cobalt (III) and chromium (III) complexes in solution", Reference **W**, **6**, 1968, 123.
24. M. Wrighter, "Photochemistry of metal carbonyls", *Chem. Rev.*, **74**, 1974, 401.
25. R.E. Bozak, "Photochemistry of the metallocenes", Reference **W**, **8**, 1971, 227.

Section 9

1. G.H. Wald, "Life and light", *Scientific American*, **201**, 1959, 92.
2. L.V. Berkner and L.C. Marshall, "History of oxygenic concentration in the earth's atmosphere", *Faraday Soc. Discussions*, **37**, 1964, 122.
3. D.C. Neckers, "Photochemical reactions of natural macromolecules—photoreactions of proteins", *J. Chem. Edu.*, **50**, 1973, 165.
4. A. McLaren and D. Shugar, *Photochemistry of Nucleic Acid and Proteins.* Oxford: Pergamon Press, 1964.
5. R.F. Steiner and I. Weinryb, *Excited States of Nucleic Acids and Proteins.* New York: Plenum Press, 1971.
6. C.P. Swanson (ed.), *An Introduction to Photobiology.* Englewood Cliffs: Prentice Hall, 1969.
7. J.C. Burr, "Advances in photochemistry of nucleic acid derivatives", Reference **W**, **6**, 1968, 193.
8. G.K. Radda and G.H. Dodd, "Luminescence in biochemistry", Reference **L**, Ch. 10.
9. H.J. Heller and H.R. Blathmann, "Some aspects of light protection of polymers", *Pure & Appl. Chem.*, **30**, 1972, 145.
10. R.F. Reinisch (ed.), *Photochemistry of Macromolecules.* New York: Plenum Press, 1970.
11. G. Gerischer, "Electrochemical cell for energy conversion", *Israel J. Chem.*, **14**, 1975, 150.
12. G. Gerischer, "Electrochemical techniques for the study of photosensitization", *Photochem. Photobio.*, **16**, 1972, 243.
13. E.I. Rabinowitch and Govindjee, *Photosynthesis.* New York: Wiley, 1969.
14. R.K. Clayton, *Light and Living Matter*, Vols. 1 and 2. New York: McGraw-Hill, 1971.
15. E.I. Rabinowitch and Govindjee, "Photosynthesis", *Scientific American*, **213**, 1965, 74.
16. M. Calvin, "Photosynthesis as a resource for energy material", *Ame. Scientist*, **64**, 1976, 270.
17. D. Oesterhelt, "Bacteriorhodopsin as an example of a light driven proton pump", *Angew Chem. Int. Edit*, **15**, 1976, 17.

18. M. Archer, "Photochemical system in solar energy", in *UK assessment*—UK, Section of the International Solar Energy Society, May 1976.
19. G. Stein, "Chemical storage of solar energy and photochemical fuel formation", *Israel J. Chem.*, **14**, 1975, 213.
20. Reference U.

Section 10

General references **F**, **H**, **I**, **J**, **L**, **P**, **S**, **T**
1. J.G. Calvert and J.A. Pitts, Jr., Reference **H**, Ch. 7.
2. W.R. Ware, "Transient luminescence measurements", Reference **T**, Ch. 5.
3. J.W. Longworth, "Luminescence spectroscopy", Reference **T**, Ch. 7.
4. H.A. Taylor, "Analytical methods and techniques for actinometry", Reference **S**, Ch. 3.
5. J.N. Demas and G.A. Crosby, "The measurement of photoluminescence quantum yields—a review", *J. Phys. Chem.*, **75**. 1971, 991.
6. E.J. Smith, "Photochemical equipment and actinometers", Reference **S**, Ch. 1.
7. G. Porter, "Flash photolysis", Science, **160**, 1968, 1299.
8. W. Demtroder, "Laser spectroscopy—spectroscopy with lasers", in *Topics in Current Chemistry*, Vol. 17. Berlin: Springer-Verlag, 1971.
9. P.M. Rentzepis, "Techniques of flash photolysis", *Photochem. Photobio.*, **8**, 1968, 565.
20. F.H. Fry, "Lasers in analytical chemistry" Reference **S**, Ch. 2.
11. S.R. Leone, "Application of lasers to chemistry", *J. Chem. Edu.*, **53**, 1976, 13.
12. L. Paterson and G. Porter, "Lasers in photochemical kinetics", *Chemistry in Britain*, **6**, 1970, 246.
13. J.R. Novak and M.W. Windsor, *Proc. Roy. Soc.*, London, **A 308**, 1968, 95.

Index